THE BEST TEST PREPARATION FOR THE

GRE

RECORD

EXAMINATION

BIOLOGY

Staff of Research & Education Association
Dr. M. Fogiel, Director

Research and Education Association
61 Ethel Road West
Piscataway, New Jersey 08854

The Best Test Preparation for the
GRADUATE RECORD EXAMINATION (GRE)
IN BIOLOGY

Printed in the United States of America

Library of Congress Catalog Card Number 98-66635

International Standard Book Number 0-87891-602-4

Research & Education Association
61 Ethel Road West
Piscataway, New Jersey 08854

REA supports the effort to conserve and
protect environmental resources by
printing on recycled papers.

CONTENTS

PREFACE

This book represents a massive effort to provide you with five complete practice tests based on the most recent GRE Biology Tests. Each test is 2 hours and 50 minutes in length, and is complete with every type of question that can be expected on the GRE Biology Test. In addition, an answer key and detailed explanations of the answers follow every test. By completing all five tests, and by studying the explanations of answers and our Biology Review, you can discover your strengths and weaknesses and thereby become well prepared for the actual test.

ABOUT THE TEST

The Graduate Record Examination (GRE) Biology Test is given three times a year by the Educational Testing Service under the direction of the Graduate Record Examinations Board. Applicants for graduate school submit GRE test results together with other undergraduate records as part of the highly competitive admission process to graduate school.

The questions for the test are composed by a committee of specialists who are recommended by the American Institute of Biological Sciences, the American Society of Zoologists, and the Botanical Society of America, and who are selected from various undergraduate and graduate faculties. The test consists of approximately 200 multiple-choice questions. Some questions are grouped together, based on particular laboratory experiments or descriptive paragraphs. Emphasis is placed on the major areas of biology as follows:

I. Cellular and Molecular Biology
        Cell Structure and Function
        Molecular Biology and Molecular Genetics

II. Organismal Biology
        Animal Structure, Function, and Organization
        Plant Structure, Function, and Organization
        Reproduction, Growth, and Development
        Diversity of Life

III. Ecology and Evolution
        Ecosystems and General Ecology
        Evolutionary Processes and Consequences

These major areas are given equal importance, and the questions are interspersed randomly throughout the test. The questions are taken from courses of study most commonly offered in the undergraduate curriculum.

ABOUT THE REVIEW

Our Biology Review provides a comprehensive summary of the main areas tested on the GRE Biology Test, and is written to help you understand these concepts. The following topics are discussed.

Microbiology

Enzymes and Cellular Metabolism
    Enzyme Structure and Function; Control of Enzyme Activity;
    Feedback Inhibition; Glycolysis; Krebs (Citric Acid) Cycle;
    Electron Transport Chain and Oxidative Phosphorylation

DNA and Protein Synthesis
    DNA Structure and Function; DNA as Transmitter of Genetic
    Information; Protein Synthesis: Transcription and Translation

Molecular Biology

Viral Structure and Life History
    Nucleic Acid (DNA and RNA) and Protein Components;
    Bacteriophage: Structure, Function and Life Cycle

Prokaryotic Cells
    Cell Structure and Function; Bacterial Life History and
    Physiological Characteristics

Fungi
    Major Structural Types; General Life History and Physiology

Nervous and Endocrine Systems

Nervous System Structure and Function
    Organization of the Vertebrate Nervous System; Sensor and
    Effector Neurons; Sympathetic and Parasympathetic Nervous
    System

Sensory Reception and Processing
    Somatic Sensors; Olfaction and Taste; Hearing; Vision

Endocrine System: Hormones and Their Sources
    Function of Endocrine System; Cellular Mechanisms of Hormone
    Action; Control of Hormone Secretion; Major Endocrine Glands,
    Their Hormones, Specificity, and Target Issues

Circulatory System, Lymphatic, and Immune Systems

Circulatory System
    Multiple Functions, Including Role in Thermoregulation; Four-
    Chambered Heart, Pulmonary, and Systemic Circulation;
    Arterial and Venous Systems, Capillary Beds, Systemic and
    Diastolic Pressure; Composition of Blood; Role of Hemoglobin in
    Oxygen Transport

Lymphatic System

Immune System
    Antigens, Antibodies, and Antigen-Antibody Reactions; Tissues
    and Cells of the Immune System

Digestive and Excretory Systems

Digestive System
    Ingestion: Structures and Their Functions; Stomach; Digestive
    Glands, Including Liver and Pancreas, Bile Production; Small
    Intestine, Large Intestine; Muscular Control of Digestion

Excretory System
    Role of Excretory System in Body Homeostasis; Kidney: Structure
    and Function; Nephron: Structure and Function; Formation of Urine;
    Storage and Elimination of Wastes

Muscle and Skeletal Systems

Muscle System
Functions; Basic Muscle Types and Locations; Nervous Control of
Muscles

Skeletal System
Bone Structure; Skeletal Structure

Respiratory and Skin Systems

Respiratory System
Function; Breathing Structures and Mechanisms

Skin System
Composition; Protection and Thermoregulation

Reproductive System and Development

Male and Female Gonads and Genitalia
The Male Reproductive System; The Female Reproductive System

Gametogenesis by Meiosis
Spermatogenesis (Sperm Production); Oogenesis (Production of
Egg Cells or Oocytes)

Reproductive Sequence
Males; Females

Embryogenesis
Embryonic Development; Germ Layers

Genetics and Evolution

Genetics
Genotype and Phenotype; Dominant/Recessive Inheritance;
Incomplete Dominance; Codominance; Sex-Linked Inheritance

Evolution
Darwin's Theory—Natural Selection; Evidence for Evolution;
Factors Responsible for Evolutionary Change; Hardy-Weinberg
Equilibrium

Biological Molecules

Amino Acids and Proteins
Carbohydrates
Lipids
Phosphorus Compounds

Oxygen Containing Compounds

Alcohols
Aldehydes and Ketones
Carboxylic Acids
Common Acid Derivatives
Ethers
Phenols

Amines

Hydrocarbons

Saturated
Unsaturated
Aromatic

Molecular Structure of Organic Compounds

σ and π Bonds
Multiple Bonding
Stereochemistry

Separations and Purification

Extraction
Chromatography
Distillation
Recrystallization

Use of Spectroscopy in Structural Identification

Infrared Spectroscopy
NMR Spectroscopy

By studying our review, your chances of scoring well on the actual exam will be greatly increased. It will teach you everything that you need to know. After thoroughly studying the material presented in the Biology Review, you should go on to take the practice tests. Used in conjunction, the review and practice tests will enhance your skills and give you the added confidence needed to obtain a high score.

TAKING THE PRACTICE TESTS

Make sure to take as many of the practice tests as possible. When taking the practice tests, you should try to make your testing conditions as much like the actual test as possible.

- Work in a quiet place where you will not be interrupted.
- Time yourself!
- Do not use any books, calculators, slide rules, rulers, compasses, or similar articles, as these materials will not be permitted into the test center.

Doing these things will help you become accustomed to the time constraints you will face when taking the exam, and will also help you develop speed in answering the questions because you will become more familiar with the test format.

SCORING THE TEST

Four separate scores are obtained from your test: 1) a total score based on all the questions in the test, 2) a subscore for the questions pertaining to the field of cellular and subcellular biology, 3) a subscore for the questions pertaining to the field of organismal biology, and 4) a subscore for the questions pertaining to the field of population biology. These subscores identify your strengths and weaknesses and may be used for guidance and placement purposes.

For each of your correct answers, you will receive one "raw score" point, while for each incorrect answer, one-fourth of a point will be deducted from your total score. You will not be penalized for answers left blank. If you are unsure of an answer, but are familiar enough with the material to eliminate one or more of the answer choices, it may be to your benefit to guess.

The following table is provided for you to convert your total "raw score" for each practice test into a total "scaled score." The table also allows you to compare your performance to the performance of other students on past exams. While it will give you a range into which your score will fall when you take the actual test, the table is meant to apply only to our practice tests. This is because the GRE Biology scaled scores can shift from administration to administration.

BIOLOGY TEST CONVERSION TABLES

TOTAL SCORES

Raw Score	Scaled Score	%*	Raw Score	Scaled Score	%*
203-210	990		90-92	590	37
200-202	980		87-89	580	34
197-199	970		84-86	570	32
195-196	960		82-83	560	29
192-194	950		79-81	550	26
189-191	940		76-78	540	25
186-188	930		73-75	530	22
183-185	920		70-72	520	19
180-182	910		67-69	510	17
178-179	900		65-66	500	15
175-177	890	99	62-64	490	14
172-174	880	99	59-61	480	12
169-171	870	99	56-58	470	11
166-168	860	98	53-55	460	9
164-165	850	98	51-52	450	9
161-163	840	98	48-50	440	7
158-160	830	97	45-47	430	6
155-157	820	96	42-44	420	5
152-154	810	95	39-41	410	4
149-151	800	94	36-38	400	4
147-148	790	93	34-35	390	3
144-146	780	91	31-33	380	3
141-143	770	89	28-30	370	2
138-140	760	87	25-27	360	2
135-137	750	85	22-24	350	1
132-134	740	83	19-21	340	1
130-131	730	81	17-18	330	1
127-129	720	78	14-16	320	1
124-126	710	75	11-13	310	1
121-123	700	72	8-10	300	
118-120	690	68	5- 7	290	
116-117	680	66	3- 4	280	
113-115	670	63	0- 2	270	
110-112	660	60		260	
107-109	650	56		250	
104-106	640	52		240	
101-103	630	50		230	
99-100	620	48		220	
96- 98	610	44		210	
93- 95	600	40		200	

SUBSCORES

Raw Score Sub 1	Raw Score Sub 2	Raw Score Sub 3	Scaled Score	Raw Score Sub 1	Raw Score Sub 2	Raw Score Sub 3	Scaled Score
67	78		99	26	33	31	59
66	77		98	25	32	30	58
65	76		97	24	31	29	57
64	75		96	23	30	28	56
63	74		95	22	29	27	55
62	73		94	21	27-28	26	54
61	71-72		93	20	26	24-25	53
60	70	65	92	19	25	23	52
59	69	64	91	18	24	22	51
58	68	63	90	17	23	21	50
57	67	62	89	16	22	20	49
56	66	61	88	15	21	19	48
55	65	60	87	14	20	18	47
54	64	59	86	13	18-19	17	46
53	62-63	58	85	12	17	16	45
52	61	57	84	11	16	15	44
51	60	56	83	10	15	14	43
50	59	55	82	9	14	13	42
49	58	54	81	8	13	12	41
48	57	53	80	7	12	11	40
47	56	52	79	6	11	10	39
46	54-55	51	78	4-5	9-10	9	38
45	53	50	77	3	8	8	37
44	52	49	76	2	7	7	36
43	51	48	75	1	6	6	35
42	50	47	74	0	5	5	34
41	49	45-46	73		4	4	33
40	48	44	72		3	2-3	32
39	47	43	71		2	1	31
38	45-46	42	70		0-1	0	30
36-37	44	41	69				29
35	43	40	68				28
34	42	39	67				27
33	41	38	66				26
32	40	37	65				25
31	39	36	64				24
30	38	35	63				23
29	36-37	34	62				22
28	35	33	61				21
27	34	32	60				20

*Percent scoring below the scaled score

GRE

BIOLOGY
REVIEW

I. Molecular Biology

1. Enzymes and Cellular Metabolism

A. ENZYME STRUCTURE AND FUNCTION

An enzyme is a protein that performs a metabolic function. Enzymes are catalysts and affect the rate but not the overall change in free energy of a chemical reaction. Most enzymes have molecular weights over 10,000 daltons.

All enzymes are complex proteins. Some proteins, for example, play only a structural role and have no metabolic function. Proteins are polypeptides, i.e., linear polymers with amino acids serving as the repeating units. Proteins are composed of one or more polypeptide chains and often contain a nonprotein moiety such as carbohydrate (as in glycoproteins), lipid (as in lipoproteins), or a metal ion (as in metalloproteins). Both noncovalent and covalent interactions can be important in stabilizing the structure of a polypeptide. Hydrogen bonds, ionic bonds, and hydrophobic interactions are important noncovalent factors. Disulfide bonds formed by cysteine side chains are a major form of covalent interaction. Each amino acid unit in a polypeptide can be any one of the twenty different naturally occurring amino acids. The sequence of amino acids in a polypeptide can, therefore, be very variable. This great variability enables proteins to assume many different structures and thereby perform many different and specific functions. The function of an enzyme is determined by its structure.

D-glucose

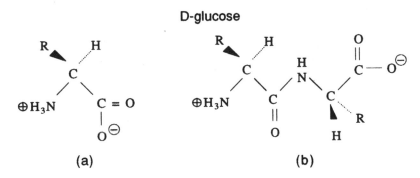

(a)   (b)

Figure 1 — (a) An Amino Acid; (b) A Peptide Bond

Individual amino acids (see Figure 1(a)) all possess an alpha-carbon that bonds to a corboxyl group ($-COO^-$), an alpha-amino group ($-NH_3^+$), a side group (R), and a hydrogen atom (H). At physiological pH the carboxyl group is negatively charged and the alpha-amino group is positively charged. With the exception of glycine (where R = H), the alpha-carbon of all amino acid is asymmetric (or chiral). All naturally occurring amino acids have L-stereochemistry.

In humans, not all the amino acids required for protein synthesis are made by the body. Eight amino acids are essential for man, i.e., valine (Val), leucine (Leu), isoleucine (Ile), lysine (Lys), phenylalanine (Phe), tryptophan (Try), threonine (Thr) and methionine (Met). Histidine (His) is required in infants but not in adults. The essential amino acids must come from dietary sources.

In a polypeptide chain the amino acid units are linked together by peptide bonds. The peptide bond is formed between the carboxyl group of one amino acid and the amino group of the next amino acid. As shown in Figure 2, the peptide bond has a partial double-bond character which restricts rotation about the C—N peptide bond. This restricted rotation limits the number of possible conformations obtainable by a polypeptide.

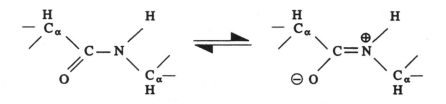

Figure 2 — Partial Double-Bond Character of Peptide Bond.

Proteins (and enzymes) can have four levels of structure. The primary structure is totally determined by the sequence of amino acids. The secondary structure is due to the formation of alpha-helices or beta-sheets. Both alpha-helices and beta-sheets arise from the periodic hydrogen bonding between peptides (see Figure 3).

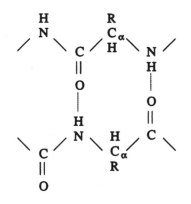

Figure 3 — Hydrogen Bonds Between Polypeptides

The tertiary structure of a protein is the complete three dimensional description of each atom in the protein. Tertiary structure is built from units of secondary structure linked together by turns in the polypeptide backbone. Quaternary structure results when protein subunits combine to form a larger structure. Hemoglobin is an example of a protein with quarternary structure since it contains two alpha-subunits (α) and two beta-subunits (β), i.e., $\alpha_2\beta_2$.

Proteins can be either water-soluble or bound to biological membranes. For water-soluble proteins, the hydrophilic amino acid side chains are generally found near the outer surface of the protein. The hydrophobic amino acid side chains are found in the interior of the protein where they are out of contact with water. Hydrophobic amino acid side chains (Cys, Val, Ile, Leu, Met, Trp, and Phe) are relatively nonpolar and do not interact with water molecules. Hydrophilic amino acid residues (Lys, Arg, His, Asp, Glu, Asn, and Gln) are polar and interact favorably with water.

Membrane proteins are associated with the lipid bilayer of biological membranes. There are two general categories of membrane proteins, i.e., intrinsic membrane proteins and extrinsic membrane proteins. Intrinsic membrane proteins are strongly associated with the biological membrane and can only be removed with a denaturing detergent such as sodium dodecylsulfate (SDS). Intrinsic membrane proteins often have a sequence containing numerous hydrophobic amino acid side chains that are strongly associated with the hydrophobic domain of the lipid bilayer. Extrinsic membrane proteins are only loosely associated with biological membranes and can be removed by alterations in the ionic strength or by a chelator such as EDTA.

PROBLEM

> Enzymes are
>
> a)   proteins.   b)   catalysts.
>
> c)   carbohydrates.   d)   Both a) and b).

Solution

d)   Enzymes are proteins that act as catalysts. Carbohydrates are not proteins, and therefore cannot be enzymes.

B. CONTROL OF ENZYME ACTIVITY

For the proper regulation of metabolism it is necessary for all organisms to exert considerable control over the location, amount and activity of enzymes. An enzyme (E) forms a complex with its substrate (S) at a well defined region called the active site. Subsequently, the enzyme-substrate complex (ES) is converted into a product (P) and the enzyme (E) is released.

$$E + S \rightleftharpoons ES \longrightarrow E + P$$

The activity of an enzyme can be regulated by changing or blocking the active site. A competitive inhibitor is one that reversibly binds to the active site and com-

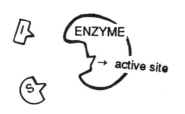

(a) Competitive inhibition
Both substrate (S) and inhibitor (I) compete for the same active site.

(b) Non-competitive inhibition.
The inhibitor (I) binds at a site different from the active site and does not prevent binding of the substrate. The inhibitor does decrease enzymatic activity.

Figure 4 — (a) Competitive Inhibition; (b) Noncompetitive Inhibition

A – 5

petes with the substrate for binding at this same locus (see Figure 4(a)). For competitive inhibition the activity of the enzymatically catalyzed reaction depends upon the concentration of both substrate and inhibitor.

Many enzymes have two (or more) alternative conformations, and the binding of ligands (substrates or other molecules) can influence which conformation the enzyme assumes. For allosteric proteins one conformation is enzymatically active and the other conformation is inactive. Allosteric enzymes are very important in the regulation of metabolic reactions (see below).

PROBLEM

Organisms can control all of the following EXCEPT

a)   location of enzymes.   b)   type of enzymes.

c)   amount of enzymes.   d)   activity of enzymes.

Solution

b)   Organisms control the activity of enzymes by changing or blocking active sites. Location and amount of enzymes may be controlled through feedback. Types of enzymes are determined genetically and cannot be altered.

C. FEEDBACK INHIBITION

Cells are required to synthesize an enormous number of essential compounds for their survival. Bacteria, although structurally simple, can use glucose to provide for their energy needs and to synthesize necessary organic components. The synthesis of these organic compounds is accomplished by specific metabolic pathways in which a precursor molecule is converted to a product by a series of enzyme-catalyzed reactions. The flow of metabolites in a metabolic pathway is often regulated by controlling the activity of key enzymes in the pathway. Usually, the first enzyme in a metabolic pathway is controlled by the end product of the pathway. This type of regulation is called feedback inhibition. In negative feedback inhibition (see Figure 5) the end product of a pathway inhibits the first enzyme in the pathway.

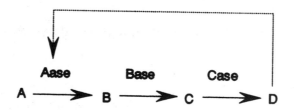

Figure 5 — Feedback Inhibition of a Metabolic Pathway.
Metabolite D inhibits Aase.

The first enzyme in a metabolic pathway is usually an allosteric enzyme. The end product of the pathway binds to a regulatory site on the enzyme, causing it to assume an inactive conformation. The regulatory site is different from the active site. In the case of negative feedback inhibition the pathway end product is usually a noncompetitive inhibitor. A noncompetitive inhibitor binds to the enzyme at a locus different from the active site (see Figure 4(b)). Furthermore, for noncompetitive inhibition the rate of the enzymatically catalyzed reaction depends only on the concentration of the inhibitor and not on the concentration of substrate. In summary, feedback regulation:

(a)   usually involves an allosteric enzyme,

(b)   is very rapid, and

(c)   can involve enzymatic inhibition or enzymatic activators (positive feedback inhibition).

Feedback regulation also provides an efficient method of conserving cellular energy and preventing the build-up of metabolic intermediates which, at high levels, could be toxic.

PROBLEM

> In feedback inhibition of metabolic pathways, which are controlled directly?
>
> | a)   End products | b)   Metabolites |
> |---|---|
> | c)   Enzymes | d)   Precursor molecules |

Solution

c)   In feedback inhibition the flow of metabolites is often regulated by controlling the activity of key enzymes in a pathway. Usually, the first enzyme in a metabolic pathway is controlled by the end product of the pathway.

D. GLYCOLYSIS

The sequence of energy producing catabolic reactions called glycolysis takes place in all living cells. Glycolysis results in the production of adenosine triphosphate (ATP), which provides cells with an efficient source of chemical energy. Catabolism is the chemical breakdown of food molecules to provide energy and building blocks for the synthesis of macromolecules. The first step in catabolism is the breakdown of macromolecular polymers to their monomeric units. Polysaccharides are broken down into sugars such as glucose. Glucose is further catabolized by the process of glycolysis.

ANAEROBIC

Glycolysis does not require the presence of oxygen. This metabolic pathway is very ancient, having evolved when the earth's atmosphere contained very little oxygen. In eucaryotes and many procaryotes, glycolysis results in the net production of two molecules of ATP, two molecules of NADPH, and two molecules of pyruvate per molecule of glucose:

$$\text{D-glucose} + 2HPO_4^{2-} + 2ADP + 2NAD \longrightarrow$$
$$2CH_3C\!-\!CO_2 + 2ATP + 2\,NADPH + 2H^+.$$

The ten steps in the glycolytic pathway are detailed below (see Figure 6). The first steps of glycolysis (steps 1-3) convert glucose to fructose 1,6-diphosphate (FDP) at the cost of two ATPs. Fructose and glucose are both six-carbon sugars (hexoses). The second stage results in the splitting (by aldolase) of fructose-1, 6-diphosphate into two three carbon sugars (trioses), i.e., dihydroxyacetone phosphate (DHAP) and D-glyceraldehyde 3-phosphate. DHAP is converted to D-glyceraldehyde 3-phosphate by triose phosphate isomerase. The third phase of glycolysis produces ATP by converting D-glyceraldehyde 3-phosphate into metabolites that can transfer phosphoryl groups to ADP. The pyruvate produced by glycolysis is a key branch-point metabolite. Under anaerobic conditions yeast converts pyruvate to ethanol and carbon dioxide and animals convert pyruvate to lactic acid. Some procaryotes utilize different pathways than the one described above. All function to generate ATP, $NADH_2^+$, and pyruvate, but some are less efficient than glycolysis.

AEROBIC

The key regulatory enzyme in glycolysis is phosphofructokinase. Its activity is inhibited by ATP, citrate and fatty acids and activated by ADP, AMP, cyclic AMP, and FDP. When cells that are undergoing anaerobic glycolysis are switched to aerobic conditions the rate of glycolysis rapidly drops. This is called the Pasteur effect. The effect is explained by the fact that under aerobic conditions the pyruvate produced by glycolysis can undergo further oxidation, via the citric acid cycle (see below). This results in the production of 18 ATP molecules per pyruvate. Thus, the energy needs of the cell are met with a considerably reduced rate of glycolysis. The decreased rate of glycolysis with higher levels of ATP is consistent with the fact that phosphofructokinase is inhibited by ATP.

PROBLEM

Glycolysis does not

a) occur in the cytoplasm.   b) require oxygen.

c) produce ATP.   d) break down glucose.

Solution

b) Glycolysis is the series of metabolic reactions by which glucose is converted to pyruvate (a 3-carbon sugar) with the concurrent formation of ATP. Glycolysis occurs in the cytoplasm of the cell and for this process the presence of oxygen is unnecessary.

E. KREBS (CITRIC ACID) CYCLE

For most eukaryotic cells and aerobic bacteria the pyruvate produced by glycolysis enters mitochondria and is completely oxidized to CO_2 and H_2O. This process, called cellular respiration, produces reducing power in the form of NADPH and $FADH_2$. NADPH and $FADH_2$ are then utilized by the electron transport system to produce ATP. Electron transport occurs in the mitochondria of eukaryotic cells and in the plasma membrane of aerobic bacteria.

In eukaryotic cells, the pyruvate produced by anaerobic glycolysis enters the mitochondrion and is decarboxylated to yield acetyl-CoA and NADH by pyruvate dehydrogenase. The primary function of the citric acid pathway is to oxidize acetyl groups to CO_2 and H_2O. The overall reaction is

$$acetyl\ CoA + 2H_2O + 3NAD^+ + FAD + GDP + HPO_4^- \longrightarrow$$
$$2CO_2 + 3NADH + FADH_2 + GTP + H^+ + CoA$$

A – 8

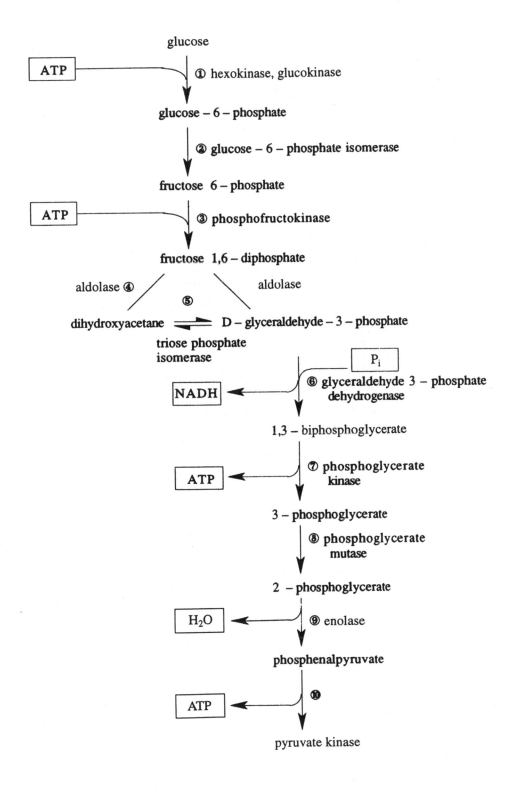

Figure 6 — Glycolysis

A – 9

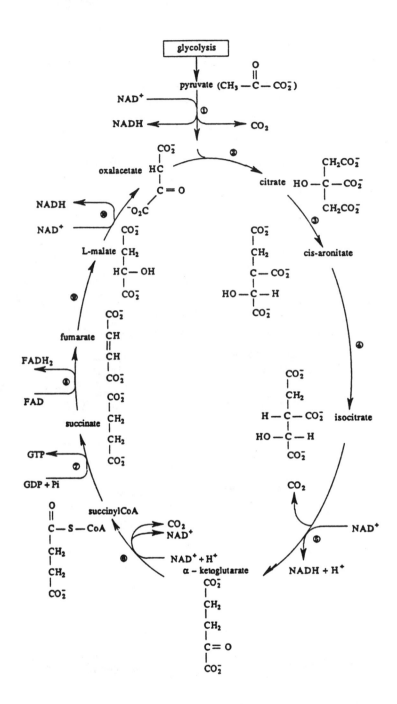

Figure 7 — The Krebs Cycle

(1) pyruvate carboxylase
(2) citrate synthetase
(3) aconitase
(4) aconitase
(5) isocitrate dehydrogenase
(6) alpha-ketoglutarate dehydrogenase
(7) succinyl CoA synthetase
(8) succinate dehydrogenase
(9) fumarase
(10) malate dehydrogenase

and the enzymatically catalyzed steps shown in Figure 7. The citric acid (or Krebs) cycle occurs in the mitochondrial matrix (see Figure 10).

It is noteworthy that molecular oxygen (O_2) does not enter the citric acid cycle. The additional oxygen atoms required for CO_2 production come from H_2O. The one GTP produced by step 7 is easily converted to ATP (GTP + ADP = GDP + ATP). The oxidation of one NADH molecule by the electron transport system produces three ATPs and, similarly, the oxidation of one $FADH_2$ produces two ATPs. The complete oxidation of glucose yields 38 ATPs.

PROBLEM

In the Krebs Cycle, all of the following occur EXCEPT

a) oxidation of succinate.

b) formation of $FADH_2$.

c) formation of NADH.

d) transformation of NADH to NAD.

Solution

d) NADH is converted to NAD during oxidative phosphorylation, which yields 3 ATP.

F. ELECTRON TRANSPORT CHAIN AND OXIDATIVE PHOSPHORYLATION

The last steps in catabolism, called oxidative phosphorylation, result in the efficient production of ATP. In these steps, electrons (e^-) are ultimately transferred to oxygen (see Figure 8) with the generation of ATP. Oxidative phosphorylation is dependent upon the structure of mitochondria.

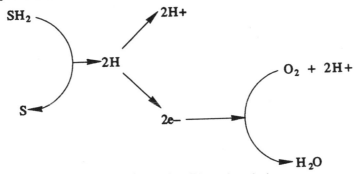

Figure 8 — Oxidative Phosphorylation.
Substrates (SH_2) contain H-atoms consisting of a proton (H^+) and an electron (e^-). The energy locked in H^+ and e^- is harnessed during oxidative phosphorylation to produce ATP.

ATP can be generated in two ways:

(1) by substrate level phosphorylation as indicated in glycolysis and the citric acid cycle (i.e., transfer of a high energy phosphoryl group to ADP to make ATP and,

(2)   by adding a phosphate group to ADP, i.e., the reverse hydrolysis of ATP

   ATP = ADP + Pi

Oxidative phosphorylation generates ATP by the second mechanism. The enzyme catalyzing this reaction is a proton (H^+) driven adenosine triphosphatase (H^+-ATPase). H^+-ATPase is a transmembrane protein embedded in the inner mitrochondrial membrane. The energy to drive ATP formation by H^+-ATPase comes from a proton gradient across the inner mitochondrial membrane. This proton gradient is generated by the movement of electrons down the respiratory chain. The respiratory chain is a series of membrane bound redox carriers with cytochrome oxidase being the terminal electron acceptor (see Figure 9).

The electrons that ultimately reach O_2 (and form H_2O) are initially carried by the hydrogen atoms of NADH and $FADH_2$ (from glycolysis and the citric acid cycle). These hydrogen atoms can be dissociated into an electron (e^-) and a proton (H^+). The electrons are transported by the respiratory chain and the protons are released into the aqueous medium. The released H^+ ions are translocated from the matrix space to the intermembrane space, and a pH gradient is established. The energy created by this gradient is trapped by the H^+-ATPase when the H^+ flow back into the matrix (see Figure 10). This process is known as chemiosmosis.

The process of oxidative phosphorylation illustrates a milestone in biochemistry, because it is an example of factorial metabolism — the coupling of metabolism with transport across a membrane.

PROBLEM

The ratio of ATP produced aerobically to anaerobically by the oxidation of one molecule of glucose is

a)   2:1.   b)   1:2.

c)   1:18.   d)   18:1.

Solution

d)   The aerobic production of ATP involves the Krebs (citric acid) cycle and the oxidation of glucose. The anaerobic production of ATP takes place during glycolysis. The citric acid cycle produces 34 ATPs and the oxidation of glucose produces two. This makes the total number of ATPs produced during aerobic processes 36. Glycolysis yields two ATPs. The net ratio of aerobic ATP to anerobic ATP is 36:2 which is equal to 18:1.

ATP yield from the complete oxidation of glucose

Reaction sequence	ATP yield per glucose
GLYCOLYSIS: GLUCOSE TO PYRUVATE (in the cytoplasm)	
Phosphoyrlation of glucose	- 1
Phosphorylation of fructose 6-phosphate	- 1
Dephosphorylation of 2 molecules of 1, 3-DPG	+ 2
Dephosphorylation of 2 molecules of phosphoenolpyruvate	+ 2

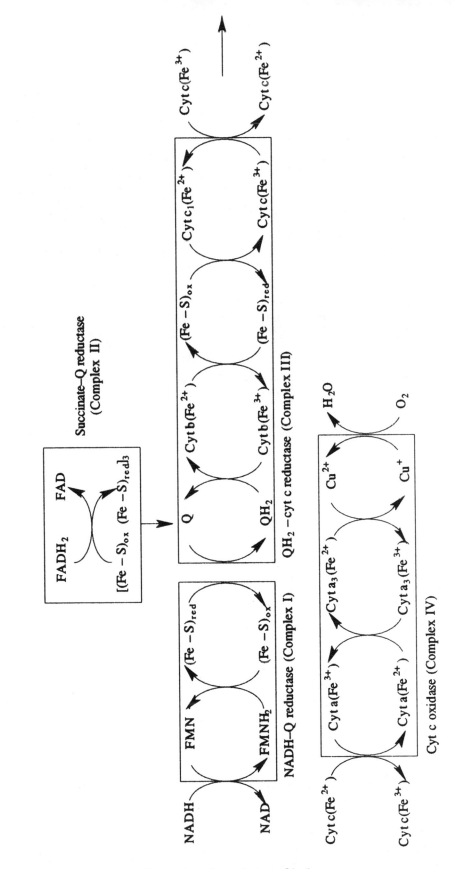

Figure 9 — Respiratory Chain

A – 13

2 NADH are formed in the oxidation of 2 molecules of
 glyceraldehyde 3-phosphate
CONVERSION OF PYRUVATE TO ACETYL CoA (inside mitochondria)
 2 NADH are formed

CITRIC ACID CYCLE (inside mitochondria)
 Formation of 2 molecules of guanosine triphosphate from 2
  molecules of succinyl CoA   +2
 6 NADH are formed in the oxidation of 2 molecules of
  succinate

OXIDATION PHOSPHORYLATION (inside mitochondria)
 2 NADH formed in glycolysis; each yields 2 ATP
  (not 3 ATP each, because of the cost of the shuttle)   + 4
 2 NADH formed in the oxidative decarboxylation of
  pyruvate; each yields 3 ATP   + 6
 2 FADH formed in the citric acid cycle;
  each yields 2 ATP   + 4
 6 NADH formed in the citric acid cycle;
  each yields 3 ATP   + 18

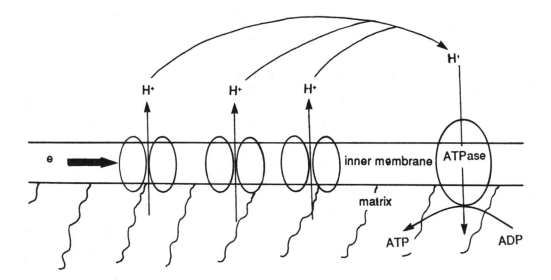

Figure 10 — Oxidative Phosphorylation.
As high energy electrons move down the respiratory chain, a proton
gradient is created. H^+-ATPase uses this gradient to make ATP.

2. DNA and Protein Synthesis

A. DNA STRUCTURE AND FUNCTION

The information determining the sequence of amino acids in a polypeptide resides in another linear polymer called deoxyribonucleic acid (DNA). DNA, and nucleic acids in general, are polynucleotides in which the repeating units are nucleotides. DNA contains the cell's genetic information. DNA has been called an "aperiodic crystal." Aperiodicity is a requirement for coding information and the "crystalline" structure of DNA provides considerable thermodynamic stability.

Nucleotides contain either a purine or a pyrimidine base that is attached to a five-carbon sugar-phosphate (see Figure 11(a)). In DNA the sugar is 2-deoxyribose (dRib) and in ribonucleic acids (RNA) the sugar is ribose (Rib). Four bases are found in both DNA and RNA. Adenine and guanine are the two purine bases present in DNA. Thymidine and cytosine are the two pyrimidines. RNA also contains adenine, guanine, and cytosine, but uracil substitutes for thymine. Nucleotides are linked to each other by phosphodiester bonds between the 3′ hydroxyl group of one nucleotide and the 5′ end of the next (see Figure 11(b)).

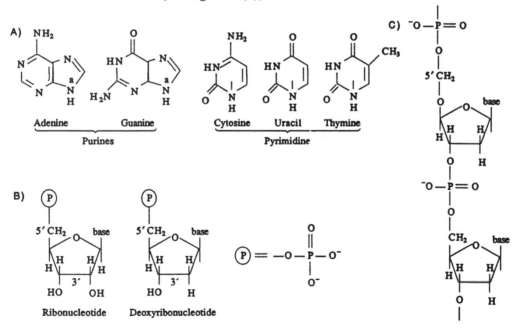

Figure 11 — Nucleic Acid Structure.
(a) Structure of bases in nucleic acids. Purines attach to ribose (or deoxyribose) at the 9-position and pyrimidines at the 1-position. (b) Structure of a ribonucleotide and deoxyribonucleotide. (c) Structure of single chain deoxyribonucleic acid.

The composition of DNA provides important clues with regard to both its function and structure. Chargaff found that:

(1)   the base composition of DNA in different tissues from the same species was identical,

(2)   the base composition of DNA from similar species was similar and the base composition of DNA from widely divergent species was dissimilar,

(3)   in DNA from all species the number of adenine bases equalled the number of thymine (A = T) bases and the number of guanine bases equalled the number of cytosine bases (G = C).

The Watson-Crick model of DNA provides an immediate and simple explanation for the fact that A = T and G = C. The Watson-Crick model proposes that DNA is a double stranded helix with the two strands running in opposite directions, i.e., antiparallel. The purines and pyrimidine bases are stacked on top of each other forming the inside of the double helix. The planes of the bases are essentially parallel to one another and perpendicular to the long axis of the DNA molecule. Adenine, on one strand, forms a specific base pair with thymine on the other, antiparallel strand. The AT base pair is stabilized by two hydrogen bonds. Guanine and cytosine also form a specific base pair (GC), but it is stabilized by three hydrogen bonds. The complementary base pairing for double stranded DNA is illustrated in Figure 12. In addition to hydrogen bonding and charge separation between phosphates along the helix, the structure of DNA is stabilized by hydrophobic interactions. The stacked bases are removed from contact with water.

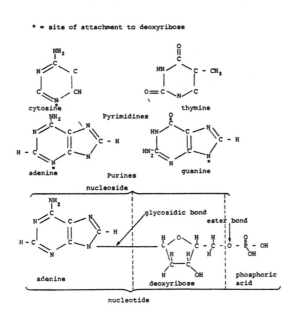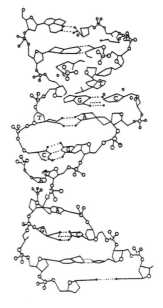

Figure 12 — Complementary Base Pairing in DNA.
(a) Four base pairs; (b) DNA double helix.

B. DNA AS TRANSMITTER OF GENETIC INFORMATION

The unique structure of DNA enables it to serve two template functions. One function is to serve as a template for its own replication. The other is to provide a template for its own replication.

DNA replication involves

(1)   strand separation, and

(2)   the synthesis, via DNA polymerase, of a complementary daughter strand from each parent strand.

The biochemical details of this process are complex, involving numerous proteins. DNA replication involves a replication form in which both nascent strands are synthesized in a 5′ to 3′ direction. This results in continuous synthesis for the leading strand but discontinuous synthesis for the lagging strand (see Figure 13). Discontinuous synthesis results in Okazaki fragments, which are later joined to form a continuous strand.

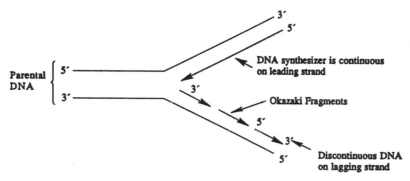

Figure 13 — DNA replication fork with Okazaki fragments.

C. PROTEIN SYNTHESIS: TRANSCRIPTION AND TRANSLATION

Transcription

A second function of DNA is to provide a template for the synthesis of messenger RNA (mRNA), ribosomal RNA (rRNA) and transfer RNA (tRNA), all of which are single stranded ribonucleic acid polymers. Messenger RNA is the intermediary polymer used to transmit information about the sequence of amino acids in protein from DNA. The synthesis of RNA from a DNA template is called transcription. In most cells the flow of genetic information is:

$$\text{DNA} \xrightarrow{\text{transcription}} \text{mRNA} \xrightarrow{\text{translation}} \text{protein}$$

In eukaryotic cells DNA is found almost exclusively in the cell nucleus, and most of the RNA is found in the cytoplasm where protein synthesis occurs. Both rRNA and tRNA are involved in the biosynthesis of proteins (translation) but do not carry any information coding for the sequence of amino acids in a protein. For RNA synthesis the bases on one strand of DNA (the "sense strand") are matched with complementary ribonucleotide triphosphates and polymerized into an RNA molecule.

In prokaryotes a multi-subunit RNA polymerase is responsible for the synthesis of mRNA, tRNA and rRNA from the DNA template. RNA polymerase cannot, however, utilize an RNA template (double stranded or single stranded) or an RNA/DNA hybrid. Transcription involves three steps:

(1)   initiation,

(2)   elongation, and

(3)   termination.

Initiation starts at specific sites on the DNA termed promoters. One subunit of RNA polymerase, the sigma subunit, recognizes these promoters. RNA chain elongation proceeds in a 5′ $\longrightarrow$ 3′ direction until a termination signal is encountered on the DNA template. Some termination signals require a protein called Rho and are referred to as Rho-dependent. After the polymerase encounters the signal, Rho designates it from the template. Other termination signals are Rho-independent.

Transcription in eukaryotic cells is mechanistically very similar to prokaryotic transcription but more complex. Eukaryotic cells contain three different polymerases, denoted RNA polymerase I, II, and III. Each transcribes a different set of genes. RNA polymerase I transcribes most of the ribosomal RNA. RNA polymerase II transcribes mRNA and most of the snRNP RNAs. RNA polymerase III transcribes small RNAs such as tRNA and the 5S ribosomal RNA. Each polymerase is comprised of ten or more subunits — considerably more complex than *E. coli* RNA polymerases. The three polymerases share common subunits, although each also has private subunits. Unlike *E. coli* RNA polymerase, eukaryotic polymerases cannot bind to promoter sequences. They require that other protein factors first bind the DNA.

The base sequence of an mRNA molecule (or of the DNA gene itself) encodes the information for the amino acid sequence of a protein. Since proteins vary greatly in molecular weight, it follows that mRNAs must be heterologous in length. Each amino acid in a polypeptide is determined by a three base codon. Three bases allow for 64 different codons. However, proteins are comprised of only twenty different amino acids. Therefore, either some codons are not used or more than one codon can code for an amino acid (i.e., the code is degenerate). In fact only three codons (UAA, UAG and UGA) do not code for amino acids. They provide the termination signal for translation and are referred to as "stop codons." The genetic code used in a wide variety of organisms is identical with only minor exceptions.

In prokaryotic cells, the sequence of bases coding for a given polypeptide (the gene or cistron) are continuous. In marked contrast, the coding regions for polypeptides in eukaryotic cells can be discontinuous. The newly synthesized mRNA (the primary transcript) in eukaryotic cells contains regions called introns that are not expressed in the synthesized protein product. These intron regions are removed and the regions that are expressed (exons) are spliced together to form the final functional mRNA molecule.

Translation

The process for translating an mRNA molecule into a polypeptide is very complex. Ribosomes and tRNA are of primary importance, but numerous other proteins are also required. Polypeptides are synthesized by sequentially adding amino acids to the carboxyl end of the growing polypeptide chain. The mRNA contains the codons specifying the sequence of amino acids but cannot itself associate with them. An intermediary RNA molecule called tRNA performs this task. At least one unique tRNA exists for each amino acid. The amino acid is activated and attached to its specific tRNA by an ATP driven process utilizing the enzyme aminoacyl tRNA synthetase. A tRNA carrying its cognate amino acid is said to be "charged."

$$\text{amino acid} + \text{ATP} + \text{tRNA} + H_2O \longrightarrow \text{aminoacyl-tRNA} + \text{AMP} + 2Pi$$

Aminoacyl-tRNA synthetase enzymes are very selective in attaching each amino acid to its cognate tRNA. Without this strict selectivity, the process of polypeptide synthesis would be compromised.

Each tRNA molecule has an anticodon consisting of three bases complementary to the bases of a codon on an mRNA molecule. The process of assembling a polypeptide from aminoacyl-tRNAs and an mRNA takes place on ribosomes. A ribosome is composed of both RNA and protein. It has two binding sites for charged tRNA: a P or peptidyl site and A or aminoacyl site. Polypeptide synthesis involves initiation, elongation, and termination.

Prokaryote Translation

Initiation results when the mRNA and the initiator tRNA (formylmethionyl-tRNA) bind to 30S ribosome subunit in either order. The charged initiator tRNA

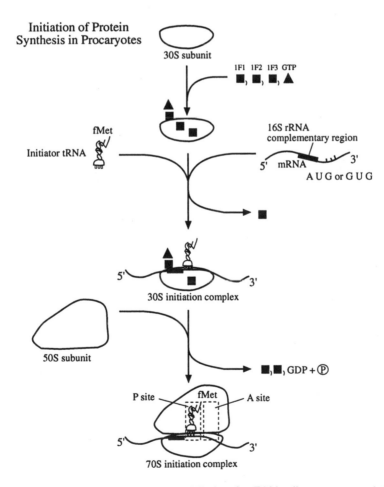

Figure 14 (a) — Protein Synthesis and Role of mRNA, ribosomes and tRNA

binds to the P side. (See Figure 14(a).) mRNA has a purine rich sequence called the ribosome binding site (RBS) followed by the AUG codon. The RBS promotes the binding of the mRNA to RNA of the 30S ribosomal subunit. The 50S ribosomal subunit then binds to the 30S subunit giving the 70S initiation complex. One GTP is spent to form the 70S initiatition complex. Many non-ribosomal proteins, termed initiation factors; are involved in establishing the 70S unit complex.

Elongation proceeds by binding of a aminoacyl-tRNA to the A site. Formation of a peptide bond between the amino group of incoming amionacyl-tRNA and the carboxyl group of the adjacent fmet-tRNA costs one GTP and causes the release of uncharged fmet-tRNA. The translocation of the peptidyl-tRNA from the A site to the P side moves the ribosome to the next codon and expends a second GTP. Each elongation step requires the hydrolysis of two GTPs.

Termination of polypeptide synthesis occurs when a stop codon is encountered. The stop codon is read by one of two protein release factors which cause release of the polypeptide chain from the ribosome.

As with transcription, eukaryotic and prokaryotic translation are similar but have several differences. For one, eukaryotic ribosomes are relatively large. Both the small subunit (40S) and the large subunit (60S) are larger than their prokaryotic equivalents. Together they form an 80S ribosome. Also, eukaryotic initiation is nota-

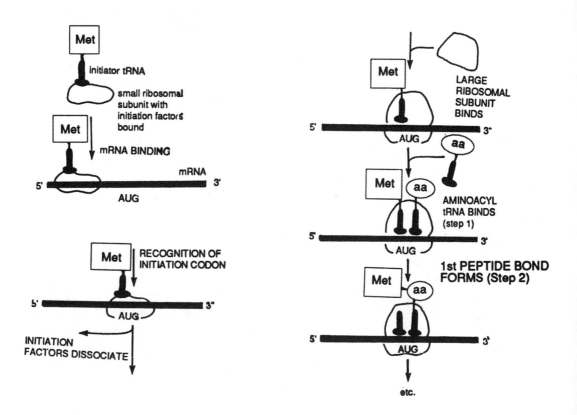

Figure 14 (b) — Initiation of Eukaryotic Protein Synthesis

bly more complex (see Figure 14(b)). It involves more initiation factors, many of which are themselves multi-subunited. The initiator tRNA is a special methoinine tRNA (termed Met-tRNA^Met), but it is not formylated. Also, the initiator tRNA always binds to the 40S subunit before the mRNA does, rather than in either order (as in prokaryotes). Furthermore, binding of the mRNA to the 40S subunit requires the hydrolysis of one ATP. With the exception of the mRNA of a few viruses, eukaryotic mRNA does not possess a ribosome binding site but instead requires a 5′ cap for efficient initiation (see Figure 14(c)). This cap is a 7-methylguanosine linked at its 5′ end to the 5′end of the mRNA via a triphosphate bridge. Finally, one release factor (RF) recognizes all three stop codons.

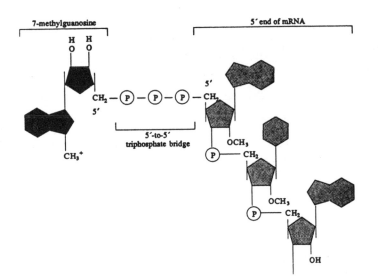

Figure 14 (c) — Structure of Eukaryotic mRNA Cap

PROBLEM

Nitrous acid converts cytosine to uracil by deamination. This type of conversion in one DNA strand would lead to a change in the complementary base in the other strand to

a)   adenine.

b)   cytosine.

c)   thymine.

d)   guanine.

Solution

a)   Cytosine normally binds to guanine and uracil normally binds to adenine. A conversion of cytosine to uracil would lead to a conversion of guanine to adenine in the complementary strand. Thus a CG to AU (or AT) event has occurred.

II. Microbiology

Knowledge of the properties and characteristics of viruses and bacteria and fungi is essential to understanding the cause of many diseases as well as the therapeutic approaches used to alleviate these diseases. Modern molecular approaches are proving to be particularly useful in this regard.

1. Viral Structure and Life History

Viruses are important because they:

(1) provide insight into evolution,

(2) are important tools for understanding the molecular biology of normal cells,

(3) are important in many diseases such as AIDS,

(4) they may provide powerful molecular tools for the many diseases.

Some viruses have been shown to cause cancer in animal models. Viruses have been obtained in homogeneous state and some viruses have been crystallized and their three dimensional structure fully determined by x-ray crystallography. Recently, much emphasis has been placed on the possibility of using viruses to transmit selected genetic information into eukaryotic cells to correct defective genes, i.e., gene therapy.

A. NUCLEIC ACID (DNA AND RNA) AND PROTEIN COMPONENTS

Viruses are the simplest supramolecular complex capable of initiating replication. They contain nucleic acids (either DNA or RNA but not both) with a surrounding protein coat called the capsid that protects the encapsulated nucleic acid from damage. Some animal viruses also have an envelope of lipid and glycoprotein surrounding the capsid. An extracellular viral particle (or virion) cannot independently reproduce itself and requires a host cell for this function. It accomplishes this task by diverting the biosynthetic machinery of the host cell to synthesize its own components. In some RNA viruses the viral mRNA preferentially bind to the host ribosomes. Hence, synthesis of viral proteins is favored over synthesis of host proteins.

There are four classes of RNA eukaryotic viruses that are distinguishable by the relationship of their viral RNA to their mRNA (see Table 1). mRNA is designated as (+) RNA and, its complementary RNA, as (-) RNA. Class I viruses contain (+) RNA which, in turn, is the template (+) mRNA. The parental RNA also functions as mRNA since it is capable of polymerizing ribonucleotides from an RNA template. For class I, as well as class II and class III viruses, this is accomplished by a viral RNA-directed-RNA polymerase (or RNA replicase).

Class II viruses contain (-) RNA which is transcribed into monocistronic mRNAs by a viral RNA transcriptase contained in the virion. One of these mRNAs codes for an RNA replicase which generates double stranded RNA from the parental (-) RNA. The RNA replicase also synthesizes progeny (-) RNA strands from the double stranded RNA.

Class III viruses contain double-stranded RNA, and the (-) strand provides the template for (+) mRNA. Class IV viruses are particularly important because the flow of genetic information is from (+) RNA to DNA and then back to DNA (see Table 1). Class IV viruses are called retroviruses and they code for a RNA-directed DNA polymerase (or reverse transcriptase). The HIV virus, which causes AIDS, is a retrovirus.

An important property of some RNA retroviruses (class IV RNA viruses) is their ability to induce tumors in animal models. Some DNA viruses (i.e., Simian virus 40 and polyoma virus) can also cause tumors. Cancer causing viruses (i.e., oncogenic viruses) transform their host cells by inserting their viral specific genes into the host chromosome. Normal cells stop multiplying when in close contact with one another, i.e., contact inhibition. Transformed cells no longer exhibit contact inhibition and, therefore, grow continuously.

Class	Viral RNA	Flow of Genetic Information
I	(+) RNA ⟶	(-) RNA ⟶ (+) mRNA
II	(-) RNA ⟶ (±) RNA ⟶ RNA	(+) mRNA
III	(±) RNA ⟶	(+) mRNA
IV	(+) RNA ⟶	(-) DNA ⟶ (±) DNA ⟶ (+) mRNA

Table 1 — Classes of RNA Viruses

In the DNA or RNA viruses the DNA provides the template for the synthesis of mRNA molecules which preferentially use the host ribosomes to synthesize viral specific proteins and the enzymes necessary for viral DNA synthesis.

Viruses contain very few genes (between 3 and 240) and, therefore, construct much of their molecular machinery from identical protein subunits. For example, the protein coat of the TMV (Tobacco mosaic virus), which contains only 6 genes, is made up of 2,130 identical protein subunits. Coat protein subunits usually arrange themselves into either rods or spheres or a combination of these shapes.

PROBLEM

> Viruses differ from other living organisms because
>
> a)   viruses possess no bounding membrane.
>
> b)   viruses lack all metabolic machinery.
>
> c)   viruses lack all reproductive machinery.
>
> d)   All of the above.

d)   Viruses differ from living things in many ways. They do not have any membranes because they have no need to take in or expel material. Viruses lack all metabolic machinery and do not produce ATP because they do not perform energy-requiring processes. Viruses do possess either DNA or RNA, but cannot independently reproduce. They must rely on host cells for reproductive machinery and components.

B. BACTERIOPHAGE: STRUCTURE, FUNCTION AND LIFE CYCLE

Bacteriophages (or phages) are bacterial viruses that have either RNA or DNA genomes. Figure 15 illustrates the structure of a typical bacteriophage which has a head, tail and tail fibers. Infection of a bacterium (1 to 10 μm in length) begins when a phage (100-300 nm) attaches its tail fibers to a surface receptor on the bacterium. The DNA, which is tightly packed in the phage head, is subsequently injected through the cell wall and the cell membrane into the bacterium (see Figure 16). In only a few minutes all the metabolism of the infected bacterium is directed towards the synthesis of a new phage particles. About 30 minutes after infection the bacterium undergoes lysis and hundreds of completed bacteriophages are released.

The complex coordination of phage life cycle is a result of different phage genes being expressed at different times. The early phage genes are expressed before phage DNA synthesis begins. For many phages some of these gene products shut down the biosynthetic capacity of the bacterium. One of early phage gene products that helps to shut down the metabolism of the host cell is a nuclease specific for bacterial DNA but not the phage DNA.

The late gene products are associated with the synthesis of viral DNA, capsid formation, packaging of the viral DNA into preformed heads, and the synthesis of lysozyme to degrade the bacterial cell wall thus causing lysis. Not all phages cause immediate lysis of the infected bacterium. In some cases the phage DNA incorporates itself into the bacterial chromosome and is only replicated when the host chro-

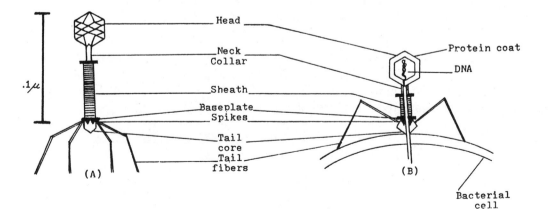

Figure 15 — The Structure of a Typical Bacteriophage

A – 24

mosome is replicated. This process is called lysogeny. Viruses that exhibit this state are called temperate or moderate viruses.

The viral DNA incorporated into the host chromosome is called a provirus, or prophage. In the case of bacteriophages this prophage can be induced to become virulent and lyse its host bacterium. The resulting infectious phage often carry small amounts of bacterial chromosome which can be transferred to newly infected bacteria. The process whereby DNA is transferred from one bacteria to another by a phage is called transduction.

PROBLEM

Moderate viruses may

a)   replace DNA only when the host replicates.

b)   induce tumors.

c)   cause immediate lysis of infected bacteria.

d)   have both DNA and RNA.

Solution

a)   In moderate viruses the phage DNA is incorporated directly into the host chromosome, and thus replicates only when the host does. RNA retroviruses may induce tumors. Most viruses, with the exception of moderate ones, cause immediate lysis of infected bacteria. Viruses may contain either DNA or RNA, not both.

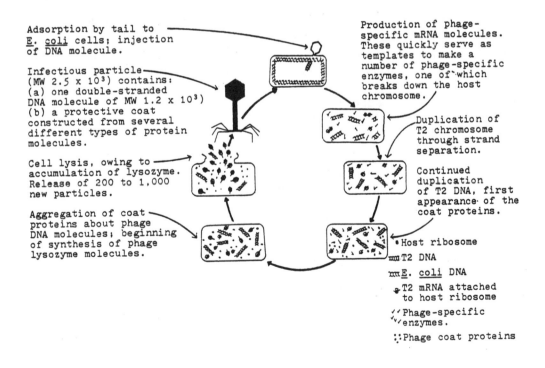

Figure 16 — The Life Cycle of a Bacteriophage

2. Prokaryotic Cells

A. CELL STRUCTURE AND FUNCTION

All living organisms have a cell structure that can be classified as either eukaryotic or prokaryotic. The prokaryotic cell is distinguished by the absence of a membrane bound nucleus. Prokaryotic cells (1-10 μm in length) are much smaller than eukaryotic cells (10-100 μm in length). Both types of cells can have flagella for motility but these structures are relatively simple in prokaryotic cells. Prokaryotic cells include eubacteria, archaebacteria, blue-green algae, spirochetes, rickettsia, and mycoplasma.

In prokaryotic cells such as bacteria (see Figures 17(a) and 17(c)) the single chromosome is a large single circular double stranded DNA molecule that is not separated from the cytoplasm by a nuclear membrane. The prokaryotic chromosome lies in the nuclear zone. Prokaryotic cells contain less DNA than more advanced eukaryotic cells. Furthermore, in prokaryotic cells the processes of transcription and translation occur simultaneously.

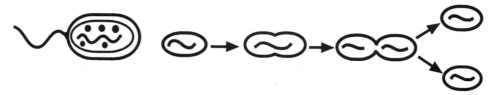

Figure 17 — (a) The Structure of a Bacterium; (b) Bacterial Reproduction

The only two membraneous structures in prokaryotic cells are the plasma membrane and the outer membrane which are lipid bilayers with associated intrinsic and

	Prokaryotic	Eukaryotic
DNA		
	No nuclear membrane	DNA contained in nucleus with surrounding nuclear membrane
	No histones	DNA associated with histones
membranes		
	plasma membrane	plasma membrane and other
	outer membrane	membranous organelles such as mitochondria, endoplasmic reticulum, Golgi complex, peroxisomes, and lysosomes

Table 2 — Differences Between Prokaryotic and Eukaryotic Cell

extrinsic membrane proteins. In contrast, eukaryotic cells contain membraneous organelles such a mitochondria and a membrane bound nucleus (see Table 2). Ribosomes and cytosol are present in both cell types. The cytosol contains water soluble enzymes, metabolic intermediates and inorganic ions.

PROBLEM

Which of the following structures is found in a bacterial cell?

a) Golgi apparatus

b) Nuclear membrane

c) Ribosomes

d) Mitochondrion

Solution

c) Unlike eukaryotes, bacterial cells lack Golgi apparatus, endoplasmic reticulum, mitochondria and a nuclear membrane. The lack of an endoplasmic reticulum means that the ribosomes are free (not bound to rough endoplasmic reticulum).

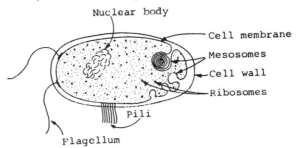

Figure 17 (c) — Typical bacterial cell

B. BACTERIAL LIFE HISTORY AND PHYSIOLOGICAL CHARACTERISTICS

Bacterial growth is the result of cellular division. For bacteria the process of cell division is straightforward, i.e., the cell doubles in size and then divides in two. This type of growth is exponential. The time it takes for a doubling of a number of bacteria is called the mean generation time, which is typically less than one hour. During cellular division, the nuclear body first replicates (see Figure 17(b)), the resulting homologous chromosomes separate, a cross wall forms between the chromosomes, the cell divides and separates.

A considerable store of fundamental information about molecular biology has come from microbiological studies. Strong evidence for DNA being the genetic material comes from studies in which DNA from one bacterial strain is transferred to another. The transfer of the donor DNA is accompanied by the transfer of some donor phenotype(s) (such as virulence) to the recipient strain. DNA can be transferred from one bacterium to another by transformation, conjugation, and transduction.

Transduction is the transfer of a fragment of the bacterial genome from a donor strain to a recipient strain of bacteria using a bacteriophage as the vector. In transformation, a DNA fragment isolated from a donor strain is directly taken up by the recipient strain. In bacterial conjugation the male and female cells adhere, chromosome or episome replication occurs in the male cell and one copy is injected into the female cell. No transfer of DNA occurs from the female to the male bacterium.

Bacterial cells are noted for their metabolic versatility and their highly efficient regulation of metabolic and catabolic activities.

Under adverse conditions some bacteria shift from their normal vegetative state to a dormant state, i.e., they undergo sporogenesis. Sporogenesis is a form of cellular differentiation resulting in a metabolically dormant structure such as an endospore which are formed by the Gram-positive bacteria of the genera Bacillus and Clostridium. Under favorable conditions the spore can undergo germination to return the cells to a vegetative state.

PROBLEM

In transduction, a

a)   male chromosome is injected into a female cell.

b)   female chromosome is injected into a male cell.

c)   bacteriophage transfers genetic material between bacteria.

d)   DNA fragment from a donor strain is directly taken up by a recipient strain of bacteria.

Solution

c)   a) refers to bacterial conjugation. b) is an impossibility. d) refers to transformation.

3. Fungi

A. MAJOR STRUCTURAL TYPES

All fungi are eukaryotic organisms having at least one nucleus with a nuclear membrane, an endoplasmic reticulum and mitochondria. They lack chloroplasts and chlorophyll. Fungi are spore-bearing organisms with absorptive nutrition. They reproduce sexually and asexually. The primitive plant body formed by fungi is called a thallus but it has no true roots, leaves, stems, or vascular tissue. Although there are over 100,000 species of fungi, only about 100 are important in human diseases. Ringworm (dermatophytoses) is, however, a very common infectious disease caused by a fungus. Fungi are further divided into yeasts and molds.

B. GENERAL LIFE HISTORY AND PHYSIOLOGY

Yeasts are unicellular forms of fungi with a spherical shape (3 - 15 μm in diameter). Yeast reproduce by budding or by binary fission. Molds grow in multicellular tubular colonies called hyphae. During growth these hyphae bunch together to form a mycelium.

Fungi and myxobacterium have the ability to form "fruiting bodies" which are an effective adaptation to a land environment. The fruiting bodies serve to disperse spores or cysts. Asexual spores formed from the body (or thallus) of a fungus are called thallospores and asexual spores formed from specialized structures are called conidia.

Fungi are also capable of sexual reproduction. In all sexual reproduction there is an alteration in chromosome number. At fertilization two haploid nuclei join to form a diploid nucleus. The diploid cells eventually give rise to haploid cells by meiosis. In lower fungi the visible organism often exists primarily in the haploid state (haplophase) and only transiently in the diploid state (diplophase). Sexual reproduction in fungi follows this sequence:

(1) compatible haploid nuclei are brought together in the same cell of the thallus;

(2) two genetically different nuclei fuse to form a diploid nucleus;

(3) meiosis occurs to form haploid nuclei which develop into sexual spores.

PROBLEM

All of the following are true of fungi EXCEPT they do not

a) reproduce sexually.   b) reproduce asexually.

c) produce spores.   d) produce seeds.

Solution

d)   Fungi may produce spores sexually or asexually. Seeds are produced by plants, not fungi.

III. Generalized Eukaryotic Cell

1. Plasma Membrane: Structure and Function

A. COMPOSITION, STRUCTURE, AND MOVEMENT OF PROTEINS AND LIPIDS

The plasma membrane surrounds the cell and separates the inside (intercellular) from the outside (extracellular) of the cell. Structurally, the plasma membrane is composed of a lipid bilayer and membrane bound proteins (see Figure 18(a)). Lipid bilayers are also present in other organelles of eukaryotic cells, e.g., mitochondria and endoplasmic reticulum. Phospholipid (PL) molecules are the primary lipid constituents of most lipid bilayers (see Figure 18(b)). Cholesterol and glycolipids are also present in many biological membranes. PL molecules are amphipathic molecules, i.e., they have a polar head group and two nonpolar hydrocarbon "tails." PL molecules self aggregate to form a lipid bilayer because in this molecular arrangement their head groups remain in contact with water and their tails are removed from contact with water. The lipid bilayer structure has two important properties:

(1) It is a permeability barrier for charged molecules. Charged, hydrophilic molecules, cannot move through the lipid bilayer because they would have to give up thermodynamically favorable interactions with polar water molecules. Water molecules, although permeable to the bilayer, have an extremely low concentration in the hydrophobic domain of the bilayer. For charged molecules to pass through the bilayer specific transport proteins must be present. Hydrophobic molecules such as O_2 as well as small uncharged polar molecules (H_2O, CO_2 and urea) are, however, membrane permeable.

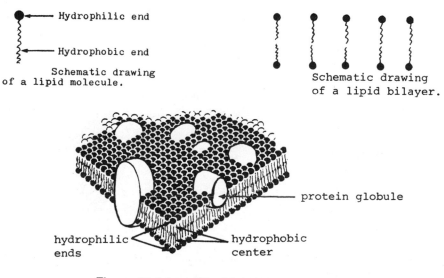

Figure 18 (a) — The Lipid Bilayer

(2)   The individual PL molecules move rapidly in the plane of the lipid bilayer. Proteins associated with the lipid bilayer also can have rapid lateral motion. The bilayer, therefore, acts as two-dimensional fluid and this fluidity is necessary for diffusion of membrane bound enzymes and receptor molecules.

The plasma membrane and the membranes of other subcellular organelles are asymmetric with respect to the head groups found on the inner and outer monolayers. In addition, the proteins associated with biological membranes are also embedded in the bilayer in an asymmetric manner. For example, glycoproteins (as well as glycolipids) in the plasma membrane usually have their carbohydrate moities facing the extracellular space.

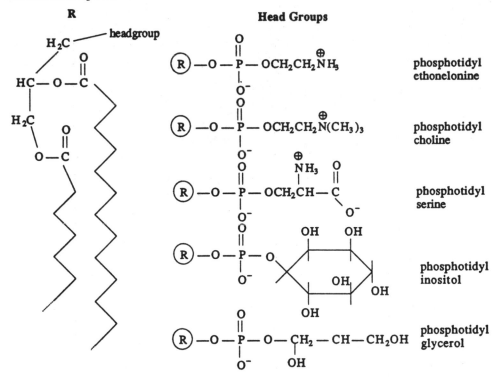

Figure 18 (b) — The Structure of PL Molecules

PROBLEM

Cell membranes are generally composed of a

a)   double layer of phospholipids with proteins dispersed throughout the membrane.

b)   double layer of phosphoproteins with glucose dispersed throughout the membrane.

c)   double layer of nucleic acids.

d)   double layer of proteins with phospholipids dispersed throughout the membrane.

a)   The plasma membrane contains about 40 percent lipid and 60 percent protein by weight although there is considerable variation between different cell types. The lipid molecules of the plasma membrane are polar. One end is hydrophobic, the other end is hydrophilic. The lipid molecules are arranged in two layers so that the hydrophobic ends are near each other and the hydrophilic ends face outside. The individual lipid molecules can move laterally, so the bilayer is actually fluid and flexible. Protein molecules of the plasma membrane may be arranged at various sites and imbedded to different degrees. The highly selective permeability of the plasma membrane is dependent upon the specific types and amounts of proteins and lipids present.

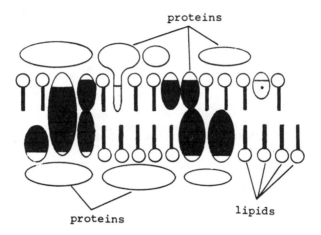

Figure 18 (c)

B. MEMBRANE TRANSPORT — PASSIVE AND ACTIVE TRANSPORT

The transport of charged biomolecules across biological membranes is dependent upon transport molecules most of which are proteins. Transport proteins can achieve great specificity and permit only one class of molecules to be transported (e.g., sugars or amino acids) or one specific molecule in a class. Some transport molecules simply permit a solute to reversibly diffuse from one side of the membrane to the other. This process is called passive transport. The direction of transport for a solute will be influenced by a concentration gradient across the membrane (i.e., from high to low concentration), as well as the electric charge across the membrane, i.e., the membrane potential.

The combination of the chemical and electrical gradient is called the electrochemical gradient. Plasma membranes are more negatively charged on the cytoplasmic side than the extracellular side and this hinders the passive transport of positively charged ions. In some cases, the passive transport of a solute is through an aqueous pore created by the transport protein. This type of transport protein is called a channel protein or porin. Transport of a solute through a channel protein is not saturable (see Figure 19). In other cases, the solute molecule binds to a transport protein which then facilitates its translocation to the other side of the membrane. This process, called facilitated diffusion, is similar to a substrate binding to the active site of an enzyme, which is a saturable process.

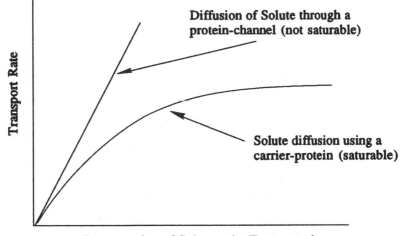

Figure 19 — Passive Transport

In order to transport a solute against an unfavorable electrochemical gradient it is necessary to expend energy, usually ATP. This type of transport is called active transport.

PROBLEM

Porins are important in

a)   facilitated diffusion.   b)   substrate binding.

c)   saturable processes only.   d)   passive transport.

Solution

d)   Passive transport may occur when a solute passes through an aqueous pore created by a channel protein, or porin. Transport of a solute through a porin is not saturable. Facilitated diffusion occurs when a substrate binds to the active site of an enzyme.

C. THE Na⁺ - K⁺ PUMP AND MEMBRANE POTENTIAL

The membrane potential of plasma membranes is generated by two important transport proteins: the Na^+, K^+-ATPase and the K^+-channel. Na^+,K^+-ATPase uses ATP to pump Na^+ ions out of the cell and K^+ ions into the cell (see Figure 20). This is an example of active transport because the concentration of Na^+ outside the cell is higher than inside. The reverse is true for K^+ ions.

The K^+-channel permits K^+ ions to diffuse out of the cell and this loss of positive ions causes the inside of the cell to become more negative than the outside. Eventually increasing negative charge inside the cell retards the outflow of K^+ ions (i.e., the negative charge inside the cell attracts the positively charged K^+ ions) and equilibrium is achieved when the inflow of K^+ ions equals the outflow. The end result is a plasma membrane potential between - 20 and - 70 mV depending on the cell type.

PROBLEM

In most cells, the concentration of Na^+ is _____ in the cell than outside it, because of _____.

a) higher ... active transport

b) higher ... passive transport

c) lower ... active transport

d) lower ... passive transport

Solution

c) The $Na^+ - K^+$ pump is a form of active transport where energy is used to pump Na^+ ions out of the cell and K^+ ions into the cell.

D. OSMOTIC EFFECTS AND CELL VOLUME

The Na^+,K^+–ATPase along with the K^+-channel primarily controls the level of ions inside the cell. Thus, these elements are also the main controllers of intracellular osmotic pressure and cellular volume. The charged macromolecules inside the cell require counterbalancing ions like Na^+, K^+ and Cl^-. This creates an osmotic pressure causing cellular swelling from the influx of water. Counterbalancing this intracellular osmotic pressure is the osmotic pressure caused by the ions in the extracellular fluid, which are primarily Na^+ and Cl^-. These ions tend to move down their concentration gradient and into the cell. Were it not for the Na^+, K^+-ATPase pumping Na^+ out and consequently preventing Cl^- from leaking in (by maintaining a negative membrane potential) the cell would swell and burst.

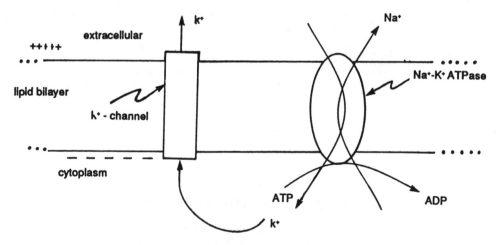

Figure 20 — Membrane Potential and Active Transport

PROBLEM

Which ions must be pumped out of the cell to prevent its rupture?

a) Na^+

b) K^+

c) Cl^-

d) All of the above.

Solution

d)   The macromolecules inside the cell require counter balancing ions outside the cell, such as Na^+, K^+, and Cl^-. These counterbalancing ions create an osmotic balance that prevents the influx of water into the cell, thus preventing cell rupture.

E. MEMBRANE RECEPTORS

The plasma membrane contains a wide variety of protein receptors to which ligands can bind. A major function of these receptors is to receive signals from the extracellular environment. Neurotransmitters and hormones are examples of ligands that bind to protein receptors on target cells and influence the behavior of the cell.

Receptor proteins (see Figure 21) are usually transmembrane proteins that have an extracellular domain where signals are received, a hydrophobic domain going through the lipid bilayer, and a cytoplasmic signal transducing domain. The initial binding of the signal molecule (i.e., the first message) alters the conformation of protein receptor and this activates an intracellular signal pathway. The intracellular signal is often transmitted by a second class of small and rapidly diffusable molecules called second messengers. Calcium and cyclic AMP (see Figure 22) are two important second messengers. Alternately, the cytoplasmic domain of the receptor may have protein kinase activity which is activated upon ligand binding. Thus, the receptor molecule itself can activate or inactivate certain intracellular substrates via phosphorylation.

The second messengers can then regulate a wide variety of biochemical and physiological processes. For example, the release of fatty acids from adipocytes (fat cells) is regulated by catecholamines. When catecholamines (the first message) bind to a surface receptor on the adipocyte plasma membrane this causes the receptor molecule to activate an adenlyate cyclase enzyme which catalyzes the production of intracellular cAMP from ATP (see Figure 21). The increased cAMP (the second messenger) activates a protein kinase which, in turn, phosphorylates the hormone

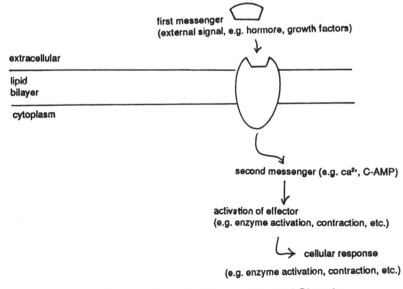

first messenger
(external signal, e.g. hormone, growth factors)

extracellular

lipid
bilayer

cytoplasm

second messenger (e.g. ca^{2+}, C-AMP)

activation of effector
(e.g. enzyme activation, contraction, etc.)

cellular response
(e.g. enzyme activation, contraction, etc.)

Figure 21 — Cell Receptors and Signals

sensitive lipase enzyme. The phosphorylation activates hormone sensitive lipase, and it then hydrolyzes triglyceride into fatty acids (see Figure 23).

Some receptors serve to bind very large molecules that are brought into the cell as a source of nutrients. Most cells, for example, have receptors for low density lipoprotein (LDL) which is a very large lipid-protein complex. The LDL receptor is called the apoB,E receptor and it recognizes the apoB protein moiety of LDL. After binding to the apoB,E receptor, LDL is internalized by endocytosis (see Figure 24) and provides the cell with an external source of cholesterol and other lipids.

PROBLEM

Ligands bind to			
a)	target cells.	b)	neurotransmitters.
c)	protein receptors.	d)	hormones.

Solution

c) Neurotransmitters and hormones are examples of ligands that bind to protein receptors on target cells.

F. EXOCYTOSIS AND ENDOCYTOSIS

Macromolecules are too large to be transported through the plasma membrane by specific transport proteins. The transport of these macromolecules is accomplished by the processes of exocytosis and endocytosis. In exocytosis an intracellular vesicle is transported to the plasma membrane where it fuses with the plasma membrane. The fusion process releases the contents of the vesicle to the extracellular space. Endocytosis is essentially the reverse of this process.

PROBLEM

Which is most likely to be transported by exocytosis?			
a)	Urea	b)	Lipoprotein
c)	Na+	d)	Hormones

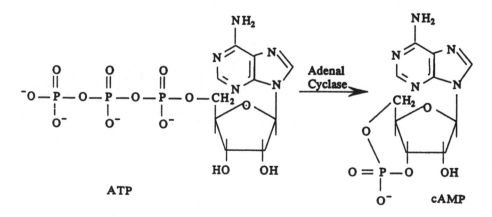

Figure 22 — Structure of cyclic AMP

b) Exocytosis is used for the transport of the largest molecules, macromolecules. Lipoproteins have larger molecules than any of the other materials listed.

G. CELLULAR ADHESION

Cells in a tissue are in contact with a network of molecules called the extracellular matrix. This matrix plays a major role in promoting cell-cell adhesion. In addition, cells that are in direct contact with each other can form cell junctions between specialized regions of their plasma membranes.

2. Membrane-Bound Organelles

A. MITOCHONDRIA

Mitochondria are the primary site for the production of ATP. Mitochondria appear to be associated with the microtubules of the cytoskeleton. Mitochondria contain their own genome. Proteins from both the mitochondrial genome and the nuclear DNA are required for mitochondrial replication. In mammals, mitochondrial genes are maternally inherited.

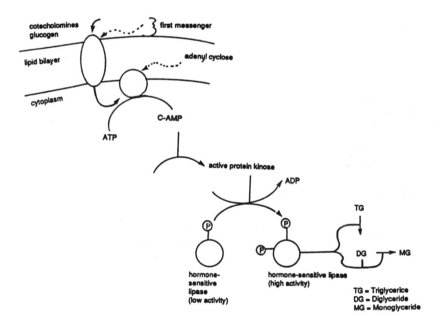

Figure 23 — Hormone Sensitive Lipase

PROBLEM

> Which of the following is responsible for the majority of cellular ATP production?
>
> a) Endoplasmic reticulum   b) Lysosomes
>
> c) Golgi apparatus   d) Mitochondria

Solution

d)   The mitochondria are responsible for 95 percent of all ATP produced in the cell. For this reason the mitochondria are commonly referred to as the "powerhouse" of the cell. Mitochondria are membrane-bound organelles that are distributed throughout the cell. Mitochondria tend to be most concentrated in regions which require large amounts of energy, such as muscle.

B. ENDOPLASMIC RETICULUM

Eukaryotic cells contain an endoplasmic reticulum (ER) which represents about one-half of all the cellular membrane (see Figure 25 on page 40). Prokaryotic cells do not contain an ER. The ER is a primary site for lipid biosynthesis. It also serves as a delivery site for proteins that are to be excreted from the cell or to be delivered to other intracellular organelles. Structurally, the ER is thought to be a single, highly convoluted, membrane sheet enclosing a single space called the ER lumen. The cytoplasm is separated from the ER lumen by a single membrane (the ER membrane). The ER membrane is continuous with the outer nuclear membrane.

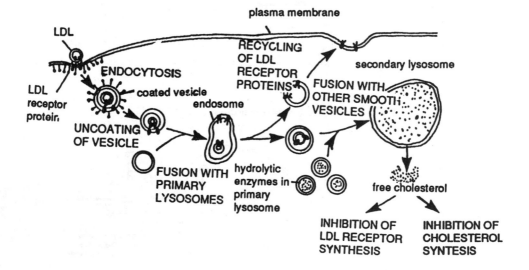

Figure 24 — LDL Uptake by Receptor Mediated Endocytosis

Proteins that are to be secreted by the cell or sent to other intracellular organelles are delivered to the lumen of the rough ER. Rough ER has attached ribosomes. The polypeptides being translated on these ribosomes are transported from the cytoplasmic side of the ER membrane into the ER lumen. The ribosomes attached to the rough ER are identical to ribosomes that are not attached to the rough ER. Attachment of some ribosomes to the ER is directed by small sequence of amino acids at the amino end of the polypeptide being translated (i.e., the signal sequence). The signal sequence is removed once the polypeptide has been delivered to the ER lumen. Many polypeptides undergo "core glycosylation" in the ER. Smooth ER has no attached ribosomes.Transport vesicles carrying newly synthesized lipids and proteins bud off the smooth ER for transport to the Golgi apparatus. Many important detoxification and lipid metabolism reactions take place on the smooth ER.

PROBLEM

Prokaryotic cells contain

a) an endoplasmic reticulum.

b) ribosomes.

c) a nuclear membrane.

d) both an endoplasmic reticulum and ribosomes.

Solution

b) Prokaryotic cells contain ribosomes, but they do not have an endoplasmic reticulum or nuclear membrane.

C. GOLGI APPARATUS

The Golgi apparatus is composed of flattened membrane-bound sacs surrounded by a swarm of smaller membrane-bound vesicles called "coated vesicles." Proteins associated with the ER are transported to the Golgi apparatus by these small vesicles which are coated with a protein called clathrin. Glycoproteins are received by the convex side of Golgi apparatus and undergo "terminal glycosylation" in the Golgi apparatus. The sugar moieties of glycoproteins are extensively modified by enzymes in the Golgi apparatus and the modified glycoproteins are sorted and delivered to either other organelles or to the plasma membrane where they can be secreted into the extracellular fluid. It should be noted that the luminal side of both the ER and the Golgi apparatus correspond to the extracellular side of the plasma membrane. Furthermore, two membranes separate the lumen of the ER from the lumen of the Golgi apparatus.

PROBLEM

The Golgi apparatus primarily functions in

a) packaging protein for secretion.

b) synthesizing protein for secretion.

c) packaging protein for hydrolysis.

d) synthesizing protein for hydrolysis.

Solution

a)   The Golgi apparatus is an organelle that is responsible only for the packaging of protein for secretion.

D. LYSOSOMES

Lysosomes (250 - 750 nm in diameter) are membrane-bound vesicles found in the cytoplasm. These organelles are responsible for the intracellular digestion of macromolecules. A primary lysosome is a newly synthesized vesicle and contains a wide variety of hydrolytic enzymes (all are acid hydrolases) such as proteases phospholipases, and nucleases. These hydrolytic enzymes are almost all glycoproteins and have optimal enzymatic activities at pH 5.0, the pH inside the lysosomes. The primary lysosome arises from budding of specialized regions of the Golgi apparatus. A secondary lysosome is a lysosome that is actively digesting a substrate (see Figure 24). The substrate can be a foreign pathogen such as a bacterium or an endogenous macromolecule such as LDL.

PROBLEM

Lysosomes contain

a)   glycogen stores.

b)   lipids.

c)   acid hydrolases.

d)   ATP.

Solution

c)   Lysosomes are cell organelles found in the cytoplasm. They are vesicles surrounded by a single membrane and contain enzymes, mostly acid hydrolases. These hydrolases are released when the membrane bursts, permitted the digestion of cellu-

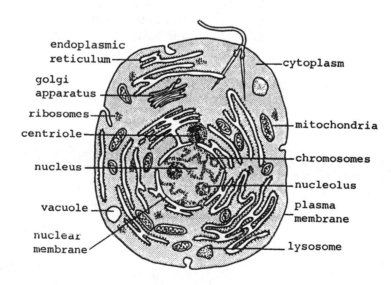

Figure 25 — The Generalized Eukaryotic Cell

lar structures and macromolecules. During the normal metabolism of the cell, enzyme release is carefully controlled by mechanisms which are still very poorly understood.

3. Cytoskeleton

A. MICROFILAMENTS, MICROTUBULES AND INTERMEDIATE FILAMENTS

The cytoskeleton of eukaryotic cells plays a key role in maintaining the cellular structure and cellular motility. Microfilaments and microtubules are composed of cytoskeletal filaments. These filaments are polymers of soluble subunits and can rapidly assemble and dissemble. The assembly process is energy dependent and requires ATP or GTP. A third type of filament is designated as intermediate filament because it has a diameter in between that of microfilaments and microtubules. Intermediate filaments are found in most animal cells. They are of a more permanent nature than either microfilaments or microtubules. The molecular mechanisms involved in the contraction actin and myosin filaments are discussed in the section on muscle tissue.

PROBLEM

Intermediate filaments are

a)   found in animal cells.

b)   more permanent than microfilaments.

c)   intermediate in diameter between microfilaments and microtubules.

d)   All of the above.

Solution

d)   Intermediate filaments, found in animal cells, are more permanent than, and intermediate in diameter between, microfilaments and microtubules.

B. CILIA, FLAGELLA AND CENTRIOLES

Cilia are hair-like projections (0.25 µm) that extend from the surface of many animal cells. These structures are used for cell movement (as in protozoa) or to move fluid (such as mucus) at the surface of the cell. Ciliated epithelial cells are found in the respiratory tract. Ciliary movement is dependent upon movement of the axoneme which is primarily composed of microtubules. It is a relatively permanent structure. The soluble subunit used to construct a microtubule is called tubulin. Ciliary motion requires ATP hydrolysis which generates a sliding movement of microtubules.

In order for microtubules to perform their functions they must be attached to other parts of the cell. Cilia end in a structural unit, the basal body, located at the base of the ciliary axoneme. The cytoplasmic microtubules observed in interphase cells (period between mitoses) are attached to centrioles. Basal bodies and centrioles have very similar structures each having a nine fold array of triplet microtubules.

The centrosome, which is present in most animal cells, has a centriole pair at its center. Higher plants do not have centrosomes. The centrosome, also called the cell center, is adjacent to the cell nucleus. It serves to organize microtubules and plays a major role in cell division.

Flagella in eukaryotic cells have a structure very similar to cilia and generate movement using the same principle detailed for cilia. Sperm cells and protozoa are examples of flagellated eukaryotic cells. The flagella of bacteria differ markedly from those of eukaryotic cells.

4. Nucleus: Structure and Function

The nucleus of the cell contains the nuclear DNA encoding the genetic information required for cellular replication, differentiation and functions. DNA replication and RNA synthesis occur in the nucleus. The RNA in the nucleus can be processed (e.g., RNA splicing) before being transported to be cytoplasm.

Nuclear DNA associates with histone proteins to form nucleosomes, the unit particles of chromatin. Chromatin, in turn, is packaged to form very compact structures called chromosomes. Other proteins, called nonhistone proteins, are also associated with nuclear DNA. Most of the DNA in the nucleus does not code for protein.

A. NUCLEAR ENVELOPE AND NUCLEAR PORES

The nucleus is bound by an envelope that is made up of two membranes, i.e., an inner and an outer nuclear membrane. The inner and outer nuclear membranes are fused at points called nuclear pores. The nuclear pores contain a nuclear pore complex thought to permit the selective transport (in and out) of macromolecules. For example, DNA and RNA polymerases which are synthesized in the cytoplasm must be transported into the nucleus through nuclear pore complexes.

PROBLEM

Nuclear pore complexes

a) transport micromolecules.

b) transport the nucleus.

c) fuse the nuclear and cell membranes.

d) fuse the inner and outer nuclear membrane.

Solution

d) Nuclear pore complexes fuse the inner and outer nuclear membrane to permit the transport of macromolecules to and from the nucleus.

B. NUCLEOLUS

The nucleolus is a highly ordered structure specially designed to produce the rRNA required for ribosomes. The synthesized rRNA immediately complexes with ribosomal proteins made in the cytoplasm and is transported through the nuclear pores into the nucleolus. The final maturation of ribosomes occurs, however, in the cytoplasm.

Solution

b)   rRNA is synthesized in the nucleolus and combined with ribosomal proteins.

5. Mitosis

A. MITOTIC PROCESS, PHASES OF THE CELL CYCLE

Cells are continuously subjected to various kinds of stress that can result in cell death. For an organism to grow and survive, cells must reproduce themselves. For cellular division to occur a cell must first double its contents, divide its nucleus and then divide its cytoplasm. The process of nuclear division is called mitosis. A cell that is not undergoing active division is said to be in interphase. The interphase period had been further delineated on the basis of when DNA synthesis occurs (see Figure 26). The period of active DNA synthesis and replication is called S-phase. The gap period before S-phase is called G1 phase, and the gap period after S-phase is called G2 phase. The mitotic phase, designated M-phase, begins after the G2 phase.

PROBLEM

A mitotic cell produces

a)   two cells with half of the chromosomes of the first cell.

b)   two cells each with the full chromosome complement of the original cell.

c)   four cells with half of the chromosome complement of the original cell.

d)   four cells with the full chromosome complement of the first cell.

Solution

b)   Mitosis refers to the process by which a cell divides to form two daughter cells, each with exactly the same number and kind of chromosomes as the parent cell.

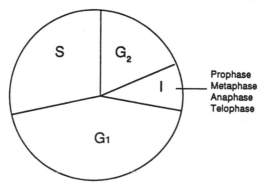

Figure 26 — Cell Cycle Phase

B. MITOTIC STRUCTURES

CHROMATIDS AND CENTROMERE

Most of the synthetic events necessary for cell division occur during interphase. In animal cells the initiation of DNA replication (i.e., the start of S-phase) is closely associated with the replication of the cell's pair of centrioles. The centrioles are the microtubule organizing center of the cell. Each centriole pair forms the spindle pole during mitosis. Once DNA synthesis is initiated, it continues until all the DNA is replicated. As the DNA replicates, new histones are attached and chromatin is formed. Each chromosome is duplicated in S-phase forming two sister chromatids joined by a centromere.

CENTRIOLES, ASTERS, SPINDLES

G2-phase begins at the end of the S-phase. During G2-phase cells prepare for mitosis by constructing much of the macromolecular machinery used in the mitotic spindle (see below). M-phase begins at the end of G2-phase. The M-phase has been further divided into prophase, prometaphase, metaphase, anaphase and telophase (see below). During prophase, chromatin condenses into chromosomes. In addition the mitotic spindle is formed. The mitotic apparatus consists of the two centrioles, a set of microtubules and two pair of centrioles. The microtubules form a radial array called the aster around each pair of centrioles. Some of the microtubules eventually connect each pair of centrioles. These microtubules comprise the spindle. The microtubules are responsible for the movement of chromosomes during mitosis.

KINETOCHORE

Prometaphase starts with the dissolution of the nuclear envelope. Microtubules subsequently become attached to chromosomes at a locus called the kinetochore. During metaphase the chromosomes become aligned at a plate halfway between the spindle poles. At anaphase the chromosomes are broken apart by the microtubules attached to the kinetochores. During telophase the daughter chromosomes arrive at opposite spindle poles, and the kinetochore microtubules dissociated. Furthermore, a nuclear envelope appears around each set of new chromosomes and nucleoli reappear. This completes the process of mitosis. The subsequent division of the cytoplasm is called cytokinesis.

PROBLEM

Kinetochore microtubules dissociate during	
a) prophase.	b) metaphase.
c) telophase.	d) anaphase.

Solution

c)   Kinetochore microtubules attach during prometaphase. An anaphase they break apart the chromosomes. During telophase the kinetochore microtubules dissociate.

IV. Specialized Eukaryotic Cells and Tissues

1. Neural Cells and Tissues

A. STRUCTURES (CELL BODY, AXON, DENDRITES, MYELIN SHEATH, SCHWANN CELLS, AND NODES OF RANVIER)

The neuron is the key cell type in the brain and the peripheral nervous system. A neuron receives information from other neurons or from sensory receptors and transmits the information to either other neurons or to muscles. The structure of typical nerve cell is shown in Figure 27.

Information is transmitted by neurons either by an action potential or by synaptic transmission. The action potential is an all-or-none response and it is a time dependent change in the transmembrane potential of the neuronal plasma membrane. The action potential is carried (see below for details) away from the cell body by the axon which usually branches and has many termini. When the action potential comes to an axon terminal it contacts the synapse which is a knob-like structure. The synapse is the junction between the end of the axon and the dendrites of an adjacent neuron.

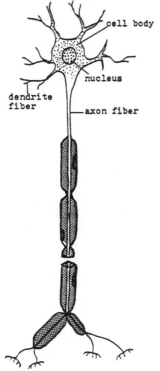

Figure 27 — The Structure of a Typical Vertebrate Neuron

In vertebrates many axons are insulated by layers of myelin which serve to increase the speed at which an action potential is transmitted along the axon. The myelin sheath (in peripheral neurons) is formed by glial cells called Schwann cells (see Figure 28). Between one Schwann cell and the next the there is a small region where the axon has no sheathing — this region is called the node of Ranvier and this region is very rich in Na+-channels.

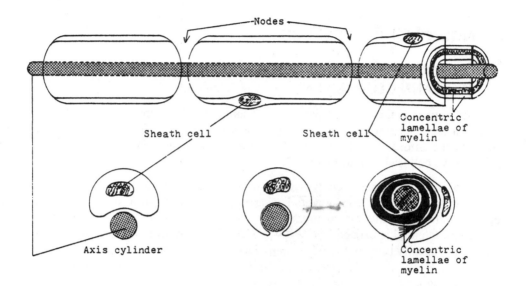

Figure 28 — The Structure of Schwann Cells and the Nodes of Ranvier

PROBLEM

The mylein sheath of many axons is produced by the

a) node of Ranvier.   b) nerve cell body.

c) Schwann cell.   d) astrocytes.

Solution

c)   Schwann cells are the myelin-forming cells of the peripheral nervous system. Each Schwann cell forms a single myelin internodal segment around a portion of an axon. Schwann cells may also surround unmyeinated axons, without producing myelin.

B. SYNAPSE

The structure of a synapse is shown in Figure 29. The presynaptic cell is separated from the postsynaptic cell by the synaptic cleft. The electrical signal from the axon triggers the release of neurotransmitter substances form storage vesicles (synaptic vesicles). The release of the neurotransmitter causes an electrical change in the postsynaptic cell. The postsynaptic cell sums up electrical signals induced by the release of the neurotransmitter and when a critical total signal level is reached an action potential is generated by the postsynaptic cell.

PROBLEM

The correct sequence for signal transmission in a synapse is

a) presynaptic cell, postsynaptic cell, synaptic cleft.

b) presynaptic cell, synaptic cleft, postsynaptic cell.

c) synaptic cleft, presynaptic cell, postsynaptic cell.

d) None of the above.

Solution

b) Electrical signals in the presynaptic cell cause release of neurotransmitter substances which flow through the synaptic cleft to the postsynaptic cell. The postsynaptic cell creates an electrical signal in response to the received neurotransmitter.

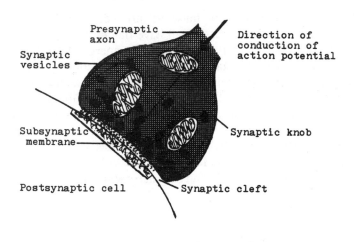

Figure 29 — The Structure of a Typical Synapse

C. RESTING POTENTIAL AND ACTION POTENTIAL

The resting potential of a neuron is established by the Na^+,K^+-ATPase and the K^+-channel as described above (see text on p. 33 and Figure 20 on p. 34). The action potential is generated by a voltage gated Na^+-channel. A voltage gated channel is one whose permeability can increase or decrease depending upon the level of the membrane potential. The resting membrane potential of a neuron is about -70mV. When the axon receives a nerve impulse this electrical signal reduces the membrane potential to about 0 mV (see top panel of Figure 30). This initial increase is called membrane depolarization. Concomitant with the membrane depolarization there is an opening of a voltage gated N^+-channel which permits Na^+ to flow into the cell (recall that Na^+ is high outside the cell and low inside the cell). This influx of positive charge causes the membrane potential to become even more positive and this is accompanied by a closing of the Na^+-channel and a subsequent drop in membrane potential until the resting potential is reestablished (see lower panel of Figure 30).

Referring to Figure 30, membrane polarization occurs

a)	before 0 ms.	b)	at 0 ms.
c)	at 1 ms.	d)	at 4 ms.

Solution

b)   The resulting neuron potential of a neuron is about − 70 mV. When the axon receives a nerve impulse the potential reduces to about 0 mV, resulting in a membrane depolarization.

2.   Contractile Cells and Tissues

A. STRIATED, SMOOTH, AND CARDIAC MUSCLE

Vertebrates have three types of muscles: striated, smooth, and cardiac. The specialized muscle cells in muscle tissues all have the ability to contract using actin and myosin filaments. ATP hydrolysis provides the energy for muscle contraction. Striated muscles are under voluntary control and they connect bones in a limb. Striated muscles (or skeletal muscles) are used for complex activities such as walking. Skeletal muscles are made of long muscle fibers (myofibers) and each fiber is considered a large single cell that is formed by the fusion of many separate cells. Each myofiber has many nuclei and bundles of myofibrils. As shown in Figure 31, a myofiber (1-40 mm in length and 10-50 μm in width) has a striated appearance. The striated appearance is due to bundles of aligned myofibrils which have dark bands (A bands) alternating with light bands (I bands). A narrow line bisects each I band and is called the Z-disc.

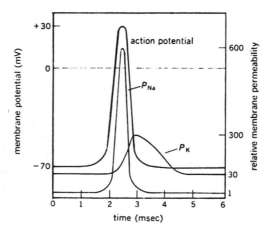

Figure 30 — Action potential

Smooth muscles are under involuntary control by the central nervous system. Smooth muscles are involved with movements of the small and large intestine, the bladder, and they also control the diameter of blood vessels. Smooth muscle cells are not striated and have one nucleus per cell. Cardiac muscle, or heart muscle, is very similar to striated muscle but is under involuntary control. Cardiac muscles produce the synchronous contraction of the heart (i.e., the heartbeat).

PROBLEM

Which of the following statements is false?

a) Cardiac muscle is uninucleate, striated and controlled by the autonomic nervous system.

b) Skeletal muscle is multinucleate, striated and controlled by the somatic nervous system.

c) Smooth muscle is uninucleate, non-striated and controlled by the autonomic nervous system.

d) None of the above.

Solution

d) All of the statements regarding muscle types are true. Skeletal muscle is responsible for most voluntary movements, and it is controlled by the somatic nervous system. It contains striations due to the ordered arrangement of thick and thin filaments and has many nuclei. Smooth muscle lines the stomach, intestinal tracts and blood vessels whose involuntary movements are controlled by the autonomic nervous system. Smooth muscle is uninucleate and does not have striations. Cardiac muscle contains features of both types of muscle. It is striated, uninucleate, and is not under voluntary control.

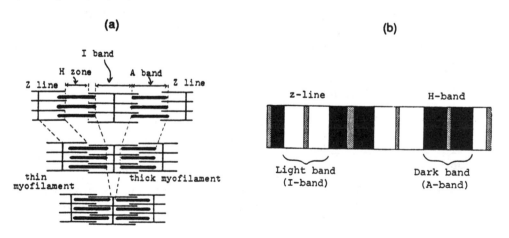

Figure 31 — (a) The Structure of Two Adjacent Myofibrils, and (b) Diagrammatic Structure of a Sarcomere.

B. SARCOMERE

The sarcomere (see Figure 31(b)) is the basic unit of contraction in striated muscles and it is the region between one Z-disk and the next. Myofibrils are made up of repeating sarcomere units. The sarcomere has thick myosin filaments as well as thin actin filaments. The thin filaments are attached to the Z-disk. The movement of thick and thin filaments between within each sarcomere leads to muscle contraction.

PROBLEM

During muscular contraction,

a) sarcomeres move between myofibrils.

b) myofibrils move between Z-disks.

c) myosin and actin filaments move between sarcomeres.

d) Z-disks are made of myofibrils.

Solution

c) The sarcomere has myosin and actin filaments. The movement of these filaments between each sarcomere leads to muscular contraction. Actin filaments attach to the Z-disk. Myofibrils are made of sarcomeres.

C. CALCIUM REGULATION OF CONTRACTION

The movement of striated muscles is under voluntary control and muscle contract is initiated by an electrical impulse from a neuron. The electrical impulse from the neuron triggers an action potential in the plasma membrane of the myofiber and is rapidly spread by transverse (or T) tubules (from the plasma membrane) to the Z-disk of the myofibrils. The signal is then transmitted to the sarcoplasmic reticulum. The sarcoplasmic reticulum surrounds each myofibril and when activated by the electrical impulse it releases Ca^{+2} ions which causes all the myofibrils in the myofiber to simultaneously contract.

PROBLEM

Which is required for muscular contraction?

a) Electrical impulse   b) Ca^{+2} ions

c) Na^+ ions   d) Both a) and b)

Solution

d) Muscular contraction is initiated by electrical impulse from a neuron. The impulse makes the sarcoplasmic reticulum release Ca^{+2} ions, which cause myofibrils to contract.

3. Epithelial Cells and Tissues

A. Simple Epithelium and Stratified Epithelium

Epithelial cells line the inner and outer surfaces of the body. These cells have many specialized shapes and functions. Epithelial cells adhere to each other and to the basal lamina. Simple squamous epithelium cells (see Figure 32(a)) form a thin layer of cells that cover the inner lining of most blood vessels. Simple cuboidal epithelium (Figure 32(b)) also consist of a single layer of tightly fitting cells but they have a cube-like shape. Cuboidal epithelium cells line the ducts of many glands. A layer of elongated simple columnar epithelial cells form the lining of the stomach, the cervix and the small intestine (Figure 32(c)). Goblet cells are found in simple columnar epithelium and they secrete mucus. Stratified epithelium are several cell layers thick and they form the surface of the mouth, the esophagus, and the vagina. Skin is also formed from stratified squamous epithelial cells that have undergone a process of keratinization. Intestinal epithclial cells are very specialized and the cell surface facing the lumen of the small intestine has many microvilli that are important in the absorption of nutrients. The microvilli contain actin filaments which help maintain their rigidity.

A major function of all epithelial cells is to provide a boundary between different cell types. The basal lamina provides a distinct boundary between the epithelial cells and the cells that underlay the basal lamina.

PROBLEM

Epithelial tissues perform many functions. Which of the following is a function of this tissue?	
a) Absorption	b) Protection
c) Secretion	d) All of the above.

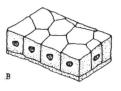

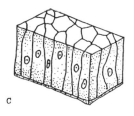

Figure 32 — Types of Epithelial Cells
(a) Simple squamous epithelium; (b) Simple cuboidal epithelium,
and (c) Simple columnar epithelium

Solution

d) Epithelium exhibits a multiplicity of structural forms, all with the common role of covering the outer surface and lining the inner surfaces of the body. In some cases, such as skin, the function is mainly protective. In many other instances, however, epithelial tissue carries out functions such as surface and transepithelial transport, absorption, and secretion.

4. Connective Cells and Fiber Types

A. MAJOR CELL AND FIBER TYPES

Connective tissues are characterized by the presence of relatively few cells with a large complement of extracellular matrix. The extracellular matrix is primarily composed of collagen fibers which are secreted by fibroblasts. Cartilage is formed by chondroblast and bone is formed by osteoblasts.

At least four distinct types of collagen fibers have been identified. Type IV is found exclusively in the basal lamina. Types I and III are found in skin whereas bone contains only type I. The extracellular matrix also has polysaccharides which are primarily glycosaminoglycans. These polysaccharides are often cross-linked with protein to form proteoglycans. Bone and teeth are composed of extracellular matrix with a secondary deposition of calcium phosphate crystals.

PROBLEM

Collagen fibers are found in

a)	skin.	b)	blood.
c)	bones.	d)	Both a) and c)

Solution

d) Collagen is found primarily in connective tissue, such as skin and bones. Blood is not a connective tissue.

B. LOOSE VS. STRONG CONNECTIVE TISSUE

Connective tissue is classified as either "loose" or "strong." Loose connective tissue forms the bed for epithelial cells and many glands. Blood vessels are found in loose connective tissues. Strong connective tissue is found in bone, cartilage and tendons.

PROBLEM

Which of the following is not a connective tissue?

a)	Bone	b)	Tendons
c)	Cartilage	d)	Muscle

Solution

d) Connective tissues function to support and hold together structures of the body. Bone, cartilage, tendons, ligaments, and fibrous connective tissues are all different types of connective tissue. The cells of these tissues characteristically secrete a large amount of noncellular material, called matrix. The nature and function of each kind of connective tissue is determined primarily by its matrix. Most of the connective tissue volume is made up of matrix.

C. CARTILAGE

Cartilage is a component of rigid connective tissue and it provides support, framework and protection. There are three types of cartilage, i.e., hyaline, elastic and fibrocartilage. Each contains a different kind of extracellular matrix. Hyaline cartilage is the most abundant and it occurs in many joints and bone ends. During embryonic development skeletal components are first formed from hyaline cartilage which is subsequently replaced by bone. Elastic cartilage is more flexible than hyaline cartilage and it forms the external structure of the ears. Fibrocartilage is mechanically very strong and it serves a protective role by functioning as a cushion between bones in the knees and the pelvic girdle. Cartilage cells are found in small chambers called lacunae which are surrounded by extracellular matrix.

PROBLEM

Elastic cartilage is found			
a)	on bone ends.	b)	in ears.
c)	in the pelvis and knees.	d)	Both a) and c)

Solution

b) Elastic cartilage is found in ears. Hyaline cartilage coats bone ends. The pelvis and knees contain fibrocartilage.

D. EXTRACELLULAR MATRIX

The extracellular matrix is primarily composed of collagen and tropocollagen is the basic structural unit from which collagen is constructed. Collagen fibers are extremely strong. Tropocollagen has a unique triple helix structure and each polypeptide strand is called an alpha-chain. Intramolecular hydrogen bonds link each alpha-chain to the other two alpha-chains. The amino acid sequence of each alpha-strand is given by:

(gly-pro-X)n

where every third residue is glycine and X can be any amino acid. Pro is proline. In collagen fibers, the tropocollagen molecules are aligned along their long axis but are displaced by about 64 nm. The adjacent tropocollagen molecules are also cross linked to one another. This cross-linking greatly enhanced the mechanical strength of the collagen fibers.

PROBLEM

The extracellular fibers found in all connective tissues are composed mainly of			
a)	collagen.	b)	calcium.
c)	elastin.	d)	glycans.

Solution

a)   The connective tissues are defined as the complex of cells and extracellular materials which provide the supporting and connecting framework for all other body tissues. The connective tissues consist of extracellular fibers, amorphous ground substance, and connective tissue cells. The fibers are composed mainly of the protein collagen. The ground substance occupies the spaces between the cells and fibers and contains proteoglycans, glycoproteins, and other molecules secreted from the cells.

V. Nervous and Endocrine Systems

1. Nervous System Structure and Function

Neurons are the fundamental cell type of the nervous system. Neurons can transmit information from inside and outside the body to processing centers in the brain and spinal column. The processed signals can evoke responses, also transmitted by neurons, by muscles and glands. The coordination and integration of these events leads to behavioral adaptation to environmental changes and helps maintain a stable internal environment.

A. ORGANIZATION OF THE VERTEBRATE NERVOUS SYSTEM

The structure of the neuron, action potentials and synaptic transmission has already been discussed (see pp. 45-48). The organs of the brain and spinal column form the central nervous system. The nerves that connect the central nervous system to other body parts are called the peripheral nervous system.

PROBLEM

The central nervous system is composed of the

a)   brain and spinal column.

b)   spinal column and nerves.

c)   neurons, synapses, and spinal column.

d)   sense organs, spinal column, and brain.

Solution

a)   The organs of the brain and spinal column form the central nervous system. The nerves that connect the central nervous system to other body parts are called the peripheral nervous system.

B. SENSOR AND EFFECTOR NEURONS

The sensory function of the nervous system is achieved by sensors at the ends of peripheral nerves. Neurons with a sensory function are called afferent or sensory neurons. The dendrites of these neurons have either sensors at their terminals or their dendrites are in close association with specialized sensor cells. Most sensor neurons have a unipolar structure (see Figure 33(a)).

The information from sensory neurons is transmitted, in the form of a nerve impulse, over peripheral nerves to the central nervous system. After integration of the sensory information a response can be transmitted by the peripheral nerves to effectors, i.e., muscles and/or glands. Interneurons form linkages between neurons within the brain (or spinal cord) and are involved with processing and integration. The interneurons are multipolar (see Figure 33(b)).

Motor neurons, also called efferent neurons, transmit nerve impulses from the brain or spinal column to effectors. Motor neurons are usually multipolar.

Neurons are bundled together to form nerve fibers. Some nerve fibers contain only motor neurons (motor nerves), some only sensory neurons (sensory nerves). Most nerve fibers have, however, both motor and sensory neurons (mixed nerves).

PROBLEM

Bundles of neurons are known as

a)   interneurons.   b)   association areas.

c)   nerve fibers.   d)   effectors.

Solution

c)   Neurons are bundled together to make nerve fibers. Interneurons link neurons to the brain. Effectors are muscles or glands that respond to stimuli.

C.   SYMPATHETIC AND PARASYMPATHETIC NERVOUS SYSTEM

The autonomic nervous system is not under voluntary control and functions independently. The contraction of smooth muscles, blood pressure regulation, temperature regulation, and the secretory function of most glands are under the control of

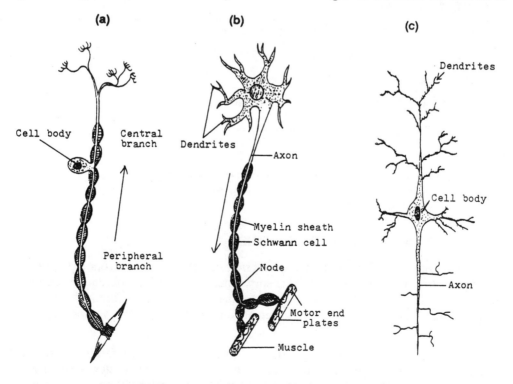

Figure 33 — Neuronal Structures: (a) Unipolar Neuron;
(b) Multipolar Neuron, and (c) Bipolar neuron.

A – 56

the autonomic nervous system. The autonomic nervous system has been further divided into the sympathetic and parasympathetic nervous systems.

In general, the sympathetic subdivision serves to prepare an organism for energy expenditure while the parasympathetic system restores and maintains an organism in a resting state. Organs are innervated with nerve fibers from both the sympathetic and parasympathetic divisions. The sympathetic nervous system, for example, increases heart rate, and decreases intestinal secretions. These physiological adaptations are restored by the parasympathetic nervous system.

Most of the nerve fibers in the autonomic system are composed of motor neurons and two neurons are used to connect the brain or spinal cord to the effector. The preganglionic axon comes from a neuron in the brain or spinal cord and forms a synapse with a ganglion outside the brain and spinal cord. The postganglionic axon comes from this second neuron and it goes to the effector.

Sympathetic nerves are adrenergic and secrete the neurotransmitter called norepinephrine at the end of their postganglionic fibers. Parasympathetic nerves are cholinergic and secrete acetylcholine at the ends of their postganglionic fibers.

PROBLEM

> Which of the following is characteristic of stimulation of the sympathetic nervous system?
>
> a) Elevated heartbeat
>
> b) Increased saliva excretion
>
> c) Elevated gastric secretion
>
> d) All of the above.

Solution

a) In general, the sympathetic nervous system produces the effects which prepare an animal for emergency situations, such as quickening of the heart and breathing rates and dilation of pupils.

2. Sensory Reception and Processing

A. SOMATIC SENSORS

Somatic sensors transmit information from the nonspecialized parts of the body. The specialized senses refer to smell, taste, hearing, equilibrium and sight. The somatic sensors can be further divided into exteroreceptive, proprioceptive, visceral, and deep sensations. The exteroreceptive sensations arise from the surface of the body. The proprioceptive sensations arise from muscles and tendons as well as body position, the visceral sensations from the internal organs, the deep sensations from "deep" tissues (e.g., bones).

The somatic sensors can be:

(1) mechanoreceptors, which respond to mechanical movement.

(2) thermoreceptors, which respond to hot and cold.

(3)   pain (or nociceptors) receptors which signal tissue damage.

A wide variety of mechanoreceptors exist. These include:

(1)   free ends of sensory nerve fibers which are found predominantly in epithelial cells which respond to touch and pressure.

(2)   Meissner's corpuscles which respond to light touch.

(3)   Pacinian corpuscles which respond to deep pressure and tissue vibrations.

PROBLEM

Somatic sensors could not detect a	
a) bright light.	b) hot stove.
c) stomachache.	d) sunburn.

Solution

a)   Somatic sensors transmit information from non-specialized parts of the body. They do not transmit smell, taste, hearing, equilibrium, or sight.

B.   OLFACTION AND TASTE

The specialization sensation of smell (olfaction) and taste rely on chemoreceptors. Chemoreceptors are also present on internal tissues where they can detect changes in oxygen levels, glucose levels and pH. In general, chemoreceptors require a threshold level of stimulation in order to generate a receptor potential.

The neurons in the superior part of the nasal cavity that detect odors are called olfactory receptors. These bipolar (see Figure 33(c)) neurons lie in a surrounding matrix of columnar epithelial cells. Bowman's glands, which secrete the mucous necessary for receptor functioning, are also embedded in the columnar epithelial cells. The mucosal ends of the olfactory neurons have many cilia which are the primary receptor sites for gaseous molecules dissolved in the mucosal fluid.

The precise mechanism whereby different gaseous molecules are distinguished is not yet clearly known. When the cilia are stimulated a receptor potential is generated which triggers a nerve impulse in the olfactory nerve fibers. This signal is transmitted to the central nervous system. The olfactory receptors undergo a progressive adaptation; with time that diminishes their response to a stimulus. The taste cells undergo a similar adaptation.

The sense of taste is generated by taste buds located in the tongue, and to a lesser extent, on the roof of the mouth. It is thought that taste consists of different combinations of four primary tastes, i.e., sour, salty, sweet and bitter. The taste receptors are microvilli that protrude from taste cells that are specialized epithelial cells. The outer surface of the taste bud in covered with stratified squamous epithelial cells and the microvilli from the taste cells protrude from a pore on this surface. The taste cells are replaced about every ten days. After stimulation the taste cells generate a receptor potential that, in turn, triggers a nerve impulse that is transmitted to the central nervous system.

> Following exposure to a strong odor over a long period of time, olfactory receptors exhibit a diminished response. This is due to
>
> a)   receptor stress.   b)   progressive adaptation.
>
> c)   receptor death.   d)   lack of mucosal fluid.

Solution

b)   Progressive adaptation to strong stimuli reduces the response of taste and olfactory receptors.

C.   HEARING

EAR STRUCTURE

The ear, which functions in both hearing and balance, has external, middle and internal components. The external ear consists of the auricle, which is funnel shaped, and the auditory meatus, which is tube shaped (see Figure 34). These structures serve to funnel sound waves into the ear where they produce pressure oscillations on the eardrum. The middle ear is in the tympanic of the temporal bone. The eardrum or tympanic membrane separates the external and middle ear.

Three small bones in the tympanic cavity transmit the vibration of the eardrum to the inner ear (see Figure 35). The malleus or hammer is attached to the eardrum.

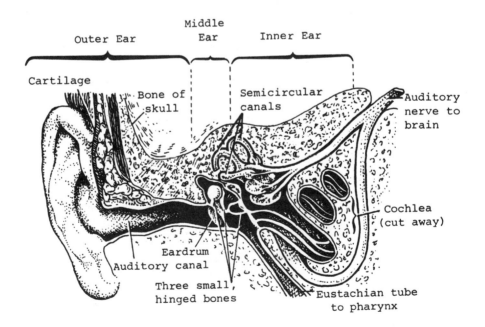

Figure 34 — The Structure of the Ear

The malleus causes the incus (or anvil) to vibrate and this movement is then transmitted to the stapes (or stirrup). It is the movement of the stapes that causes movement of fluid in the inner ear. The stapes is connected to an opening in the middle ear called the oval window.

The inner ear contains the labyrinth (see Figure 36), a complex set of interconnecting and coiled tubes. The labyrinth includes the cochlea and three semicircular canals. The cochlea contains a fluid that is moved by the impact of the stapes. The surface of the basilar membrane inside the cochlea contain the organ of Corti. The organ of Corti has the hair cells that function as the receptors for sound oscillations.

The eustachian tube connects the middle ear to the throat and permits pressure equilibration between the ear and the outside of the body.

MECHANISM OF HEARING

The organ of Corti generates a receptor potential when stimulated by the vibrations of the basilar membrane. The hearing receptor cells have cilia or hair-like structures that project into the endolymph of the cochlear duct. The sensitivity of the

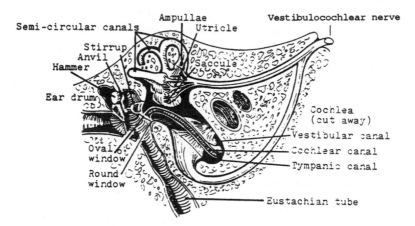

Figure 35 — Structures of the Middle Ear

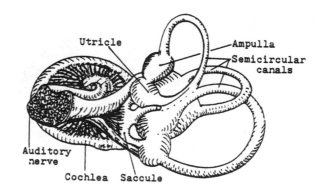

Figure 36 — The Labyrinth of the Inner Ear

ear to different sound frequencies depends upon the differential sensitivity of the hair cells. The movement of the hairs causes a receptor potential to be generated which is transmitted to the cochlear nerve fibers. Some of the nerve impulses from each ear reach both sides of the brain.

PROBLEM

Which of the following items is not part of the human ear?

a) Malleus

b) Cochlea

c) Hyoid

d) Oval window

Solution

c)   Only the hyoid, which is a very small bone near the base of the tongue, is not a part of the human ear. The tectorial membrane is part of the cochlea which is in the inner ear. The oval window is a membrane which separates the middle ear and the inner ear. The malleus is one of the small bones in the middle ear which conducts sound.

D. VISION

EYE STRUCTURE

The light receptors in the eye are extremely sensitive. A retinal rod cell can detect a single photon. The structure of the human eye is shown in Figure 37. The cornea is transparent and helps focus light and provides mechanical protection to the other underlying tissues. The anterior chamber contains aqueous humor and it lies between the cornea and the lens. The lens focuses light on the retina. The iris controls the diameter of the pupil and helps control the intensity of the light impinging on the retina. The vitreous humor in the eye cup helps control the internal pressure of the eye.

The retina contains the photoreceptor cells which are either specialized for color vision (i.e., the cones) or for night vision (i.e., the rods). In the human retina there is a specialized region called the fovea which has a high density of cone cells. Before

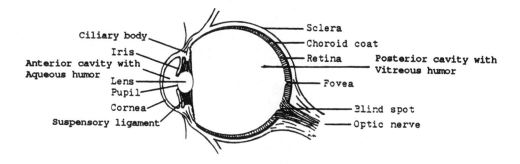

Figure 37 — The Structure of the Human Eye

light reaches the photoreceptor cells it must pass through a number of other retinal layers. The rod photoreceptor cells (see Figure 38) are adjacent to the retinal pigment epithelium. The rod cells shed their tips each day and these tips are phagocytized by the retinal pigmented epithelium. Blood supply to the retina is by way of the choroid or by retinal blood vessels. The choroid is posterior to the retinal pigment epithelium.

The outer segment of the rod photoreceptor cell has numerous disc membranes that are not in direct contact with the plasma membrane. The photosensitive pigment, rhodopsin, is an intrinsic membrane protein found in the disc membranes.

LIGHT RECEPTORS

The rhodopsin molecules in the disc membranes are covalently linked with 11-cis-retinal. Retinal is a aldehyde form of vitamin A. Light causes an isomerization of the cis-retinal to the all-trans-retinal form. This isomerization triggers a change in the conformation of rhodopsin. The light induced conformational change in rhodopsin causes Na^+-channels on the photoreceptor plasma membrane to close. In the dark, the photoreceptor cells are depolarized. This depolarization is due to open Na^+-channels that permit a constant influx of Na^+-ions. The result of a light stimulus is to close the Na^+-channels and thereby cause the receptor cell to become hyperpolarized. This action potential causes a decreased release of inhibitory neurotransmitter.

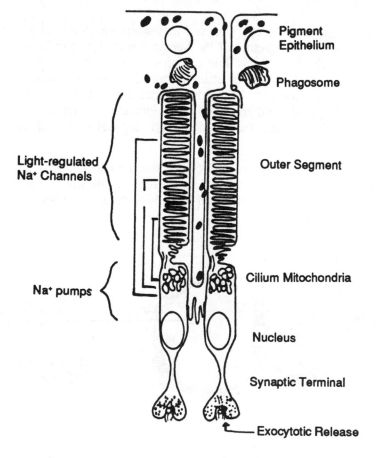

Figure 38 — Structure of the Rod Photoreceptor Cell

The part of the eye which regulates the amount of incoming light is the

a)   retina.   b)   lens.

c)   iris.   d)   cornea.

Solution

c)   Light entering the eye passes through the cornea and enters the lens via a small opening called the pupil. The size of the pupil can be changed by a diaphragm-like muscular structure, the iris, so that the amount of incoming light can be regulated. The iris may contain various colored pigments. The light then falls on a light-sensitive region, the retina, which is located at the rear of the eye.

3.   Endocrine System: Hormones and their Sources

A.   FUNCTION OF ENDOCRINE SYSTEM

Bodily functions are controlled by both the nervous system and the endocrine system as well as the interaction between these two systems. The nervous system, as detailed above, relies on electrical signals. The endocrine system utilizes chemical signals and the signal molecules are called hormones. The endocrine system refers to the set of glands, tissues and cells that secrete hormones directly into bodily fluids.

Hormones regulate a wide variety of metabolic functions and transport functions, as well as development, growth, and reproduction. Hormones are structurally diverse and can exert physiological effects on their target tissues at very low concentrations. Hormones can be peptides, proteins, glycoproteins, biological amines, or steroids.

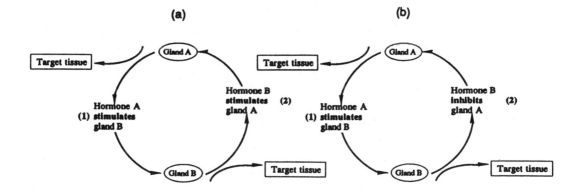

Figure 39 — (a) Positive and (b) Negative Feedback Regulation.

A – 63

Hormones do not regulate		
a) growth.	b)	reproduction.
c) digestion.	e)	temperature.

Solution

d)   Temperature regulation is controlled by the autonomic nervous system. Growth, reproduction, and digestion are controlled, in part, by hormones.

B.   CELLULAR MECHANISMS OF HORMONE ACTION

Many hormones exert their action by first binding to specific receptors on the cell surface. Some of the overall molecular characteristic of this type of signalling process has been discussed above (see the section on the eukaryotic plasma membrane).

Many hormones activate a cyclic AMP system. The first event is the binding of a hormone to a surface receptor on the plasma membrane of the target cell. The receptor-hormone complex then activates adenyl cyclase which produces cAMP from ATP in the cytoplasm. The cAMP is a "second messenger" which relays the initial extracellular signal from the hormone (the first messenger) to an intracellular signal (increased levels of cAMP). The increased cAMP then, in turn, can activate a wide variety of physiological responses. Often a protein kinase is activated which can phosphorylate specific enzymes and thereby regulate their enzymatic activity.

Steroid hormones do not utilize a cAMP system. The steroid hormones are freely permeable to the plasma membrane and do not have surface receptors. Instead, they bind to cytoplasmic receptors and the receptor-hormone complex then initiates a series of events leading to the activation of specific genes in the cells' nucleus.

PROBLEM

Steroid hormones are unlike other hormones because they
a)   utilize the cyclic AMP system.
b)   have cytoplasmic receptors.
c)   have surface receptors on the plasma membrane.
d)   are secreted directly into bodily fluids.

Solution

b)   Steroids have cytoplasmic receptors. Most other hormones have surface receptors on the plasma membrane and utilize the cyclic AMP system. All hormones are secreted directly into bodily fluids.

C.   CONTROL OF HORMONE SECRETION

The secretion of hormones into circulating blood is a very regulated process. Both negative and positive feedback loops help regulate this process (see Figure 39). In a negative feedback loop gland A secrets hormone A which stimulates gland B to produce hormone B which then can inhibit the secretion of hormone A by gland A. In a positive feedback loop gland A produces hormone A which stimulates gland B to produce hormone B which further stimulates gland A to secret hormone A.

Hormonal secretions can also be controlled by the nervous system. For example, the adrenal medulla (see below) secretes catecholamines in response to nerve impulses and not by the influence of other hormones or any other stimulus.

PROBLEM

Hormone secretion is controlled by

a)   negative and positive feedback.

b)   the nervous system.

c)   negative feedback only.

d)   Both a) and b).

Solution

d)   Hormone secretion can be controlled by positive feedback, negative feedback, or the nervous system.

D.   MAJOR ENDOCRINE GLANDS, THEIR HORMONES, SPECIFICITY, AND TARGET ISSUES

The major endocrine glands, their hormone products, their target tissues and their functions are detailed below.

PITUITARY GLAND

Pituitary Gland — lies at the base of the brain and is connected to the hypothalamus. The pituitary gland is divided into the anterior and posterior pituitary gland.

ANTERIOR PITUITARY GLAND

Anterior pituitary gland — all the major hormones produced by the anterior pituitary gland influence other glands.

Adrenocorticotropic hormone (ACTH) — a protein hormone whose target tissue is the adrenal cortex. ACTH controls the secretion of some adrenocortical hormones and thereby influences the metabolism of glucose, fats, and proteins.

Follicle stimulating hormone (FSH) — a protein hormone whose target tissue is the ovary. FSH stimulates the growth and reproductive activities of the gonads.

Growth hormone (GH) — a protein hormone that promotes body growth and has a major impact upon formation of body protein. It increases:

(1)   the transport of amino acids through cell membranes and,

(2)   the synthesis of proteins by ribosomes.

It also decreases the rate of protein catabolism.

Luteinizing hormone (LH) — a protein hormone whose target tissue is the ovary. LH stimulates the growth and reproductive activities of the gonads.

Prolactin (PRL) — a protein hormone that stimulates growth of the mammary gland and production of milk.

Thyroid stimulating hormone (TSH) — a protein hormone whose target tissue is the thyroid gland. TSH controls the synthesis of thyroxine in the thyroid gland. Thyroxine, in turn, controls many metabolic reactions.

POSTERIOR PITUITARY GLAND

Antidiuretic hormone (ADH) — this peptide hormone, also called vasopressin, causes a decreased secretion of water by the kidneys, i.e., antidiuresis.

Oxytocin — a peptide hormone whose target tissues include the uterus and the mammary gland. Oxytocin is thought to play a key role in the birthing process, causing contraction of the uterus. Oxytocin also stimulates the expression of milk from the mammary gland in response to suckling.

THYROID GLAND

Thyroid gland — this gland is located below the larynx and on both sides of the trachea.

Thyroxine (T4) — an iodinated amino acid derivative that increases the overall metabolic rate and, in children, promotes growth. In particular, T4 increases protein synthesis, it increases the number and size of mitochondria, and stimulates both carbohydrate and fat metabolism. Secretion of T4 is controlled by TSH from the anterior pituitary gland.

Triiodothyronine (T3) — an iodinated amino acid derivative whose functions are similar to those detailed for thyroxine.

PARATHYROID GLANDS

Parathyroid glands — these glands are located on the posterior surface of the thyroid gland. The parathyroid hormone is the only hormone secreted by the parathyroid gland.

Parathyroid hormone (PTH) — this protein hormone causes an absorption of calcium and phosphate from bone. Moreover, the PTH causes a dramatic increase in the secretion of phosphate by the kidney. The overall result of increased levels of PTH in plasma is an increase in calcium, but a decrease in phosphate levels. PTH promotes the conversion of vitamin D into 1,25-dihydroxycholecalciferol which, in turn, helps promote calcium transport through cell membranes. 1,25-Dihydroxycholecalciferol is the active form of vitamin D. High levels of plasma calcium decrease the secretion of PTH.

ADRENAL GLANDS

The adrenal glands lie at the top of the kidney. The adrenal consists of two distinct glands that secrete different hormones. The exterior part of the adrenal is called the cortex and the central region the medulla. The cells of the medulla are modified postganglionic cells. The cells of the adrenal medulla are in contact with the sympathetic division of the autonomic nervous system.

Adrenal medulla — Nerve impulses from the sympathetic nerve fibers are the stimulus for the secretion of epinephrine and norepinephrine.

Epinephrine — this hormone, also called adrenalin, is a biological amine. Both epinephrine and norepinephrine are catecholamines. Epinephrine prepares the body for a "fight or flight" response, i.e., heart rate, metabolic rate, and systemic blood pressure increases. The liver converts glycogen into glucose, the airways dilate and the force of cardiac muscle contraction increases.

Norepinephrine — this biological amine has a structure similar to that of epinephrine and its biological effects are very similar.

ADRENAL CORTEX

Adrenal cortex — this gland secretes a group of hormones called corticosteroids that are all synthesized from cholesterol. Corticosteroids are further divided into glucocorticoids, mineralocorticoids and androgenic hormones. The glucocorticoids increase blood glucose, the mineralocorticoids affect electrolytes. The androgenic hormones are similar to testosterone.

Aldosterone — is the primary mineralocorticoid and causes sodium ions to be retained and potassium ions to be excreted. This hormone also reduces urinary output, promotes water retention, and increases extracellular fluid volume. Aldosterone exerts its effects on the tubules of the kidney. The secretion of aldosterone is controlled by many factors such as the potassium concentration in extracellular fluid, the renin-angiotensin system, body sodium and adrenocorticotropic hormone.

Cortisol is the primary glucocorticoid and this hormone has the liver as its primary target. Cortisol influences carbohydrate, protein and fat metabolism. One effect of cortisol is to stimulate gluconeogenesis, i.e., the synthesis of glucose from noncarbohydrates, particularly from amino acids. Increased gluconeogenesis, in turn, causes an increased formation of glycogen in the liver. In addition, cortisol causes an increased release of fatty acids from fat cells (adipocytes). The secretion of cortisol is first stimulated by the hypothalamus (of the brain) which secretes *corticotropin-releasing hormone* (CRH). CRH causes the anterior pituitary to secrete ACTH and ACTH causes the adrenal cortex to release cortisol. Stress of almost any kind will cause the release of ACTH which is rapidly followed by secretion of cortisol. Cortisol also exerts an anti-inflammatory effect on tissues damaged by injury.

THE PANCREAS

The *pancreas* lies behind the stomach and is connected to the duodenum. The secretory cells of the pancreas play a role in both the endocrine and exocrine system. The exocrine part of the pancreas secretes digestive enzymes into the small intestine (duodenum). The role of the pancreas in digestion will be discussed below. The endocrine part of the pancreas is due to the islets of Langerhans which contain alpha-, beta-, and delta-cells. These cells secrete their products directly into the blood stream. The alpha-cells secrete glucagon, the beta-cells insulin, and the delta-cells somatostatin. Humans with diabetes have beta-cells that are incapable of secreting insulin. Insulin and glucagon work in concert to control many metabolic activities.

Insulin — a protein hormone that influences carbohydrate, fat, and amino acid metabolism. Insulin decreases the release of fatty acids and fat cells and promotes the utilization of glucose. The high levels of blood glucose (e.g., after a meal) stimulates the secretion of insulin. Insulin promotes the uptake and storage of glucose by almost all tissues in the body, particularly those of the liver and muscles. In liver and muscle tissue, glucose is stored as glycogen. The glycogen in the liver is used to supply the blood with glucose when the dietary supply of glucose decreases. Insulin also causes the liver to convert glucose into fatty acids which are subsequently stored in fat cells as triglycerides. Insulin also promotes the transport of amino acids into many tissues. Low levels of insulin causes fatty acids and glycerol to be released from adipocytes into plasma. The increased plasma levels of nonesterified fatty acids stimulates the liver to synthesize triglycerides, cholesterol esters, phospholipids and cholesterol. These lipids are secreted by the liver in the form of very low density lipoprotein. In addition, high levels of plasma fatty acids also stimulates liver mitochondrial fatty acid oxidation producing ketone bodies (i.e., betahydroxybutyrate and acetoacetate). Humans with the inability to secrete insulin often have very high levels of very low density lipoprotein and also develop premature atherosclerosis.

Glucagon — this protein hormone counteracts many of the metabolic effects of insulin. In particular, glucagon promotes an increase in blood glucose levels by causing a breakdown in glycogen, i.e., glycogenolysis. The secretion of glucagon is regulated by blood glucose levels, i.e., low levels of blood glucose stimulate glucagon secretion.

OVARY

See the Section on Reproductive System and Development.

THE TESTES

See the Section on Reproductive System and Development

PROBLEM

The adrenal medulla is most closely associated with	
a) insulin.	b) epinephrine.
c) chorionic gonadotropin.	d) vasopressin.

Solution

b)   Epinephrine is a secretory product of the adrenal medulla. It causes a breakdown of glycogen to glucose in the liver and skeletal muscle with a consequent rise in blood glucose levels. Epinephrine elevates the blood pressure and heart rate. It also constricts cutaneous blood vessels and dilates skeletal muscle vessels. In addition, it causes the organs of the digestive tract to experience vasoconstriction.

VI. Circulatory System, Lymphatic, and Immune Systems

1. Circulatory System

A. MULTIPLE FUNCTIONS, INCLUDING ROLE IN THERMOREGULATION

The circulatory system has a major role in maintaining the stability of the body's internal environment, i.e., homeostasis. The fluid in the body can be divided into intracellular fluid and extracellular fluid which have different compositions. The extracellular fluid can be further divided into interstitial fluid and the fluid of the circulatory system, i.e., plasma. The circulatory system is responsible for the movement and mixing of the extracellular fluid.

Some major roles of the circulatory system are:

(a)   the delivery of oxygen and required nutrients,

(b)   the removal of metabolic waste products,

(c)   the transport of regulatory molecules such as hormones,

(d)   the transport of protective chemicals and enzymes. Vitamin E is an example of a protective chemical which inhibits free radical damage to cell membranes and macromolecules,

(e)   the transport of molecules and cells essential to the immune system.

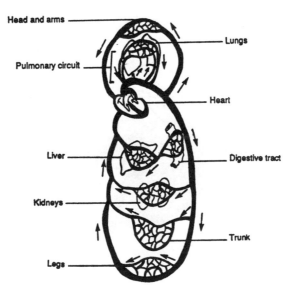

Figure 40 — The Circulatory System

> The circulatory system does all the following EXCEPT
>
> a) deliver oxygen.   b) regulate blood pressure.
>
> c) transport enzymes.   d) remove waste products.

Solution

b)   The circulatory system transports materials to and from tissues, it does NOT regulate blood pressure. Blood pressure is controlled, in part, chemically.

B. FOUR-CHAMBERED HEART, PULMONARY, AND SYSTEMIC CIRCULATION

The overall organization of the circulatory system is shown in Figure 40. In essence, the four-chamber heart is two pumps: one pumps blood to the lungs and the other to the systemic circulation.

The four-chambered heart (see Figure 41) is composed of two atria and two ventricles. The atria are filling chambers and pump blood to the ventricles which provide the main contractile force needed to move blood through the circulatory system.

Deoxygenated venous blood from the venae cavae continuously flows into the right atrium and then directly into the right ventricle (before contraction). Atrial contraction then fills the right ventricle. The contraction of the right ventricle (and the closing of the tricuspid valve) pumps the blood into the pulmonary circulation where the blood is oxygenated and carbon dioxide is lost to the atmosphere. The oxygenated blood is returned to the heart via the pulmonary veins and enters the left atrium. With the contraction of the left atrium the blood enters the left ventricle. The contraction of the left ventricle closes the bicuspid valve and blood is pumped into the systemic circulatory system.

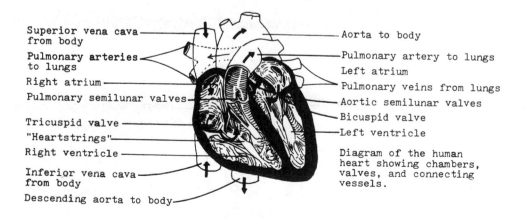

Figure 41 — The Heart

A – 70

The heart is supplied with oxygenated blood by two branches of the aorta, i.e., the right and left coronary arteries. A major cause of cardiovascular disease is the accumulation of atherosclerotic plaques in the coronary arteries. This severely narrows the lumen of these arteries and a small blood clot can clog these arteries resulting in a cut-off of oxygenated blood to the heart. This results in a coronary heart attack.

The cardiac cycle is the period from the end of contraction and the end of the next contraction. The period of relaxation is called the diastole and the period of contraction is the systole.

PROBLEM

Which of the following is not a true statement?

a) Blood enters the heart through the superior (anterior) vena cava or through the inferior (posterior) vena cava.

b) The pulmonary artery carries oxygenated blood.

c) Oxygenated blood first enters the left atrium of the heart.

d) The systemic circulation contains oxygenated blood.

Solution

b) Blood enters the right atrium of the heart through the superior or inferior vena cava. When this chamber is filled, the blood is forced through the tricuspid valve and into the right ventricle. From there, this deoxygenated blood travels through the pulmonary artery to the lungs where it exchanges carbon dioxide for oxygen. Once oxygenated, the blood travels to the left atrium through the pulmonary veins. It travels through the bicuspid valve to the left ventricle, out through the aorta, and is then distributed throughout the body.

C.   ARTERIAL AND VENOUS SYSTEMS, CAPILLARY BEDS, SYSTEMIC AND DIASTOLIC PRESSURE

Arteries carry oxygenated blood away from the heart under high pressure. Arteries have strong vascular walls which pulsate in synchrony with heart pulsations. The maximum pressure reached during the arterial pulse is called the systolic pressure and the lowest pressure is called the diastolic pressure. The arteries end in arterioles which effectively control the flow of blood into the capillary beds. The capillaries have a very permeable membrane that permits the exchange of nutrients, hormones, electrolytes, and other substances between blood and the interstitial spaces between cells. The deoxygenated blood from the capillary beds collects in the venous system and returns to the heart. The venous system is under low pressure and is thin walled.

The arterial pulse pressure is influenced primarily by the stroke volume output of the heart and by the compliance of the arterial vasculature. The stroke volume of the heart is the amount of blood pumped out of the heart with each heartbeat. The greater the stroke volume output, the greater the arterial pulse pressure. The compliance of the arterial vasculature refers to the distendability of the arteries to a pressure load. The greater the arterial compliance, the lower the arterial pulse pressure.

> The only artery in the human body which carries deoxygenated blood is the
>
> a)   pulmonary artery.   b)   right coronary artery.
>
> c)   left coronary artery.   d)   carotid artery.

Solution

a)   The pulmonary artery carries blood to the lungs to be cleaned of its carbon dioxide. All other arteries carry oxygenated blood.

D.   COMPOSITION OF BLOOD

Whole blood can be separated by low speed centrifugation into a cell free fluid called serum (or plasma if a blood anticoagulant is present) and a pellet containing cells and platelets. Plasma is about 92% water and contains electrolytes, lipoproteins, proteins, hormones, other nutrients and vitamins. The lipoproteins are lipid-protein complexes. Lipoproteins are the primary transport molecules for lipids and also transport vitamin E and beta-carotene (provitamin A). Lipoproteins are further divided into very low density lipoprotein, low density lipoprotein and high density lipoprotein. High plasma levels of low density lipoprotein are associated with atherosclerosis and cardiovascular disease. In contrast, high plasma levels of high density lipoprotein are thought to protect against atherosclerosis.

The primary proteins found in plasma are albumin, globulins, and fibrinogen. Albumin is the most abundant plasma protein (about 60%) and is a carrier molecule for nonesterified fatty acids. Albumin also plays a role in maintaining the osmotic pressure of blood. The globulins are further divided into alpha-, beta-, and gamma-globulins. The gamma-globulin fraction contains molecules that function as antibodies in the humoral immune system (see below). Fibrinogen functions in clot formation.

The red blood cell (or erythrocyte) is the primary cell found in blood. This unique cell has a plasma membrane but no other membranous organelles and does not have a cell nucleus. The primary function of red blood cells is oxygen transport to tissues and the removal of carbon dioxide. The oxygen carrying molecule in the red blood cell is hemoglobin (see below). The red blood cell has a biconcave shape and is extremely deformable and able to move through very small capillaries. In anemia the number of red blood cells in a given volume of blood is low resulting in a decreased ability to deliver oxygen to tissues. Nutritional and/or genetic factors can contribute to anemia.

Blood also contains white blood cells and platelets. White blood cells (or leukocytes) include monocytes, lymphocytes, neutrophils, eosinophils, and basophils. Neutrophils, eosinophiles, and basophils (all three are also called granulocytes) as well as monocytes are phagocytic cells. The role of these phagocytic cells in the immune system will be discussed below. Lymphocytes also play a key role in the immune system (see below). Platelets function in clot formation.

E. ROLE OF HEMOGLOBIN IN OXYGEN TRANSPORT

Hemoglobin is the primary molecule found in red blood cells and its primary function is in the transport of oxygen. The three dimensional structure of hemoglobin is known in detail from X-ray crystallographic studies. Hemoglobin is a tetramer (alpha$_2$beta$_2$) with two identical alpha subunits and two identical beta subunits. Both the alpha and beta subunits have a structure similar to myoglobin. Myoglobin is the monomeric oxygen binding protein of muscle. Each of the hemoglobin subunits has a heme group containing iron in the ferrous (Fe^{+2}) state. Each heme group can bind a single oxygen molecule. Oxygen binding does not change the oxidation state of the heme iron.

Hemoglobin is an allosteric protein. The binding of oxygen to hemoglobin is regulated by other molecules such as protons (H^+), carbon dioxide (CO_2) and 2,3-diphosphoglycerate (DPG). These molecules exert their influence on oxygen binding by binding to sites that are distinct from the oxygen binding sites. A key feature of oxygen binding to hemoglobin is the cooperative nature of this binding (see Figure 42). Cooperative binding occurs when the binding of each oxygen molecule facilitates the binding of the next oxygen molecule. This cooperative binding results in a characteristic sigmoidal dissociation curve as shown in Figure 42. In contrast, the binding of oxygen to myoglobin is not cooperative (and the dissociation curve is a hyperbola) but myoglobin does have a stronger affinity for oxygen than does hemoglobin.

The cooperative binding of oxygen to hemoglobin plays an important physiological role in the delivery of oxygen to tissues. Hemoglobin is almost fully saturated with oxygen at the partial pressure of oxygen found in the lung ($pO_2 = 100$ mm Hg). Oxygen is readily dissociated from hemoglobin and delivered to myoglobin which has a stronger affinity for the oxygen. As more oxygen is dissociated from hemoglobin the affinity of the remaining oxygen is less. This follows since dissociation is just the reverse of binding. Thus, hemoglobin is able to deliver oxygen to tissues even at the low pO_2 levels found in capillaries ($pO_2 = 20\text{-}26$ mm Hg).

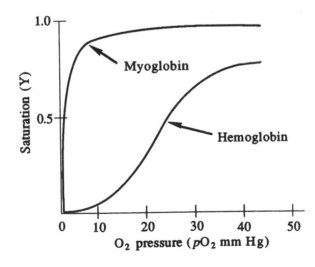

Figure 42 — Oxygen Binding to Hemoglobin and Myoglobin

CO_2, DPG and H^+ shift the oxygen dissociation curve to the right, i.e., the affinity of hemoglobin for oxygen is decreased. Tissues with a high metabolic activity, such as contracting muscle, generate large amounts of H^+ and CO_2. High H^+ and CO_2 lower the affinity of hemoglobin for oxygen and thereby increase the delivery of oxygen to these metabolically active tissues. This is called the Bohr effect.

PROBLEM

In muscles, oxygen leaves hemoglobin to bind with myoglobin because

a)   the presence of H^+ and CO_2 in muscles increases the affinity of hemoglobin for oxygen.

b)   the removal of oxygen from hemoglobin increases hemoglobin's affinity for the remaining oxygen.

c)   myoglobin has a stronger oxygen affinity than hemoglobin.

d)   the bonding of oxygen to myoglobin is cooperative.

Solution

c)   Myoglobin has a stronger oxygen affinity than hemoglobin. Myoglobin does not have the cooperative oxygen binding. The presence of H^+ and CO_2, as well as the removal of oxygen from hemoglobin, decreases the affinity of hemoglobin for oxygen.

2.   Lymphatic System

The lymphatic system provides an important link with the cardiovascular system and the immune system. It consists of lymph fluid, lymphatic vessels (see Figure 43), lymph nodes, the spleen and the thymus gland. The thymus gland plays a key role is the processing of T-lymphocytes (see below).

Almost all tissues in the body have lymphatic capillaries and these merge to form lymphatic vessels. The lymph from the upper and lower portion of the body

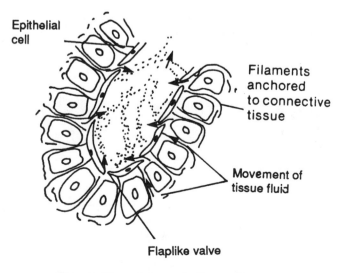

Epithelial cell

Filaments anchored to connective tissue

Movement of tissue fluid

Flaplike valve

Figure 43 — A Lymphatic capillary

flows into veins in the thorax. A vital role of the lymphatic system is to provide an alternative route for the fluid in between cells (interstitial fluid) to flow back into blood. Interstitial fluid that enters the lymph capillaries is called lymph. In addition, the lymphatic system aids in the removal of proteins and large particles from the interstitial spaces.

Lymph capillaries have many "flaps" or valves that permit the inflow of fluid and particulate matter but prevent their "back flow" into the interstitial spaces. Edema, which is the accumulation of fluid in tissues, can result if fluid flow through the lymphatic system is blocked.

Bacteria and viruses that have entered tissues cannot be directly absorbed via the blood capillaries. These pathogens enter lymph and are subsequently transported into lymph nodes which contain large numbers of lymphocytes and macrophages which defend against these microorganisms. The role of macrophages and lymphocytes in the immune system is detailed below.

PROBLEM

The lymphatic system transports all the following EXCEPT	
a) plasma.	b) interstitial fluid.
c) pathogens.	d) macrophages.

Solution

a)   Plasma travels in the circulatory system. Interstitial fluid (lymph), pathogens, and macrophages may be transported in the lymphatic system.

3. Immune System

The primary function of the immune system is to provide resistance from attack by infectious agents such as bacteria and viruses. Most immune responses in higher organisms involve the production of antibodies (i.e., acquired immunity) but other innate mechanisms for killing infectious agents also exist. Macrophages, for example, can kill invading bacteria by phagocytosis (engulfing and digesting) and this process can occur in the absence of antibodies. The acidic digestive juice of the stomach is also effective in killing infectious agents introduced by swallowing.

A. ANTIGENS, ANTIBODIES AND ANTIGEN-ANTIBODY REACTIONS

Acquired immunity depends upon the production of recognition molecules that can distinguish "self" from "nonself." The two forms of acquired immunity are:

(1)   humoral immunity and,

(2)   cellular immunity.

Cellular immunity refers to the formation of lymphocytes that are sensitized against the invading agent. Humoral immunity refers to the binding of circulating antibodies to the invading agent. Antibodies (or immunoglobulins) are remarkably specific recognition molecules.

A typical antibody consists of four polypeptide chains. There are two identical "light chains" and two identical "heavy chains" and the four chains are held together by disulfide bonds as shown in Figure 44a to form a Y-shaped molecule. As shown in Figure 44b, both the heavy and light chains of an antibody are built up from a structurally similar "domain" or polypeptide subunit of about 220 amino acids. Each light chain has two such domains: a "constant" domain and "variable" domain. Similarly, each heavy chain has three (sometimes two) constant domains and one variable domain. The variable domains are at the amino-terminal ends (see Figure 44) of both heavy and light chains and the amino acid sequence in this region is very variable. The variable regions provide the specificity which enables an antibody to bind to a very specific region of another molecule, i.e., the antigen. The constant regions of antibodies provide a mechanism for the binding of the antibody to other cells (such as macrophages) or binding to elements of the complement system (see below).

When antibodies bind to an antigen on an invading organism they mark it for destruction by either the complement system or by macrophages. The proteins of the complement system destroy an invading organism by perforating its cell membrane. Antibodies can also inactivate an invading organism by agglutination (multiple antigenic sites are bound together to form a clump), precipitation (the water soluble antigen complexes with the antibody and the complex is insoluble), or by neutralization (the antibody binds to and covers a toxic site).

PROBLEM

The complement system of immune response does NOT

   a)   agglutinate antigens.   b)   precipitate antigens.

   c)   ingest antigens.   d)   neutralize toxic sites.

Solution

c)   The complement system may agglutinate, precipitate, or neutralize antigens. Macrophages ingest antigens.

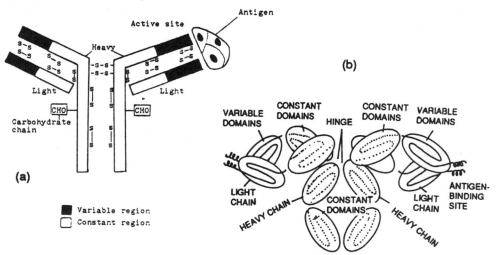

Figure 44 — The Structure of Typical IgG Antibody showing the constant and variable polypeptide segments. (a) shows position of disulfide bonds and (b) shows arrangement of domains.

B.   TISSUES AND CELLS OF THE IMMUNE SYSTEM

T-LYMPHOCYTES, B-LYMPHOCYTES, BONE MARROW, SPLEEN, THYMUS, AND LYMPH NODES

Lymphoid tissues form the "organ system" responsible for acquired immunity. Acquired immunity does not develop until after contact is made with an invading agent. Lymphoid tissues are widely distributed throughout the body and are particularly concentrated in lymph nodes, the thymus, the bone marrow and the spleen. The white cells or lymphocytes of blood are also lymphoid cells.

T-LYMPOCYTES AND B-LYMPHOCYTES

Lymphoid tissues contains two types of lymphocytes called T-lymphocytes and B-lymphocytes. The T-lymphocytes form the sensitized cells of the cellular immune system and the B-cells play a key role in the production of antibodies that provide for humoral immunity. T-cells have another class of recognition molecules called T-cell receptors which will only recognize cells that bear both "self" and "nonself" markers.

Both T- and B-lymphocytes arise from embryonic stem cells but before becoming part of lymphoid tissues they require a maturation process. For T-cells this maturation occurs in the thymus gland and for B-cells the exact site is not known but is thought to be in the bone marrow. After the maturation process the T- and B-cells migrate to and become imbedded in the lymphoid tissues.

The immunological events following an infection by an infectious agent such as a virus are as follows:

(1)   Macrophages ingest a number of viruses and display some specific viral "markers" or antigens on their surface. Some helper T-cells in circulation have the proper T-cell receptors to recognize the processed viral antigens on the macrophage surface and these T-cells become activated.

(2)   The activated helper T-cells multiply and also stimulate the multiplication of killer T-cells and activated B-cells that can also recognize the same processed viral antigens. The activated B-cells multiply and differentiate into plasma cells that produce antibodies to the viral antigen. Some of the activated B-cells become memory cells which permit a rapid response to any future infection by the virus.

(3)   The killer T-cells will destroy host cells that have become infected with the virus and thereby inhibit viral replication. The antibodies produced by the B-cells will also bind to the virus and prevent them from infecting additional host cells.

(4)   When the infection is contained, suppressor T-cells halt the immune responses and memory T-cells and memory B-cells remain in the blood and lymphatic system.

The AIDS virus is particularly damaging to the immune system because it invades and kills helper T-cells.

The functional difference between B-cells and T-cells is that

a) B-cells differentiate from stem cells and T-cells differentiate from lymphocytes.

b) T-cells differentiate from stem cells and B-cells differentiate from lymphocytes.

c) T-cells secrete antibodies in response to introduced antigens and B-cells direct the cell-mediated response.

d) B-cells secrete antibodies in response to introduced antigens and T-cells direct the cell-mediated response.

Solution

d)   T-cells and B cells are the cells involved in the immune responses of the body. Both cell types differentiate from stem cells of the bone marrow. The stem cells that migrate to the thymus become T-cells and are responsible for cell-mediated immunity. Those stem cells that migrate to the bursa or analogous structure become B-cells and are the cells of the humoral immune system.

VII. Digestive and Excretory Systems

1. Digestive System

The primary role of the digestive system is to convert food into substances that are capable of being absorbed into the body. Moreover, those substances that are incapable of absorption must be excreted. The digestive system consists of the alimentary canal and the exocrine organs that secrete digestive juices into the alimentary canal. The alimentary canal starts at the mouth and includes the pharynx, esophagus, stomach, small intestine, large intestine and the anus (see Figure 45). The salivary gland, pancreas, liver and gallbladder are organs that secrete substances into the alimentary canal.

A. INGESTION: STRUCTURES AND THEIR FUNCTIONS

The first component of the alimentary canal is the mouth which mechanically reduces the size of food materials and mixes the masticated particles with saliva. The teeth are specialized structures for breaking up food particles and increasing the surface area of the food particles. The incisors cut large food sections, the cuspids serve to grasp and tear food, and the bicuspids and molars are effective in grinding food particles.

Saliva is secreted into the mouth by the salivary glands and it increases the moisture content of the food particles and also initiates the digestion of carbohydrate. Amylase is the digestive enzyme in saliva and it splits starch and glycogen into disaccharides. In plants, glucose is stored as starch granules and in animals, glucose

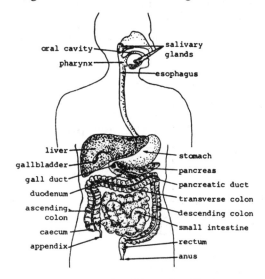

Figure 45 — The Digestive System

is stored as glycogen granules. Starch is a very heterogenous material and its two principal components are amylose and amylopectic polysaccharides.

PROBLEM

Which of the following organs is not a part of the human digestive system?

a)   Esophagus

b)   Thymus

c)   Gall bladder

e)   Pancreas

Solution

b)   All the choices are a part of the digestive system except for the thymus. The thymus is a gland that does most of its work during childhood and is almost completely inactive by puberty.

B.   STOMACH

The food mass from the mouth passes through the pharynx and the esophagus by a peristaltic wave and enters the stomach. In the stomach the food mass is mixed with gastric secretions and the digestion of protein is initiated. The gastric glands secrete pepsinogen, mucus, hydrochloric acid and intrinsic factor. Intrinsic factor is important in promoting the absorption of vitamin B12. Pepsinogen is the inactive form of pepsin.

The inactive form of an enzyme is called a zymogen. Pepsinogen is activated to pepsin by cleavage of a 44-residue peptide (from the amino terminal end) and this activation occurs spontaneously at pH 2 and is also catalyzed by pepsin. Pepsin is an acid-protease and has maximal enzymatic activity at pH 2 to 3. The hydrochloric acid in the stomach functions to maintain an acidic pH (of 1) and also denatures dietary protein to make it more susceptible to protease attack. Pepsin primarily catalyzes the hydrolysis of peptide bonds with an aromatic amino acid residue such as phenylalanine, tryptophane and tyrosine. The polypeptides produced by pepsin digestion are transported to the small intestine for further hydrolysis. The mixture of food mass and gastric juices in the stomach constitutes chyme. Chyme enters the small intestine which also receives the secretions of the liver and pancreas.

PROBLEM

Which of the following is a zymogen?

a)   Protease

b)   Tyrosine

c)   Chyme

d)   Pepsinogen

Solution

d)   Pepsinogen is a zymogen, an inactive form of an enzyme. Protease is an enzyme, tyrosine is an amino acid, and chyme is the mixture found in the stomach.

C.   DIGESTIVE GLANDS, INCLUDING LIVER AND PANCREAS, BILE PRODUCTION

The pancreas, in addition to its role in the endocrine system (see the Endocrine System, p. 63), also functions in the exocrine system. The acinar cells, which form most of the pancreatic mass, secrete pancreatic juice which travels through the pancreatic duct to the duodenum. The bile duct from the liver and gallbladder enter the duodenum at the same site. Pancreatic juice aids the digestion of carbohydrate, fat, and protein. The pH of pancreatic juice is alkaline and it neutralizes the acidic chyme from the stomach. Carbohydrate digestion is assisted by pancreatic amylase which breaks down starch and glycogen into disaccharides (see p. 79). Triglycerides are hydrolyzed into glycerol and free fatty acids by the action of pancreatic lipase.

A mixture of zymogens are released into the small intestine. These zymogens include trypsinogen, chymotrypsinogen and procarboxypeptidase and they are converted into their active forms (trypsin, chymotrypsin and carboxypeptidase, respectively). Trypsin hydrolyzes the carboxyl side of peptide bonds with argine and lysine residues while chymotrypsin acts on the carboxyl side of peptide bonds with aromatic residues (phenylalanine, tryptophane and tyrosine) as well as methionine. Carboxypeptidase A releases the carboxyl-terminal amino acids and carboxypeptidase B is restricted to peptides with an arginine or lysine carboxyl-terminal. The pancreas also secretes nucleases to breakdown nucleic acids into nucleotides.

The liver functions in the process of digestion by secreting bile salts which act like a detergent and emulsifies fat. This emulsification increases the surface area of the fat droplets. Lipases can efficiently utilize triglycerides (fat droplets) that have been emulsified by bile salts. Bile salts are synthesized by hepatic cells from cholesterol and are secreted into the common bile duct. The common bile exits to the duodenum via a sphincter muscle (the sphincter of Oddi). Between meals the sphincter of Oddi is closed and bile is stored in the gallbladder.

PROBLEM

Bile is secreted by the			
a)	stomach.	b)	liver.
c)	duodenum.	d)	gallbladder.

Solution

b)   Bile is an aqueous solution which contains various organic and inorganic solutes. Among the major organic solutes are bile salts, phospholipids, cholesterol, and bile pigments. The adult human liver produces about 15 ml of bile per kilogram body weight. The rate of synthesis and secretion is dependent mainly upon blood flow to the liver.

D.   SMALL INTESTINE, LARGE INTESTINE

The small intestine receives chyme from the stomach as well as pancreatic juice and bile from the liver. It starts at the pyloric sphincter and ends at the large intestine. The three segments of the small intestine are the duodenum, the jejunum and the ileum. The small intestine has numerous villi that project into the intestinal lumen. These villi serve to mix chyme with intestinal juices and aid in the absorption of digested nutrients. The small intestine is the primary absorbing organ of the alimentary canal. The epithelial cells of the intestinal mucosa have a variety of digestive enzymes. These include peptidases (which break polypeptides into amino acids), and enzymes that convert disaccharides into monosaccharides.

The large intestine plays almost no role in the digestion of food but does reabsorb water and electrolytes. In addition, it stores feces until defecation of undigestible food components, such as fiber.

PROBLEM

Villi are finger-like protrusions of the

a) small intestine.
b) outer ear.

c) bronchioles.
d) capillaries.

Solution

a) Villi line the lumen of the small intestine and thereby increase the intestinal surface area. Most of the nutrient absorption during digestion occurs through the villi.

E. MUSCULAR CONTROL OF DIGESTION

The smooth muscles of the alimentary canal promote both the mixing of food with gastric juices and the rhythmic wave-like (peristaltic) movements that propel food through the lumen of the digestive tract. The regulation of this muscular contraction is primarily controlled by an "intrinsic nervous system." This intrinsic system also regulates much of the secretory functions required for digestion. In addition, nerve fibers from the parasympathetic and sympathetic branches of the autonomic system also interact with the intrinsic nervous system of the gut. In general, the parasympathetic system increases the activity of the gut and the sympathetic system decreases this activity.

The vagus nerve, which is part of the parasympathetic system arises from the brain, innervates the esophagus, stomach, pancreas and the proximal half of the large intestine. Parasympathetic nerve fibers also originate from the sacral segments of the spinal cord and innervate the distal segment of the large intestine. The sympathetic innervation also regulates the gastrointestinal tract. The preganglionic fibers originate in the spinal cord and the postganglionic fibers innervate all parts of the gut. The norepinephrine secreted by the ends of the sympathetic nerves inhibits the contractions of smooth muscles in the gut and also inhibits the intrinsic nervous system.

PROBLEM

Norepinephrine inhibits all the following EXCEPT

a) smooth muscle contraction.

b) peristalsis.

c) the intrinsic nervous system.

d) the parasympathetic nervous system.

Solution

d) Norepinephrine inhibits the intrinsic nervous system and smooth muscles in the gut (which produce peristalsis). Norepinephrine does not affect the parasympathetic nervous system.

2. Excretory System

A. ROLE OF EXCRETORY SYSTEM IN BODY HOMEOSTASIS

Metabolic waste products are often toxic and must be removed from the body to prevent tissue damage. Blood and lymph are the fluids that initially receive metabolic wastes. Gaseous waste products such a CO_2 are removed from blood by the respiratory system. Salts and nitrogenous wastes are removed from the circulatory system by the urinary system. The urinary system also plays a key role in body homeostatsis by helping to regulate the volume and composition of extracellular fluid, the production of red blood cells and blood pressure. The urinary system is composed of a pair of kidneys, a pair of ureters, a urinary bladder and a urethra (see Figure 46).

PROBLEM

Wastes are removed from the blood by the

a) respiratory system.

b) urinary system.

c) digestive system.

d) both a) and b).

Solution

d)   The respiratory and urinary systems both remove wastes. The respiratory system removes gases from the blood and the urinary system removes liquid and dissolved materials from the blood. The digestive system breaks down food, but does not remove wastes.

B. KIDNEY: STRUCTURE AND FUNCTION

The kidneys (see Figure 46) are located on both sides of the spinal column and are behind the parietal peritoneum. The surface of the kidney facing the spinal column (the medial surface) is concave and contains a deep sinus through which the

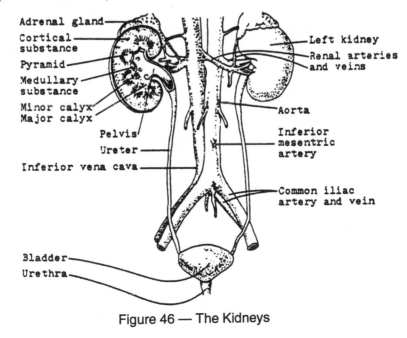

Figure 46 — The Kidneys

A – 83

renal artery (from the aorta) and the renal vein (from the inferior vena cava) enter. The ureter, which transports urine away from the kidney, also exits from the renal sinus. The extrarenal ureter is expanded into a funnel-shapped sac (the renal pelvis) at its junction with the renal sinus. The intrarenal tubes of the ureter branch to form two major calyces each of which further divides into minor calyces. Urine is delivered into the minor calyces by the renal papillae and then flows through the ureter into the bladder.

The interior of the kidney is divided into an inner renal medulla and an outer renal cortex. The renal medulla forms pyramidal structures whose apexes form the renal papillae. The renal cortex has a granular appearance due to the many small tubules of the nephrons (see Figure 47).

In addition to its role of removing metabolic wastes the kidney also secretes erythropoietin which stimulates the production of red blood cells. Renin is also secreted by the kidney and it helps regulate blood pressure. The inactive form of vitamin D (25-hydroxyl-vitamin D) is converted to the active form of vitamin D (1,25-dihydroxyvitamin D) in the kidney. The active form of vitamin D promotes Ca^{+2} absorption.

PROBLEM

> The kidney can do all the following EXCEPT
>
> a)   remove metabolic wastes.
>
> b)   help activate vitamin C.
>
> c)   help regulate blood pressure.
>
> d)   help stimulate production of red blood cells.

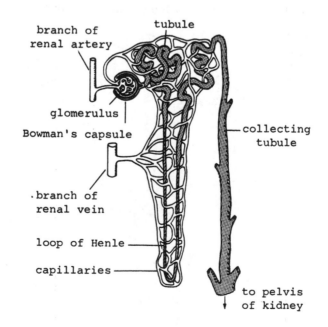

Figure 47 — A Nephron

A – 84

Solution

b)   The kidney does everything listed above but help in the activation of vitamin C. The kidney helps in the activation of vitamin D.

C.   NEPHRON: STRUCTURE AND FUNCTION

Urine is formed in the nephron which is the functional unit of the kidney. About a million nephrons are present in each kidney. The two major sections of a nephron are the corpuscle and the tubule (see Figure 47)

The corpuscle is a network of intertwined capillaries that surround a capsule (Bowman's capsule) containing the glomerulus. The end of the proximal tubule forms Bowman's capsule. The convoluted proximal tubule from Bowman's capsule forms the loop of Henle followed by the distal convoluted tubule.

PROBLEM

Which of the following is not part of the nephron in the human kidney?

a)	Proximal convoluted tubule	b)	Loop of Henle
c)	Distal convoluted tubule	d)	Major calyx

Solution

d)   The nephron is the structural and functional unit of the kidney consisting of a renal corpuscle (a Glomerulus enclosed within Bowman's capsule), and its attached tubule. The tubule consists of the proximal convoluted portion, the loop of Henle, and the distol convoluted portion. These connect by arched collecting tubules. There are approximately one million nephrons in each kidney.

The major calyx is not part of the nephron, but is, rather, a part of the intrarenal collecting system.

D.   FORMATION OF URINE

Blood enters the glomerulus through the afferent arteriole and exits through the efferent arteriole. The blood in the glomerulus is under pressure (60 mm Hg) and this pressure forces fluid into Bowman's capsule. The fluid in the Bowman's capsule flows into:

(1)   the proximal renal tubule (in the cortex of the kidney),

(2)   the loop of Henle,

(3)   the distal tubule and,

(4)   a collecting duct which, in turn, flows into the renal pelvis.

The end result of this fluid movement is the creation of urine. The solute composition of urine is, however, partly dependent upon environmental and nutritional parameters. The loops of Henle and the vasa recta provide mechanisms for regulating the osmolarity and volume of urine produced by the kidney.

The concentration of electrolytes and other substances in the fluid contained in Bowman's capsule (i.e., the glomerular filtrate) is very similar to that found in interstitial fluid. Most of the water and some of the solutes in the glomerular filtrate (under a pressure of about 18 mm Hg) is reabsorbed into peritubular capillaries which are under about 13 mm Hg of pressure, (i.e., lower than that of the loop of Henle). The solutes that are reabsorbed into circulation are nontoxic and their loss as urine would be wasteful. For example, glucose, amino acids and many electrolytes are reabsorbed. Toxic, unwanted or "excess" solutes are not reabsorbed and appear in urine. In addition to this filtration mechanism, some wastes are directly secreted, from the peritubular capillaries, into the renal tubules.

E.   STORAGE AND ELIMINATION OF WASTES

Urea is a byproduct of amino acid metabolism and it is a main constituent of urine. Uric acid which is formed from the catabolism of purines is also eliminated in urine. Urine, from the renal pelvis, flows through the ureter to the urinary bladder. The flow of urine through the ureter is promoted by peristaltic contraction of the muscular lining of the ureter. The process by which the urinary bladder empties is called micturition. Micturation occurs when the tension in the walls of the bladder reaches a threshold level caused by the increasing volume of urine. This can trigger a reflex that results in the emptying of the bladder. The micturation reflex can also be influenced by both inhibitory and stimulatory signals from the brain. In particular, the relaxation of the urethral sphincter is necessary before urination can proceed.

PROBLEM

Micturation is controlled by			
a)	reflex.	b)	signals from the brain.
c)	hormones.	d)	Both a) and b).

Solution

d)   Micturation is controlled by reflex and signals from the brain, not hormones.

VIII. Muscle and Skeletal Systems

1. Muscle System

A. FUNCTIONS

The primary function of all muscle tissue is contraction during which chemical energy is converted to mechanical energy. The contraction of muscle fibers causes tension on the body parts to which they are attached. Skeletal muscles function by applying tension to their attachment points on bones. Bones and muscles form lever systems which control body movements and help maintain posture. Muscles also function to control the movement of fluids in circulatory and excretory system and help maintain body temperature.

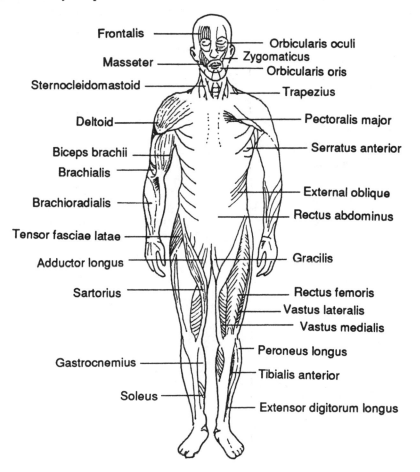

Figure 48 (a) — Anterior View of Superficial Skeletal Muscles

B.   BASIC MUSCLE TYPES AND LOCATIONS

Almost half the body is muscle mass with the vast majority being skeletal muscle. Smooth muscle and cardiac muscle account for about 5-10% of body mass. The structure of various contractile cells and the molecular mechanisms responsible for muscle contraction have been discussed in Section IV, Contractile Cells and Tissues, pp. 48-50.

SKELETAL MUSCLES

Skeletal muscle fibers are about 10-80 microns in diameter and extend the entire length of a muscle. The multinucleated cells of skeletal muscles are striated. Individual skeletal muscles are separated from each other by a surrounding fascia which can also extend beyond the muscle to become part of a tendon. The tendon functions to connect the muscle to bone.

Some major skeletal muscles are described below:

Biceps brachii — this muscle in the upper arm has two heads (immovable origins) that originate on the scapula. The muscle follows the humerus and is connected to the radius by a tendon. Contraction causes the arm to bend at the elbow.

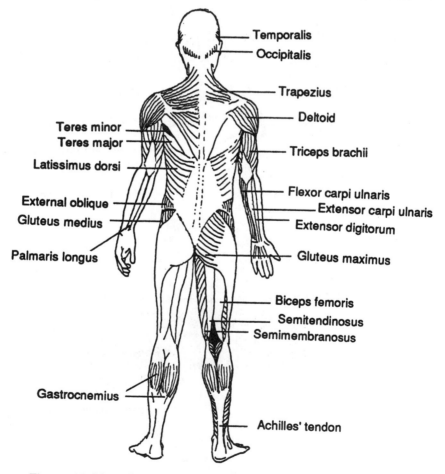

Figure 48 (b) — Posterior View of Superficial Skeletal Muscles

A – 88

pectoralis major — this large muscle of the chest connects the humerus (in the upper arm) to the bones of the thorax.

deltoid — this triangular muscle is located on the shoulder and is active in all shoulder movements.

extensor digitorum — extensor muscles act to straighten body parts away from the main body. Extensor digitorum muscles extend either the fingers or toes.

sternomastoid — these muscles connect the sternum and mastoid.

Figure 48 shows many of the major skeletal muscles.

SMOOTH MUSCLES

Smooth muscles cells form fibers that are smaller (i.e., 2-5 microns in diameter and 50-200 microns in length) than skeletal fibers and they have only a single nucleus. Smooth muscles contract and relax more slowly than skeletal muscles.

Visceral and multiunit are the two major types of smooth muscles. Visceral smooth muscles are in contact with each other (at points called gap junctions). Stimulation of one portion of a smooth muscle causes the action potential to be conducted to the surrounding fibers. Visceral smooth muscles are found in the intestines, the bile ducts, the ureters, and the uterus. Visceral smooth muscles are responsible for the peristaltic movement of the intestinal tract.

In contrast, each multiunit smooth muscle fiber acts independently and is usually innervated by a single nerve which controls its contraction. Multiunit smooth muscles are found in the walls of blood vessels and in the iris of the eye.

CARDIAC MUSCLES

The primary function of cardiac muscles is the rhythmic pumping action of the heart. The ventricles provide the primary force for pumping blood through the lungs

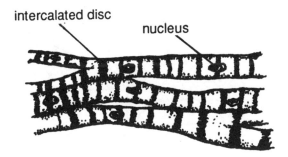

Figure 49 — The Intercalated Disc of Cardiac Muscle

and the peripheral circulatory system. Cardiac muscles exist only in the heart. Cardiac muscle cells are striated and have a single nucleus. Cardiac muscle fibers contain angular intercalated discs (see Figure 49) which are cell membranes that separate one cardiac muscle cell from the next. The three major muscle types present in the heart are: (1) ventricular muscle; (2) atrial muscle and; (3) specialized excitatory and conductive muscle fibers. The atrial and ventricular muscles contract in a manner similar to that of skeletal muscles but the specialized excitatory and conductive fibers contract only very weakly. These specialized muscles fibers do, however, provide a mechanism for the rapid transmission of excitatory impulses throughout the heart.

The intercalated discs of cardiac muscle fibers provide a very low resistance pathway for the rapid transmission of action potentials. Cardiac muscle is, therefore, a syncytium in which the action potentials rapidly propagate throughout the lattice of interconnected fibers. There are, in fact, two separate syncytium systems in the heart, i.e, the atrial syncytium and the ventricular syncytium. Although separated by fibrous tissue, an impulse can be conducted from the atrial to the ventricular syncytium via the A-V bundle.

C.   NERVOUS CONTROL OF MUSCLES

MOTOR AND SENSORY CONTROL

Motor (or efferent) neurons carry nerve impulses out from the brain or spinal cord. Each skeletal muscle fiber is connected to a myelinated motor neuron at a region called the neuromuscular junction. The specialized region of the muscle fiber membrane that forms a junction with the axon of the motor neuron is called a motor end plate (see Figure 50)

A nerve impulse reaching the neuromuscular junction will cause the release of acetylcholine which, in turn, triggers the generation of action potential in the muscle fiber. The acetylcholine released into the synaptic cleft between an axon terminal and

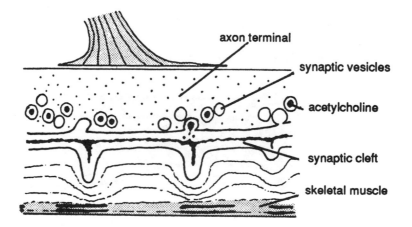

Figure 50 — The Motor End-Plate

the plasma membrane of the muscle fiber is rapidly destroyed by acetylcholinesterase.

THE MOTOR UNIT

Each motor neuron branches to form contacts with many muscle fibers. The neuron and the muscle fibers attached to it form a motor unit. When an impulse is transmitted through the motor neuron it will cause the simultaneous contraction of all muscle fibers to which it is attached. Very fine movements require the number of connections a motor neuron makes with muscle fibers to be small (e.g., about 10).

THE REFLEX ARC

Motor neurons and sensory (or afferent) neurons usually occur in the same "mixed" nerve. The simplest manner in which sensory and motor neurons are integrated to evoke behavior is in a reflex arc (see Figure 51). For example, the knee-jerk response when striking the patella stimulates a stretch receptor neuron which sends an impulse to the spinal cord. Within the gray matter of the spinal cord the axon of the sensory neuron forms a synapse with a dendrite of an anterior motor neuron. The impulse travels via the anterior motor neuron to the quadriceps femoris muscle which responds by contracting and causing an extension of the leg.

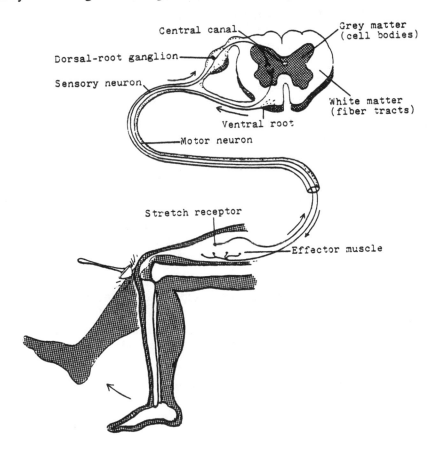

Figure 51 — A Simple Reflex Arc that Elicits a Knee-Jerk Response

Most of the sensory neurons that enter the spinal cord do not terminate on anterior neurons but rather on interneurons. Interneurons are more numerous (by a factor of about 30) than anterior motor neurons and they have numerous connections to each other and also to anterior motor neurons. These interconnections provide the basis for complex reflex responses such as the withdrawal reflex.

VOLUNTARY AND INVOLUNTARY CONTROL

Reflex behavior is unconscious and automatic but plays a critical role in homeostasis. Changes in vascular tone in response to hot or cold, sweating and some motor functions of the gut are all examples of autonomic reflexes that occur in the spinal cord.

The brain stem, which connects the cerebrum to the spinal cord, has numerous nerve pathways that help regulate the involuntary functions involved with equilibrium, respiration, cardiovascular function, eye movements, and support of the body against gravity.

"Voluntary" control of motor functions is primarily under the control of the frontal lobes of the cerebral cortex and the cerebellum of the brain. The cerebral cortex contains a "pyramidal area" that contains very large pyramid shaped cells. Motor signals from the brain originate in the pyramidal cells and travel through the brain stem and to the spinal column via the pyramidal or corticospinal nerve tract. Most of the pyramidal fibers terminate on interneurons in the cord gray matter. These interneurons, in turn, form synapses with motor neurons controlling various voluntary muscles.

In addition to the corticospinal tracts there are "extrapyramidal tracts" that also transmit motor signals from the brain. A specialized region of the frontal lobe that coordinates the muscular area involved with speech is called "Broca's area."

2. Skeletal System

A. BONE STRUCTURE

Bone, like other connective tissues, consists of cells and fibers, but unlike the others its extracellular components are calcified, making it a hard, unyielding substance ideally suited for its supportive and protective function in the skeleton.

Upon inspection of a long bone with the naked eye, two forms of bone are distinguishable: cancellous (spongy) and compact (see Figure 52). Spongy bone consists of a network of hardened bars having spaces between them filled with marrow. Compact bone appears as a solid, continuous mass, in which spaces can be seen only with the aid of a microscope. The two forms of bone grade into one another without a sharp boundary.

In typical long bones, such as the femur or humerus, the shaft (diaphysis) consists of compact bone surrounding a large central marrow cavity composed of spongy bone. In adults, the marrow in the long bones is primarily of the yellow, fatty variety, while the marrow in the flat bones of the ribs and at the ends of long bones is primarily of the red variety and is active in the production of red blood cells. Even this red marrow contains about 70 percent fat.

The ends (epiphyses) of long bones consist mainly of spongy bone covered by a thin layer of compact bone. This region of the long bones contains a cartilaginous region known as an epiphyseal plate. The epiphyseal cartilage and the adjacent spongy bone constitute a growth zone, in which all growth in length of the bone occurs. The surfaces at the ends of long bones, where one bone articulates with another are covered by a layer of cartilage, called the articular cartilage. It is this cartilage which allows for easy movement of the bones over each other at a joint.

Compact bone is composed of structural units called Haversian systems. Each system is irregularly cylindrical and is composed of concentrically arranged layers of hard, inorganic matrix surrounding a microscopic central Haversian canal. Blood vessels and nerves pass through this canal, supplying and controlling the metabolism of the bone cells. The bone matrix itself is laid down by bone cells called osteoblasts. Osteoblasts produce a substance, osteoid, which is hardened by calcium, causing

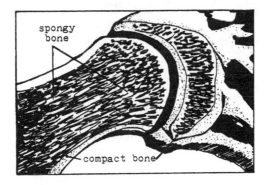

Figure 52 — Longitudinal section of the end of a long bone.

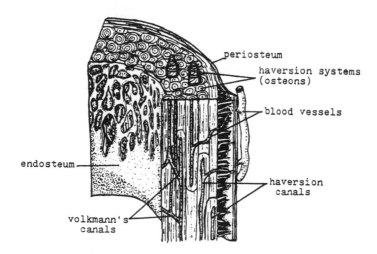

Figure 53 — Cross-section of a long bone showing internal structures

calcification. Some osteoblasts are trapped in the hardening osteoid and are converted into osteocytes which continue to live within the bone. These osteocytes lie in small cavities called lacunae, located along the interfaces between adjoining concentric layers of the hard matrix. Exchange of materials between the bone cells and the blood vessels in the Haversian canals is by way of radiating canals. Other canals, known as Volkmann's canals, penetrate and cross the layers of hard matrix, connecting the different Haversian canals to one another. (See Figure 53)

With few exceptions, bones are invested by the periosteum, a layer of specialized connective tissue. The periosteum has the ability to form bone, and contributes to the healing of fractures. Periosteum is lacking on those ends of long bones surrounded by articular cartilage. The marrow cavity of the diaphysis and the cavities of spongy bone are lined by the endosteum, a thin cellular layer which also has the ability to form bone (osteogenic potencies).

Haversian type systems are present in most compact bone. However, certain compact flat bones of the skull, such as the frontal, parietal, occipital, and temporal bones, and part of the mandible, do not have Haversian systems. These bones, termed membrane bones, have a different architecture and are formed differently than bones with Haversian systems.

B. SKELETAL STRUCTURE

The axial skeleton consists of the skull, vertebral column, ribs, and the sternum. The primary function of the vertebrate skull is the protection of the brain. The part of the skull that serves this function is the cranium. The rest of the skull is made up of the bones of the face. In all, the human skull is composed of twenty-eight bones, six of which are very small and located in the middle ear. At the time of birth, several of the bones of the cranium are not completely formed, leaving five membranous regions called fontanelles. These regions are somewhat flexible and can undergo changes in shape as necessary for safe passage of the infant through the birth canal.

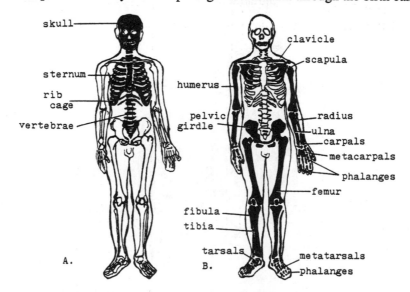

Figure 54 — Diagrams of human body showing, A, the bones of the axial skeleton and B, the bones of the appendicular skeleton.

The human vertebral column, or spine, is made up of 33 separate bones known as vertebrae, which differ in size and shape in different regions of the spine. In the neck region there are 7 cervical vertebrae; in the thorax there are 12 thoracic vertebrae; in the lower back region there are 5 lumbar vertebrae, in the sacral or hip region, 5 fused vertebrae form the sacrum to which the pelvic girdle is attached; and at the end of the vertebral column is the coccyx or tailbone, which consists of four, or possibly five, small fused vertebrae. The vertebrae forming the sacrum and coccyx are separate in childhood, with fusion occurring by adulthood. The coccyx is man's vestige of a tail.

A typical vertebra consists of a basal portion, the centrum, and a dorsal ring of bone, the neural arch, which surrounds and protects the delicate spinal cord which runs through it. Each vertebra has projections for the attachment of ribs or muscles or both, and for articulating (joining) with neighboring vertebrae. The first vertebra, the atlas, has rounded depressions on its upper surface into which fit two projections from the base of the skull. This articulation allows for up and down movements of the head. The second vertebra, called the axis, has a pointed projection which fits into the atlas. This type of articulation allows for the rotation of the head.

In man there are 12 pairs of ribs, one pair articulating with each of the thoracic vertebrae. These ribs support the chest wall and keep it from collapsing as the diaphragm contracts. Of the twelve pairs of ribs, the first seven are attached ventrally to the breastbone, the next three are attached indirectly by cartilage, and the last two, called "floating ribs", have no attachments to the breastbone.

The bones of the appendages and the girdles, which attach the appendages to the rest of the body, make up the appendicular skeleton. In the shoulder region the pectoral girdle, which is generally larger in males than in females, serves for the attachment of the forelimbs; in the hip region, the pelvic girdle serves for the attachment of the hindlimbs. The pelvic girdle, which is wider in females so as to allow room for fetal development, consists of three fused hipbones, called the ilium, ischium and pubis, which are attached to the sacrum. The pectoral girdle consists of two collarbones, or clavicles, and two shoulder blades, or scapulas. Articulating with

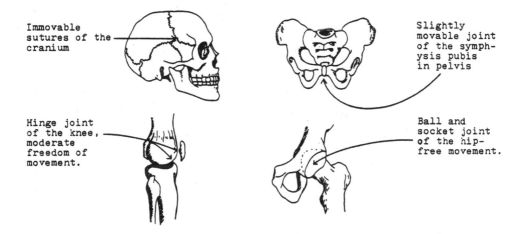

Immovable sutures of the cranium

Slightly movable joint of the symphysis pubis in pelvis

Hinge joint of the knee, moderate freedom of movement.

Ball and socket joint of the hip- free movement.

Figure 55 — Diagrams illustrating the types of joints found in the human body.

the scapula is the single bone of the upper arm, called the humerus. Articulating with the other end of the humerus are the two bones of the forearm called the radius, and the ulna. The radius and ulna permit the twisting movements of the forearm. The ulna has on its end next to the humerus a process often referred to as the "funny bone." The wrist is composed of eight small bones called the carpals. The arrangement of these bones permits the rotating movements of the wrist. The palm of the hand consists of 5 bones, known as the metacarpals, each of which articulates with a bone of the finger, called a phalanx. Each finger has three phalanges, with the exception of the thumb, which has two.

The pattern of bones in the leg and foot is similar to that in the arm and hand. The upper leg bone, called the femur, articulates with the pelvic girdle. The two lower leg bones are the tibia (shinbone) and fibula, corresponding to the radius and ulna of the arm, respectively. These two bones are responsible for rotation of the lower leg. Ventral to the joint between the upper and lower leg bones is another bone, the patella or knee cap, which serves as a point of muscle attachment for upper and lower leg muscles. This bone has no counter part in the arm. The ankle contains seven irregularly shaped bones, the tarsals, corresponding to the carpals of the wrist. The foot proper contains five metatarsals, corresponding to the metacarpals of the hand, and the bones in the toes are the phalanges, two in the big toe and three in each of the others.

The point of junction between two bones is called a joint. Some joints, such as those between the bones of the skull, are immovable and extremely strong, owing to an intricate intermeshing of the edges of the bones. The truly movable joints of the skeleton are those that give the skeleton its importance in the total effector mechanism of locomotion. Some are ball and socket joints, such as the joint where the femur joins the pelvis, or where the humerus joins the pectoral girdle. These joints allow free movement in several directions. Both the pelvis and the pectoral girdle contain rounded, concave depressions to accommodate the rounded convex heads of the femur and humerus, respectively. Hinge joints, such as that of the human knee, permit movement in one place only. The pivot joints at the wrists and ankles allow freedom of movement intermediate between that of the hinge and the ball and socket types. (Refer to Figure 55.)

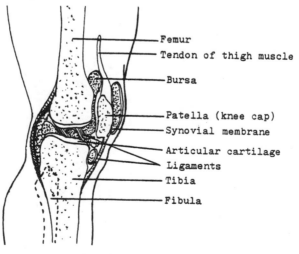

Figure 56 — The structure of a knee joint

The different bones of a joint are held together by connective tissue strands called ligaments. Skeletal muscles, attached to the bones by means of another type of connective tissue strand known as a tendon, produce their effects by bending the skeleton at the movable joints. The ends of each bone at a movable joint are covered with a layer of smooth cartilage. These bearing surfaces are completely enclosed in a liquid-tight capsule, called the bursa.

The joint cavity is filled with a liquid lubricant, called the synovial fluid, which is secreted by the membrane lining the (refer to Figure 56) cavity. During youth and early maturity the lubricant is replaced as needed, but in middle and old age the supply is often decreased, resulting in joint stiffness and difficulty of movement. A common disability known as bursitis is due to the inflammation of cells lining the bursa, and also results in restrained movement.

IX. Respiratory and Skin Systems

1. Respiratory System

A. FUNCTION

Oxygen and carbon dioxide are the important respiratory gases. While a small amount of oxygen is carried dissolved in the blood, most of the oxygen is carried by hemoglobin, an important protein found in red blood cells. Carbon dioxide (CO_2) is also dissolved in the blood and carried by hemoglobin, but most carbon dioxide is carried in the blood as bicarbonate ion (HCO_3).

There are chemoreceptors which are sensitive to changes in the chemical composition of the blood. Central chemoreceptors in the medulla oblongata are quite sensitive to levels of hydrogen ion (H^+) or CO_2. When H^+ or CO_2 increase, the receptors are excited and they in turn send signals to the breathing centers in the medulla oblongata, stimulating breathing. In addition, there are peripherally located chemoreceptors located in the aortic arch and carotid arteries. They too can stimulate the medullary breathing centers when there is an increase in H^+ ($\emptyset$ ph) or CO_2, as well as when there is a decrease in oxygen in the blood. The increase in breathing will function to increase the oxygen and/or decrease the carbon dioxide levels in the blood.

The lungs are the site of gas exchange in the pulmonary circulation. The alveoli or air sacs of the lungs are thin-walled, as are the pulmonary capillaries which supply them. Hence, the respiratory gases can easily diffuse through these walls. Gases diffuse from a region of high partial pressure to one of lower partial pressure. When one inhales, there is an increase in oxygen in the alveoli; when one exhales, there is a decrease in carbon dioxide in the alveoli. Since the blood entering the pulmonary capillaries has a low level of oxygen and a high level of carbon dioxide, oxygen will diffuse from the alveoli into the pulmonary capillaries and carbon dioxide will diffuse from the pulmonary capillaries into the alveoli. Hence the blood returning to the heart from the lungs will be replenished with oxygen, and the body tissues will be ridded of carbon dioxide, the major waste product of cellular metabolism.

B. BREATHING STRUCTURES AND MECHANISMS

The respiratory system in man and other air-breathing vertebrates includes the lungs and the tubes by which air reaches them. Normally, air enters the human respiratory system by way of the external nares or nostrils, but it may also enter by way of the mouth. The nostrils, which contain small hairs to filter incoming air, lead into the nasal cavities, which are separated from the mouth below by the palate. The nasal cavities contain the sense organs of smell, and are lined with mucus secreting epithelium which moistens the incoming air. Air passes from the nasal cavities via the internal nares into the pharynx, then through the glottis and into the larynx. The larynx is often called the "Adam's apple," and is more prominent in men than women. Stretched across the larynx are the vocal cords.

The opening to the larynx, called the glottis, is always open except in swallowing, when a flap-like structure (the epiglottis) covers it. Leading from the larynx to the chest region is a long cylindrical tube called the trachea, or windpipe. In a dissection, the trachea can be distinguished from the esophagus by its cartilaginous C-shaped rings which serve to hold the tracheal tube open. In the middle of the chest, the trachea bifurcates into bronchi which lead to the lungs. In the lungs, each bronchus branches, forming smaller and smaller tubes called bronchioles. The smaller bronchioles terminate in clusters of cup-shaped cavities, the air sacs. In the walls of the smaller bronchioles and the air sacs are the alveoli, which are moist structures supplied with a rich network of capillaries. Molecules of oxygen and carbon dioxide diffuse readily through the thin, moist walls of the alveoli. The total alveolar surface area across which gases may diffuse has been estimated to be greater than 100 square meters.

Each lung, as well as the cavity of the chest in which the lung rests, is covered by a thin sheet of smooth epithelium, the pleura. The pleura is kept moist, enabling the lungs to move without much friction during breathing. The pleura actually consists of two layers of membranes which are continuous with each other at the point at which the bronchus enters the lung, called the hilus (roof). Thus, the pleura is more correctly a sac than a single sheet covering the lungs.

The chest cavity is closed and has no communication with the outside. It is bounded by the chest wall, which contains the ribs on its top, sides and back, and the sternum anteriorly. The bottom of the chest wall is covered by a strong, dome-shaped sheet of skeletal muscle, the diaphragm. The diaphragm separates the chest region

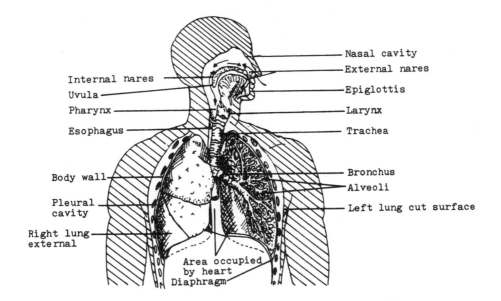

Figure 57 — Diagram of the human respiratory system

A – 99

(thorax) from the abdominal region, and plays a crucial role in breathing by contracting and relaxing, changing the intrathoracic pressure.

2. Skin System

A. COMPOSITION

Human skin is composed of a comparatively thin, outer layer, the epidermis, which is free of blood vessels, and an inner, thick layer, the dermis, which is packed with blood vessels and nerve endings. The epidermis is a stratified epithelium whose thickness varies in different parts of the body. It is thickest on the soles of the feet and the palms of the hands. The epidermis of the palms and fingers has numerous ridges, forming whorls and loops in very specific patterns. these unique fingerprints and palmprints are determined genetically, and result primarily from the orientation of the underlying fibers in the dermis. The outermost layers of the epidermis are composed of dead cells which are constantly being sloughed off and replaced by cells from beneath. As each cell is pushed outward by active cell division in the deeper layers of the epidermis, it is compressed into a flat (squamous), scalelike epithelial cell. Such cells synthesize large amounts of the fibrous protein, keratin, which serves to toughen the epidermis and make it more durable.

Scattered at the juncture between the deeper layers of the epidermis and the dermis are melanocytes, cells that produce the pigment melanin. Melanin serves as a

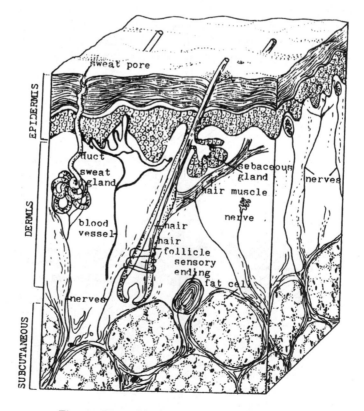

Figure 58 — Section of human skin

protective device for the body by absorbing ultraviolet rays from the sun. Tanning results from an increase in melanin production as a result of exposure to ultraviolet radiation. All humans have about the same number of melanocytes in their skin. The difference between light and dark races is under genetic control and is due to the fact that melanocytes of dark races produce more melanin.

The juncture of the dermis with the epidermis is uneven. The dermis throws projections called papillae into the epidermis. The dermis is much thicker than the epidermis, and is composed largely of connective tissue. The lower level of the dermis, called the subcutaneous layer, is connected with the underlying muscle and is composed of many fat cells and a more loosely woven network of fibers. This part of the dermis is one of the principle sites of body fat deposits, which help preserve body heat. The subcutaneous layer also determines the amount of possible skin movement.

The hair and nails are derivatives of skin, and develop from inpocketings of cells from the inner layer of the epidermis. Hair follicles are found throughout the entire dermal layer, except on the palms, soles, and a few other regions. Individual hairs are formed in the hair follicles, which have their roots deep within the dermis. At the bottom of each follicle, a papilla of connective tissue projects into the follicle. The epithelial cells above this papilla constitute the hair root and, by cell division form the shaft of the hair, which ultimately extends beyond the surface of the skin. The hair cells of the shaft secrete keratin, then die and form a compact mass that becomes the hair. Growth occurs at the bottom of the follicle only. Associated with each hair follicle is one or more sebaceous glands, the secretions of which make the surface of the skin and hair more pliable. Like the sweat glands, the sebaceous glands are derived from the embryonic epidermis but are located in the dermis. To each hair follicle is attached smooth muscle called arrector pili, which pulls the hair erect upon contraction.

B.   PROTECTION AND THERMOREGULATION

Perhaps the most vital function of the skin is to protect the body against a variety of external agents and to maintain a constant internal environment. The layers of the skin form a protective shield against blows, friction, and many injurious chemicals. These layers are essentially germproof, and as long as they are not broken, keep bacteria and other microorganisms from entering the body. The skin is water-repellent and therefore protects the body from excessive loss of moisture. In addition, the pigment in the outer layers protects the underlying layers from the ultraviolet rays of the sun.

In addition to its role in protection, the skin is involved in thermoregulation. Heat is constantly being produced by the metabolic processes of the body cells and distributed by the bloodstream. Heat may be lost from the body in expired breath, feces, and urine, but approximately 90 per cent of the total heat loss occurs through the skin. This is accomplished by changes in the blood supply to the capillaries in the skin. When the air temperature is high, the skin capillaries dilate, and the increased flow of blood results in increased heat loss. Due to the increased blood supply, the skin appears flushed. When the temperature is low, the arterioles of the skin are constricted, thereby decreasing the flow of blood through the skin and decreasing the rate of heat loss. Temperature-sensitive nerve endings in the skin reflexively control the arteriole diameters.

At high temperatures, the sweat glands are stimulated to secrete sweat. The evaporation of sweat from the surface of the skin lowers the body temperature by removing from the body the heat necessary to convert the liquid sweat into water vapor. In addition to their function in heat loss, the sweat glands also serve an excretory function. Five to ten per cent of all metabolic wastes are excreted by the sweat glands. Sweat contains similar substances as urine but is much more dilute.

X. Reproductive System and Development

1. Male and Female Gonads and Genitalia

The successful production of offspring in higher organisms is complex and reserved for mature, fully developed organisms. The organs related to the reproductive process are called genitalia. The specific organs responsible for producing the sex cells (i.e., sperm cells in males and ovum in females) are called gonads; in females this organ is the ovary and in males this organ is the testis.

A. THE MALE REPRODUCTIVE SYSTEM

In males (see Figure 59) the reproductive organs include two testes as well as accessory internal organs (i.e., the epididymides, the vas deferentia, the seminal vesicles, ejaculatory ducts, the prostate gland, the urethra, and the bulbourethral glands) and accessory external organs (i.e., the scrotum, penis). These accessory organs primarily serve to store and deliver the sperm (or spermatozoa) to the female genitalia. Various glands secrete fluids to aide in this process.

The testes respond to hormones secreted by the anterior pituitary gland (i.e., gonadotropins). The anterior pituitary, in turn, is stimulated by the hypothamamus (see section on the endocrine system, page 63). The testes also secretes hormones such as testosterone. During puberty testosterone stimulates testicular growth as well as the growth of the accessory male reproductive organs. Testosterone also helps to maintain secondary masculine sex characteristics.

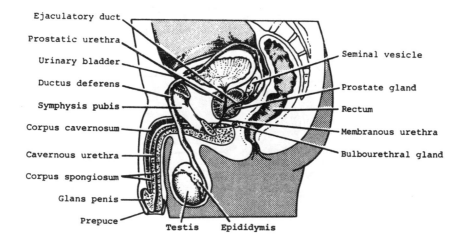

Figure 59 — The Male Reproductive System

B. THE FEMALE REPRODUCTIVE SYSTEM

The female reproductive system produces, transports and temporarily stores ovum. Ovum are produced in the ovaries which are the primary reproductive organ. Furthermore, the female reproductive system is specialized to accept sperm, to aide in the process of fertilization, to provide a highly controlled environment for long term fetal development, and to deliver the new born from an intrauterine to an extrauterine environment. Hormones are also secreted by the ovaries. Figure 60 illustrates the basic structural features of the female reproductive system. The accessory internal organs of the female reproductive tract include the fallopian tubes and the vagina. The external female accessory organs include the labia majora, the labia minora, the clitoris and the vestibule.

2. Gametogenesis by Meiosis

A. SPERMATOGENESIS (SPERM PRODUCTION)

Spermatogenesis occurs in the coiled seminiferous tubules of the testes. The seminiferous tubules have supporting columnar epithelial cells called Sertoli cells as well as spermatogenic cells. The spermatogenic cells form sperm cells. Sperm cells taken directly from the testes are neither functional nor motile. The sperm in the testes pass through ducts into the epididymis where they become motile but still lack the ability to fertilize an ovum. To be fully functional the sperm must incubate in the tubal fluid of the female.

After passing through the epididymis, sperm moves through the vas deferens which is a long muscular tube that ascends to the ejaculatory duct. The ejaculatory duct empties into the urethra.

B. OOGENESIS (PRODUCTION OF EGG CELLS OR OOCYTES)

The ovaries lie on each side of the pelvic cavity and each ovary has an exterior

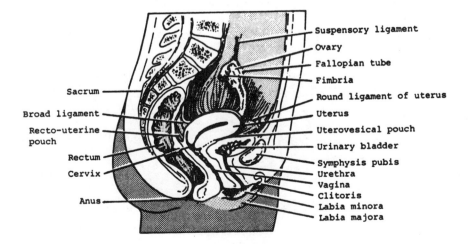

Figure 60 — The Female Reproductive System

monolayer of germinal epithelial cells that give rise to the ovum. At birth these germinal epithelial cells have already differentiated into millions of primordial follicles each of which contains an egg cell (or primary oocyte) and a surrounding single layer of follicle cells. At puberty some of the egg cells in the follicles undergo meiosis to form a secondary oocyte which contains a haploid number of chromosomes. Furthermore the primordial follicles undergo a maturation process at puberty: the oocytes enlarges and the follicular cells proliferate and form a cavity containing the oocyte and a follicular fluid. A mature follicle has an inner layer of granulosa cells and an outer layer of cells formed by ovarian stroma (the theca). The ovum is embedded in a mass of granulosa cells. During ovulation the oocyte is discharged from the mature follicle and then travels to the opening of the uterine tube. If fertilization of the egg cell does not take place the cell will die in a short period of time.

3. Reproductive Sequence

A. MALES

The reproductive sequence in males is: (a) psychic stimulation; (b) erection of the penis; (c) lubrication; (d) emission and ejaculation (orgasm). Erection is primarily a vascular event caused by dilation of arteries and constriction of veins in response to parasympathetic impulses from the sacral portion of the spinal column. The resulting high arterial blood pressure fills the erectile tissues of the penis. There are three cylindrical masses of erectile tissues (venous sinusoids) in the penis, i.e., two corpus cavernosum and one corpus spongiosum.

The parasympathetic impulses that promote erection also stimulate the secretion of a lubricating fluid from the bulbourethral gland. This fluid lubricates the end of the penis facilitating coitus (intercourse).

Emission is the movement of sperm cells (from the testes) and secretions (from the prostate and seminal vesicles) to the internal urethra to form seminal fluid. Emission is a reflex that occurs in response to sympathetic impulses from the spinal cord that results in peristaltic contractions of the smooth muscles of the epididymis, the vas deferens and the ampulla. Sympathetic impulses also trigger the contraction of smooth muscles in the prostate glands and the seminal vesicles which force the sperm down the urethra. Ejaculation is the expulsion of seminal fluid from the urethra by a reflexive contraction of the bulbocavernosus muscle (a skeletal muscle).

B. FEMALES

The reproductive sequence in females involves, in parts, a monthly (28 days is normal) sexual cycle that is under hormonal regulation. Follicle-stimulating hormone (FSH) and lutenizing hormone (LH) are secreted by the anterior pituitary at the beginning of the sexual cycle and these hormones stimulate the process of ovulation. Both FSH and LH bind to cellular receptors that, in turn, activate adenyl cyclase.

About 2 days before ovulation there is a marked increase in the secretion of LH which causes the mature follicle to rupture and release an oocyte which enters the uterine tube (at about day 14). Following ovulation, the follicular cells turn into the corpus luteum which releases increased amounts of estrogen and progesterone. The increased estrogen causes a thickening of the uterine endometrium in preparation for the potential implanting of a fertilized ovum. Similarly, progesterone causes an increased vascularization, swelling and secretory activity of the endometrium which is the innermost layer of tissue forming the uterine wall. Both estrogen and progesterone also inhibit the production of LH and FSH by the anterior pituitary gland. If fertilization does not occur, then the corpus luteum stops secreting estrogen and progesterone causing the disintegration of the uterine lining (i.e., menstrual flow). If fertilization does occur, then the placenta will secret chorionic gonadotropin which extends the life of the corpus luteum to the first 3-4 months of pregnancy.

Coitus in the female also evolves psychic stimulation, erection, lubrication, and orgasm. The clitoris (see Figure 60) contains two columns of erectile tissue called the corpus cavernosa that respond to parasympathetic impulses just as the penis does, i.e., the clitoris becomes erect. Concurrently, the Bartholin's glands located beneath the labia minor secrete a lubricating mucus.

4. Embryogenesis

A. EMBRYONIC DEVELOPMENT

Embryonic development begins when an ovum is fertilized by a sperm and ends at parturition (birth). It is a process of change and growth which transforms a single cell zygote, into a multicellular organism.

The earliest stage of embryonic development is the one-celled, diploid zygote which results from the fertilization of an ovum by a sperm. Next is a period called cleavage, in which mitotic division of the zygote results in the formation of daughter cells called blastomeres. At each succeeding division, the blastomeres become smaller and smaller. When 16 or so blastomeres have formed, the solid ball of cells is called a morula. As the morula divides further, a fluid-filled cavity is formed in the center of the sphere, converting the morula into a hollow ball of cells called a blastula. When cells of the blastula differentiate into two, and later three, embryonic

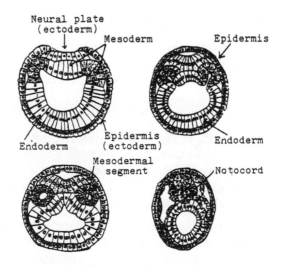

Figure 61 — Early Germ Layer Development

A – 106

germ layers, the blastula is called a gastrula. The gastrular period generally extends until the early forms of all major structures (for example, the heart) are laid down. After this period, the developing organism is called a fetus. During the fetal period (the duration of which varies with different species), the various systems develop further. Though developmental changes in the fetal period are not so dramatic as those occurring during the earlier embryonic periods, they are extremely important.

Congenital defects may result from abnormal development during this period.

B. GERM LAYERS

Early forms of all major structure are laid down during the gastrula period. These forms, called primary germ layers begin to differentiate rapidly during the fetal stage. There are three primary germ layers: ectoderm, mesoderm and endoderm.

Ectoderm gives rise to the epidermis of the skin, including the skin glands, hair, nails, and enamel of teeth. In addition, the epithelial lining of the mouth, nasal cavity, sinuses, sense organs, and the anal canal are ectodermal in origin. Nervous tissue, including the brain, spinal cord, and nerves, are all derived from embryonic ectoderm. Mesoderm gives rise to muscle tissue, cartilage, bone, and the notochord, which in man is replaced in the embryo by vertebrae.

XI. Genetics and Evolution

1. Genetics

A. GENOTYPE AND PHENOTYPE

Genes are the units of heredity and are located on chromosomes. Genes occur in various forms, or alleles, the combinations of which code for the specific expression of traits. Each individual inherits one allele of each gene from his mother and one from his father. If an individual inherits two identical alleles of a gene, he is said to be homozygous for that trait. If the alleles are different, the individual is heterozygous for that trait.

The genotype is the actual genetic constitution of the individual, but the phenotype is the expression of the genotype. For instance, there are two alleles which determine eye color. Let B represent the dominant allele, dictating brown eyes, and b represent the recessive allele dictating blue eyes. From these two alleles, there are three possible genotypes coding for eye color: BB — homozygous dominant, Bb — heterozygous, and bb — homozygous recessive. There are only two phenotypes: BB and Bb, both coding for brown eye color since B is dominant over b. Blue eyes are possible only with the genotype bb. Note that green eyes are considered as blue, both genotypically and phenotypically. The phenotype includes not only physical characteristics apparent to an observer, but all characteristics that are the result of the genotype. For instance, one's blood type is part of one's phenotype.

B. DOMINANT/RECESSIVE INHERITANCE

There are many different types of genetic inheritance patterns. The simplest is that of dominant and recessive inheritance, as exemplified by the inheritance of eye color above. Suppose two people, heterozygous for brown eyes, have children. The best way to examine the probabilities of eye color in the offspring is to use a Punnett square, where the alleles of each parent form the axes of the square. The Punnett square below shows the cross between the two heterozygotes.

Bb × Bb

	B	b
B	BB	Bb
b	Bb	bb

Phenotypically, 75% of the offspring will have brown eyes (BB and Bb), and 25% (bb) will have blue eyes. The genotypes are 25% BB (homozygous dominant), 50% Bb (heterozygous), and 25% bb (homozygous recessive). The genotypic ratio is

1BB : 2Bb : 1bb.

The Punnett square below shows the cross between a blue-eyed woman and a heterozygous brown-eyed man.

bb × Bb

	b	b
B	Bb	Bb
b	bb	bb

The proportions of the offspring are 50% heterozygous (Bb) and thus brown eyed, and 50% homozygous recessive (bb), and thus blue-eyed.

C. INCOMPLETE DOMINANCE

Some traits show incomplete dominance in which a dominant allele cannot fully mask the expression of the recessive allele. This is best exemplified in certain flowers, where color is inherited as such. Let R be the dominant allele for red flower color, and r be the recessive allele for white flower color. When a red flower (RR) is crossed with a white one (rr), the first generation will be 100% pink (Rr) as shown in the Punnett square below.

RR × rr

	R	R
r	Rr	Rr
r	Rr	Rr

If the allele for red color were fully dominant, the heterozygotes would all be red, not pink. While the result of this one cross may appear to be a blended trait, in future generations, the dominant and recessive allele can each be independently expressed again (i.e., the original traits will remerge); hence no blending has occurred. The subsequent cross between the pink flowers is shown below:

Rr × Rr

	R	r
R	RR	Rr
r	Rr	rr

The expected probabilities of phenotypic expression are 25% red (RR), 50% pink (Rr) and 25% white (rr).

A – 109

D. CODOMINANCE

In codominance, a heterozygote has two dominant alleles which are equally expressed. Codominance is best exemplified by the inheritance of blood antigens. There are multiple alleles, IA, IB, and i, possible at the locus that codes for ABO blood type. Of course, any one individual inherits only two alleles. I^A and I^B are dominant to i, but are codominant with each other.

Type A blood is expressed by the genotypes $I^A I^A$ and $I^A i$. The person has only A antigens on his red blood cells. A phenotypically type B person has the genotype $I^B I^B$ or $I^B i$. This person has only B antigens on his red blood cells. The AB phenotype is expressed by the single genotype $I^A I^B$. These alleles are codominant and the person has both A *and* B antigens on his red blood cells. The homozygous recessive genotype ii is expressed phenotypically by Type O blood. This person has neither A nor B antigens on his red blood cells.

Punnett squares can be used here as well to determine blood type probabilities for offspring. Suppose a heterozygous type A man mates with a heterozygous type B female. The Punnett square below shows that there will be an equal probability (25%) of each blood type in the offspring.

$I^A i \times I^B i$

	I^A	i
I^B	$I^A I^B$	$I^B i$
i	$I^A i$	ii

There is a 25% probability of type AB($I^A I^B$), type B($I^B i$), type A($I^A i$) and type O(ii)

E. SEX-LINKED INHERITANCE

Sex-linked inheritance is a little more complex. There are two sex chromosomes, called X and Y. A female has two X chromosomes, one inherited from each parent. A male has an X chromosome inherited from his mother, and a Y chromosome inherited from his father. The X chromosome is much larger than the Y chromosome and the X chromosome contains genes for color blindness and hemophilia on it, both of which are recessive traits.

When a male inherits an X chromosome with the recessive allele, he will fully express the trait (hemophilia or colorblindness), since his Y chromosome has no dominant allele to mask it. In contrast, when a female inherits an X chromosome with the recessive allele, she most likely will have the normal dominant allele on her other X chromosome, which will mask the recessive trait. Hence, she will merely carry the trait, but will not express it. In order for a female to be afflicted and express the trait, she must inherit two recessive alleles, one on each of her X chromosomes. That means that her father must express the trait, while her mother must either by a carrier or express the trait as well. The likelihood of a mate between two individuals each with this recessive sex-linked allele is low, unless there is mating between relatives.

Thus, the usual transmission of sex-linked traits is from a carrier mother to her son. The father cannot transmit the disease to his son since a male offspring has only one X chromosome, which must have come from his mother. The carrier mother has an equal chance of passing the gene on to either a son or a daughter; however the trait is expressed more often in the son, who lacks the dominant allele to mask it. The daughter will not express the disease unless she has also acquired the recessive gene from an afflicted father.

If h represents the recessive allele for hemophilia, then a male can be represented one of two ways: an afflicted male will have the genotype X^hY, and a normal male will have the genotype X^HY. There are three possible genotypes for a female: a normal females is X^HX^H; a carrier female is X^HX^h, and an afflicted female is X^hX^h.

Suppose that C is the dominant allele for normal color vision and c is the allele for color blindness. A male is either normal (X^CY) or color blind (X^cY); a female is either normal (X^CX^C), a carrier (X^CX^c), or colorblind (X^cX^c). As in hemophilia, the carrier genotype in a female means simply that she can transmit the condition, although she does not express the trait.

A cross between a carrier mother and a normal father is shown below:

$$X^CX^c \times X^CY$$

	X^C	X^c
X^C	X^CX^C	X^CX^c
Y	X^CY	X^cY

Note that the offspring are 25% of each of the following: normal female, carrier female, normal male, and colorblind male.

The following cross shows how a female can become afflicted. A carrier woman mates with an afflicted man.

$$X^CX^c \times X^cY$$

	X^C	X^c
X^c	X^CX^c	X^cX^c
Y	X^CY	X^cY

Hence the four possible outcomes, occurring with equal probability, are carrier female, colorblind female, normal male, colorblind male.

QUESTIONS

Let squares represent males; circles represent females; shaded are afflicted, white are not. The following questions refer to the pedigree shown in Figure 62.

PROBLEM

> All of the following statements concerning the disease in question are true EXCEPT
>
> a) it is sex-linked.
>
> b) it is caused by a recessive gene.
>
> c) it may be hemophilia.
>
> d) the afflicted boys must have received two alleles for this disease, since it is recessive.

Solution

d).

PROBLEM

> If the disease were colorblindness, the genotype of P1 must be
>
> a) $X^C X^c$.   b) $X^C X^C$.
>
> c) $X^c X^c$.   d) $X^C Y$.

Solution

d).

PROBLEM

> If F_{2-5} were to marry a woman homozygous dominant for the trait in question, the probability that they would have a a child afflicted with the disease is
>
> a) 0%.   b) 25%.
>
> c) 50%.   d) 100%.

Solution

a).

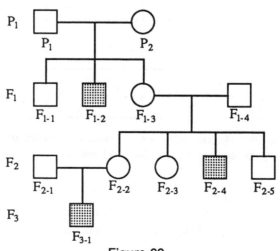

Figure 62

PROBLEM

> Different traits are characterized by certain patterns of inheritance. Which of the following are incorrectly paired?
>
> a)   Eye color — dominance/recessive inheritance
>
> b)   Red, pink, white color in plants — incomplete dominance
>
> c)   ABO blood typing — incomplete dominance
>
> d)   Hemophilia — sex-linked inheritance

Solution

   c).

2.   Evolution

A.   DARWIN'S THEORY — NATURAL SELECTION

In the 1800s, two major theories of evolution were proposed. In 1800, Jean Baptiste Lamarck proposed a theory based on the inheritance of acquired characteristics. In other words, organs, and therefore animals, evolve through use. His classic example was his explanation for why the giraffe had such a long neck: he said that giraffes stretch their necks to reach the leaves high in trees. He assumed that an animal that stretched its neck could pass a stretched and hence lengthened neck on to its offspring,

Lamarck's theory was incorrect, since characteristics acquired in life cannot be passed on to the next generation, since the information is not in the genes. For instance, if a man develops his muscles by lifting weights, his offspring would not necessarily be muscular as well.

Charles Darwin proposed a theory of evolution, largely based on his observations made during the voyage of *the Beagle*. While traveling about the Galapagos Islands off the west coast of Ecuador, Darwin observed that each of the islands had similar birds, all finches. However, these finches were of different species. The most obvious anatomical distinction among the finches was the beak. Since structure often dictates function, Darwin realized that the differences in beak structure were an adaptation to the available food sources (i.e., insects, fruit, seeds) on the various islands. The finches all were of a common ancestor, and through many generations, differentiated into distinct species particularly suited to the environment of their island. While the phrase was not coined at that time, this is a classic example of adaptive radiation, which refers to the emergence of several species from one species due to the segregation of habitats.

According to Darwin's theory, a giraffe would have a long neck because those giraffes that by chance had the selective advantage (long necks) could eat food in the trees and hence survive and reproduce. The offspring, like the parent, would have long necks. Giraffes with short necks would be selected against, and thus not be able to survive to reproduce other short-necked giraffes.

The basic premise of Darwin's theory is that individuals have differing capacities to cope with their environment. Some individuals have characteristics which are advantageous in the environment and hence, those individuals will tend to survive and reproduce. The postulates of his theory came together in his book *On the Origin of Species by Means of Natural Selection* in 1859, and are recapitulated below.

All organisms overproduce gametes. Not all gametes form offspring and of the offspring formed, not all survive. Those organisms that are most competitive (in various different aspects) will have a greater likelihood of survival. These survival traits vary from individual to individual but are passed on to the next generation, and thus over time, the best adaptations for survival are maintained. The environment determines which traits will be selected for or against, but these traits may change in time. A selected trait at one time may later be disadvantageous.

Despite all of Darwin's insight, he never could explain the mechanism whereby traits were passed on. This was the key drawback to his theory. At about the same time, unbeknownst to Darwin, Gregor Mendel was experimenting with garden peas. Mendel was the first to introduce the concept of genes (heritable factors), although he did not use the term.

B. EVIDENCE FOR EVOLUTION

Evidence for evolution comes from many scientific disciplines. Evidence from comparative biochemistry and molecular biology is based on the analysis of enzymes, proteins, nucleic acids, etc., to determine similarities in nucleotide sequences, and hence amino acid sequences. A similar molecular structure suggests a recent (by evolutionary standards) bifurcation of the original ancestral stock.

COMPARATIVE ANATOMY

Comparative anatomy distinguishes between homologous structures and analogous structures. Homologous structures are those structures that share an anatomical similarity due to a common evolutionary origin; this is exemplified by the basic bone structure in the forelimb of all vertebrates such as the flipper of a whale and the arm of a human. Analogous structures are those that serve similar functions despite differences in their anatomy, and thus have no common ancestry. The wing of an insect and the wing of a bird exemplify analogous structures.

FOSSIL RECORD

The fossil record is an important clue to evolutionary origins. While fossils can actually be the preserved remnants of life (shells, skeletons), they also include imprints and molds, such as the imprint of an animal footprint. Fossil age can be estimated by the use of carbon dating. The ratio of radioactive carbon (^{14}C) to nonradioactive carbon (^{12}C) is determined. The half life of a radioactive element is the time span in which half the atoms in a sample will have decayed. The half life of ^{14}C (to decay to ^{12}C) is 5730 years. This spontaneous rate of decay is a constant.

C. FACTORS RESPONSIBLE FOR EVOLUTIONARY CHANGE

There are four factors which bring about evolutionary change - mutation, gene flow, genetic drift and natural selection.

Genetic *mutations* are changes in the base sequence in DNA and hence genes, and as such are passed on to future generations. There are many different types of mutations. A point mutation is a substitution of a single nucleotide by another. This single base substitution changes the codon to another of the 64 possible codons. Because of the degeneracy of the genetic code and because of wobble at the third position of the codon, substitution may or may not change the amino acid encoded. If the codon does call for the incorporation of a different amino acid, there still may not be a deleterious effect, especially if an amino acid is replaced by one with similar chemical characteristics, such as replacing one nonpolar amino acid with another nonpolar one. However, if the base substitution calls for an amino acid with distinct chemical characteristics, such as a difference in the charge on the amino acid, it is more likely to be deleterious. The site of the mutation is also important. For instance if the new amino acid occurs at the functional site of a protein, such as the active site of an enzyme, this is more likely to adversely affect protein function.

Deletion or addition of a multiple of three bases results in the deletion or insertion of several amino acids, which may or may not affect protein function. Deletion or addition of one or two bases (or any other non-multiple of three) causes a reading-frame shift. All the codons beyond the mutation will be incorrectly read, and the wrong amino acids will be incorporated. Furthermore, an amino acid encoding codon may be converted to a stop codon. leading to a truncated version of the protein.

GENE FLOW

Gene flow refers to the change in allele frequency due to migration. Migration can be immigration (entrance) or emigration (exit) of individuals (and their genes) to or from a population.

GENETIC DRIFT

Genetic drift refers to a shift in allele frequency due to random fluctuation. this chance event is of special concern in small populations.

NATURAL SELECTION

Natural selection and differential reproduction are basically synonymous. This differential ability to survive and reproduce is the key factor in evolutionary change. Each individual of the population has a unique genotype (except identical twins, which have a common genotype) responsible for determining its phenotype. The environmental conditions at the time dictate which phenotypes are advantageous for survival. Those individuals with the phenotypes that can cope best in the environment will differentially reproduce and their genes will increase in frequency in the next generations.

D. HARDY-WEINBERG EQUILIBRIUM

The Hardy-Weinberg equilibrium refers to an artificial state in which the proportion of alleles at a given locus remain constant. The four factors required to maintain the equilibrium are: (1), no mutations, (2), isolation (and hence no migration and no gene flow), (3), large population size (and hence no genetic drift), and (4), equal viability and fertility of all genotypes, i.e., random reproduction (and hence no natural selection). In real populations, these factors are not usually all met and changes in allele frequency can occur.

The Hardy-Weinberg rule states that allele frequencies and genotype frequencies will be infinitely stable. For a trait which has only two alleles, p and q, the frequencies are expressed by the equation:

$$p + q = 1$$

where p is the frequency of one allele at the given locus and q is the frequency of the alternative allele. The sum of the frequencies must equal one.

Another mathematical equation holds true under a Hardy-Weinberg equilibrium:

$$(p + q)^2 = p^2 + 2pq + q^2$$

If p is the frequency of the dominant allele, then p^2 is the frequency of homozygous dominants in the population. If q is the frequency of the recessive allele, then q^2 is the frequency of homozygous recessives in the population. Thus $2pq$ is the frequency of heterozygotes in the population. Note also that

$$p^2 + 2pq + q^2 = 1.$$

Suppose a teacher does a statistical analysis of the eye color in students in a junior high school. The analysis shows that of the 1000 students, 910 have brown eyes, while only 90 have blue eyes (or green eyes). Five years later, the analysis is repeated, as the students in the first survey have graduated, and other teenagers are now in the junior high school.

The results now show that of the 1000 students, 840 have brown eyes and 160 have blue eyes. The table below summarizes the data:

Year	Brown eyes	Blue eyes	Total
1986	910	90	1000
1991	840	160	1000

In the original sample, 91% of the students have brown eyes and 9% have blue eyes. Since 9% (0.09) is equal to q^2, q must equal 0.3. Therefore p must equal 0.7 (i.e., $p = 1 - 0.3$). The 910 brown-eyed students consist of those that are homozygous and those that are heterozygous for the dominant allele. The homozygous dominant population represents 49% of the total population ($p^2 = 0.7^2$), while the heterozygous make up 42% of the total population ($2pq = 2 \times 0.7 \times 0.3$). Thus 490 students are homozygous dominant and 420 are heterozygous, giving a total of 910 brown-eyed students.

The data for 1991 is significantly different. The change in allele frequencies can be accounted for by migration (immigration and emigration). In this sample, blue eyes account for 16% of the population (160/1000). Thus $q^2 = 0.16$ and q equals 0.4. To verify this, note that p must equal 0.6 (i.e., $p = 1 - 0.4$). Thus, the homozygous dominant population would account for 36% ($p^2 = 0.6 \times 0.6$) or 360 students, and the heterozygous population would account for 48% ($2pq = 2 \times 0.6 \times 0.4$) or 480 students. Indeed, there are 840 (360 + 480) brown-eyed students in the 1991 population.

A – 116

PROBLEM

Analysis of the protein insulin shows that there is only one amino acid difference between porcine and human insulin. The scientific discipline which would make these evolutionary conclusions is

a) comparative anatomy.   b) the fossil record.

c) ^{14}C dating.   d) molecular biology.

Solution

   d).

PROBLEM

Gene flow is a factor which brings about evolutionary change due to

a) natural selection.   b) migration.

c) chance events.   d) point mutations.

Solution

   b).

PROBLEM

An equation which correctly describes a Hardy-Weinberg equilibrium is

a) $p^2 + q^2 = 1.$ b) $2p + 2pq + 2q = 1.$

c) $p^2 + 2pq^2 + q^1 = 1.$ d) $p^2 + 2pq + q^2 = 1.$

Solution

   d).

PROBLEM

If the dominant allele p has a frequency of 0.2, what is the percentage of heterozygous in a population under a Hardy-Weinberg equilibrium?

a) 16%   b) 4%

c) 64%   d) 32%

Solution

   d).

XII. Biological Molecules

1. Amino Acids and Proteins

Amino acids, as the name implies, have properties of both amines and carboxylic acids. The general formula for an amino acid is

^+H_3N - CHR - COO$^-$,

where R can be any side chain. As can be seen from the structure, amino acids are *zwitterions*, meaning that they have both negative and positive charges at neutral pH. Under acidic conditions they will react as:

^+H_3N - CHR - COO$^-$ + $H_3O^+ \longrightarrow$ ^+H_3N - CHR - COOH + H_2O.

Under basic conditions they will react by:

^+H_3N - CHR - COO$^-$ + OH$^- \longrightarrow$ ^+H_2N - CHR - COO$^-$ + H_2O.

The simplest amino acid, glycine, has a single hydrogen for its R group. Others, such as alanine and valine, have long aliphatic chains for their R group. Lysine, arginine, and histidine all have an additional amino group, and hence are basic. Serine and threonine have aliphatic hydroxyl side chains. Phenylalanine, tyrosine, and tryptophan have aromatic side chains. Aspartate and glutamate have an additional carboxylic acid group, and hence are acidic. Asparagine and glutamine have amide side chains. Cysteine and methionine contain sulfur in their side chain. Often, a one letter symbol is used to identify these amino acids. Amino acids that cannot be

Amino Acid	Three Letter Abbreviation	One Letter Symbol	Formula
Aliaphatic Amino Acids			
Glycine	Gly	G	$^+H_3N - \overset{\overset{\textstyle H}{\textstyle \|}}{\underset{\underset{\textstyle H}{\textstyle \|}}{C}} - COO^-$
Alanine	Ala	A	$^+H_3N - \overset{\overset{\textstyle H}{\textstyle \|}}{\underset{\underset{\textstyle CH_3}{\textstyle \|}}{C}} - COO^-$
Valine*	Val	V	$^+H_3N - \overset{\overset{\textstyle H}{\textstyle \|}}{\underset{\underset{\textstyle CH}{\textstyle \|}}{C}} - COO^-$

$CH_3 \quad CH_3$

Amino Acid	Three Letter Abbreviation	One Letter Symbol	Formula
Leucine*	Leu	L	$^+H_3N - \overset{\overset{\displaystyle H}{\mid}}{\underset{\underset{\displaystyle CH}{\mid}}{\underset{\underset{\displaystyle CH_2}{\mid}}{C}}} - COO^-$, $CH_3 \quad CH_3$
Isoleucine*	Ile	I	$^+H_3N - \overset{\overset{\displaystyle H}{\mid}}{\underset{\mid}{C}} - COO^-$, $H - \overset{\mid}{\underset{\mid}{C}} - CH_3$, CH_2 , CH_3

Aliphatic Hydroxyl Side Chains

Serine	Ser	S	$^+H_3N - \overset{\overset{\displaystyle H}{\mid}}{\underset{\mid}{C}} - COO^-$, $H - \overset{\mid}{\underset{\mid}{C}} - OH$, H
Threonine*	Thr	T	$^+H_3N - \overset{\overset{\displaystyle H}{\mid}}{\underset{\mid}{C}} - COO^-$, $H - \overset{\mid}{\underset{\mid}{C}} - OH$, CH_2 , CH_3

Aromatic Side Chains

Phenylalanine*	Phe	F	$^+H_3N - \overset{\overset{\displaystyle H}{\mid}}{\underset{\underset{\displaystyle CH_2}{\mid}}{C}} - COO^-$ (with phenyl ring)

Amino Acid	Three Letter Abbreviation	One Letter Symbol	Formula
Tyrosine	Tyr	Y	
Tryptophan*	Trp	W	

Basic Amino Acids

Lysine*	Lys	K	
Arginine*	Arg	R	

Tyrosine:

$$^+H_3N - \overset{\displaystyle H}{\underset{\displaystyle CH_2}{C}} - COO^-$$

(CH₂ connected to a benzene ring with OH)

Tryptophan:

$$^+H_3N - \overset{\displaystyle H}{\underset{\displaystyle CH_2}{C}} - COO^-$$

(indole ring system: benzene fused to five-membered ring containing N–H, with C, CH)

Lysine:

$$^+H_3N - \overset{\displaystyle H}{C} - COO^-$$
$$| $$
$$CH_2$$
$$| $$
$$CH_2$$
$$| $$
$$CH_2$$
$$| $$
$$CH_2$$
$$| $$
$$NH_3^+$$

Arginine:

$$^+H_3N - \overset{\displaystyle H}{C} - COO^-$$
$$| $$
$$CH_2$$
$$| $$
$$CH_2$$
$$| $$
$$CH_2$$
$$| $$
$$N - H$$
$$| $$
$$C = NH_2^+$$
$$| $$
$$NH_2$$

A – 120

Amino Acid	Three Letter Abbreviation	One Letter Symbol	Formula
Histadine*	His	H	

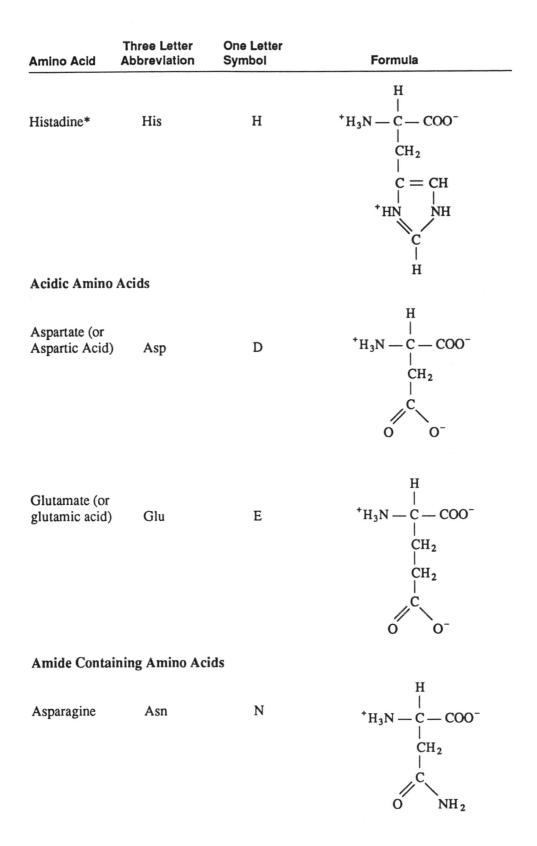

Acidic Amino Acids

| Aspartate (or Aspartic Acid) | Asp | D | |

Glutamate (or glutamic acid) Glu E

Amide Containing Amino Acids

Asparagine Asn N

Amino Acid	Three Letter Abbreviation	One Letter Symbol	Formula
Glutamine	Gln	Q	

Sulfur Containing Amino Acids

Cysteine	Cys	C	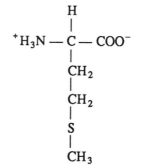
Methionine*	Met	M	

Other Amino Acids

Proline	Pro	P	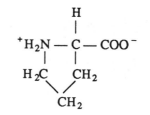

produced by the organism in question, and must be supplied by an external source, are called "essential amino acids." The table on pages 118-122 lists the abbreviations, symbols, and R groups for amino acids. Those marked with an asterisk are essential acids for *homo sapiens*.

The primary method of linking amino acids together to form peptides is by a peptide bond. The amino end of one amino acid is attracted to the carboxylic acid end of another amino acid. This bonding results in a dipeptide and water, as can be seen:

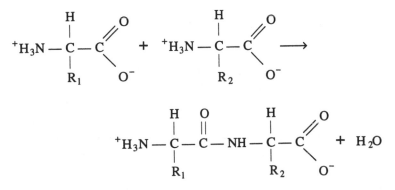

Breaking such a bond requires water; hence, it is known as hydrolysis. The sulfur containing amino acids can also bond with one another by sulfide bonds. This is important because an amino acid in one peptide chain can bond with an amino acid in another by sulfide bonds, as is illustrated below.

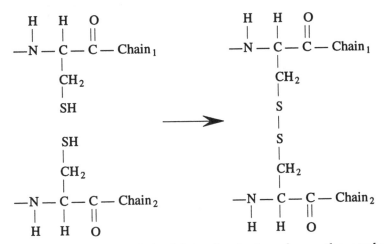

Protein molecules are made by joining hundreds and even thousands of amino acids by peptide and sulfide bonds. Each protein has a unique sequence of amino acids. This sequence of amino acids, along with the location of sulfide bonds, is known as the "primary structure" of the protein. The "second structure" refers to the seric relationship of amino acids that are close to one another in the linear sequnce.

"Tertiary structure" refers to the steric relationships of amino acids that are far apart in the linear sequence. (Remember, proteins are not straight chains, but these chains are folded, making a three-dimensional structure.) The dividing line between secondary and tertiary structure is somewhat arbitrary.

"Quaternary structure" of a protein refers to the way the various peptide chains are packed together. Each chain is referred to as a "subunit." The "isoelectric point" is the pH at which there is no net electrical charge on a protein. This is a different value for various proteins. Proteins are important to living things, as they constitute the vast majority of an organism's structure (fibrous proteins) and enzymes involved in the metabolish (globular proteins).

2. Carbohydrates

Carbohydrates all have the empirical formula

CH_2O.

Carbohydrates are polyhydroxy aldehydes, polyhydroxy ketones, or compounds that can be hydrolyzed to them. A carbohydrate that cannot be hydrolyzed to a simpler compound is called a "monosaccharide." If it can be hydrolyzed into two monosaccharides, it is called a "disaccharide." A "polysaccharide" can be hydrolyzed into many monosaccharides. Upon further classification, if a carbohydrate contains an aldehyde group, it is known as an "aldose"; if it contains a ketone group, it is known as a "ketose."

Monosaccharides are also classified by the number of carbon atoms they contain. A triose, tetrose, and pentose carbohydrate would have three, four, and five carbon atoms, respectively. An aldopentose would be a five carbon monosaccharide with an aldehyde group. A ketohexose would be a six carbon monosaccharide containing a ketone group. A ketohexose would be a six carbon monosaccharide containing a ketone group. A carbohydrate that reduces Fehling's, Benedict's, or Tollen's reagent is known as a "reducing sugar." All monosaccharides are reducing sugars. Most disaccharides are reducing sugars (sucrose is one of the exceptions). Carbohydrates can exist either in an open chain form or a closed ring. It is easy to change from one to the other. Below is a figure that shows the two forms for D-glucose. The carbon atoms are numbered in each form for reference. The figure is actually α–D-glucose. Were the hydrogen and hydroxy groups about carbon 1 to exchange places, it would be β-D-glucose. such diasteromers that differ only about carbon 1 are called "anomers."

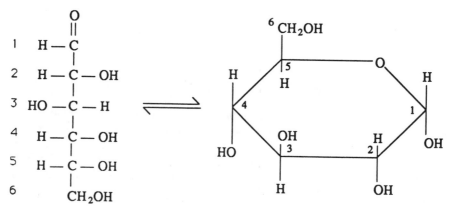

Among the aldoses, any diastereomers that differ only about the configuration of carbon 2 are called "epimers." Note that any carbohydrate can have an anomer,

A – 124

but only an aldose can have an epimer. Aldoses can be oxidized by Fehling's or Tollen's reagent, bromine water, nitric acid, and periodic acid. Ketoses can also be oxidized by Fehling's or Tollen's reagent, but not any of the other listed oxidants. The oxidation of aldoses are shown by bromine water and nitric acid in the figure below.

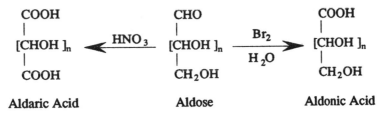

A glycoside linkage connects monosaccharides to form di- or polysaccharides. The breakage of such a linkage is shown below, changing the disaccharide, maltose, to two glucose molecules. Note that since water is involved, it is a type of hydrolysis.

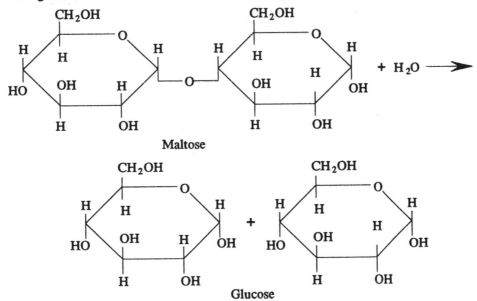

3. Lipids

Chemically, fats are carboxylic esters derived from glycerol, and are known as glycerides. The general formula for a triglyceride is shown below, where R can be

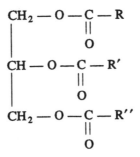

any fatty acid. Fatty acids are generally straight chained compounds from three to eighteen carbons long. As a rule, living organisms tend to produce fatty acids containing an even number of carbons. Were these fatty acids not associated with trigylcerols, they would be considered carboxylic acids. An example of a fatty acid would be linoleic acid,

$$CH_3(CH_2)_4CH=CHCH_2CH=CH(CH_2)_7COOH \text{ (cis, cis-isomer).}$$

Cholesterol is a precursor to the five major classes of steroid hormones; progestagens, glucocorticoids, mineralocorticoids, androgens, and estrogens. On page 127 is a diagram that shows cholesterol, and some of the steroids that are derived from it.

4. Phosphorus Compounds

The previous section mentions triglycerides with ester linkages to fatty acid chains. There are a group of compounds known as "phospholipids" where there are two fatty acids and a phosphate group attached to the triglycerides. These are important since phospholipids make up cell membranes, adenosine triphosphate is involved in energy exchange in all known life forms, and nucleic acids, which contain the blueprints of life, contain organic phosphates. Below is a figure that shows phosphoric acid and phosphatidic acid, a phosphoglyceride. R and R´ can be any fatty acid.

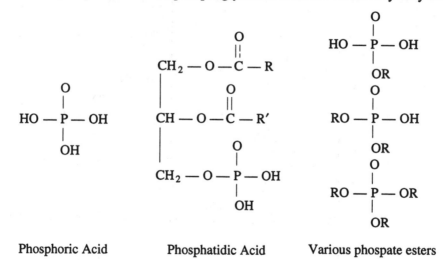

Phosphoric Acid   Phosphatidic Acid   Various phospate esters

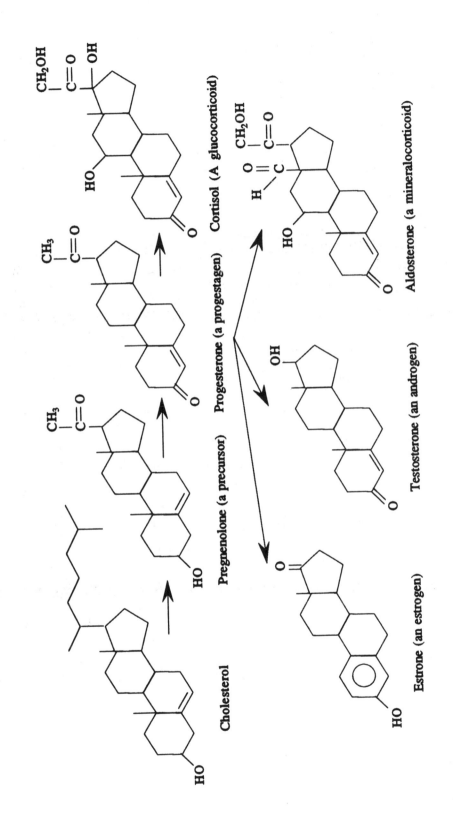

Cholesterol

Pregnenolone (a precursor)

Progesterone (a progestagen)

Cortisol (A glucocorticoid)

Aldosterone (a mineralocorticoid)

Testosterone (an androgen)

Estrone (an estrogen)

XIII. Oxygen Containing Compounds

1. Alcohols

Alcohols contain an -OH group, and the name of an alcohol ends in "-ol" (e.g., propanol, butanol, pentanol, etc.). An alcohol is further classified according to how many other carbon atoms are attached to the carbon atom that has the -OH group. Below are shown the differences between primary (1°), secondary (2°), and tertiary (3°) alcohols.

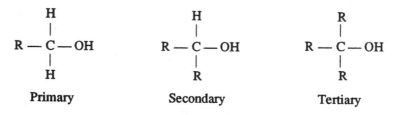

Alcohols can be dehydrated using hot acids. The ease of dehydration depends upon the type of alcohol: 3° > 2° > 1°. The steps of an alcohol dehydration are shown below for ethanol.

1) The alcohol unites with a hydrogen ion (from the acid) to form a protonated alcohol.

2) The protonated alcohol dissociates into water and a carbonium ion (hence the name "dehydration," since water is removed).

3) The carbonium ion then loses a hydrogen ion (regenerating the acid), and forms an alkene, in this case, ethene.

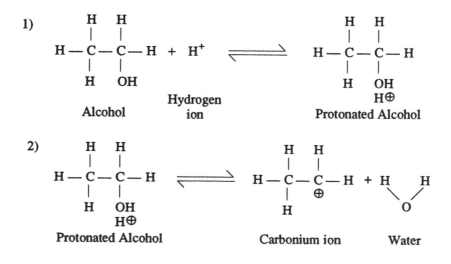

3)

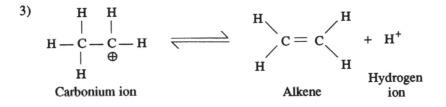

Carbonium ion   Alkene   Hydrogen ion

Alcohols will also react with hydrogen halides to yield alkyl halides and water. The order of reactivity of alcohols toward hydrogen halides is 3° > 2° > 1° > methanol. All alcohols except methanol and most 1° alcohols will react by what is termed an "S_N1" (substitution-1) reaction, whereby the halide substitutes for the -OH group in its exact location, as is shown in the example below.

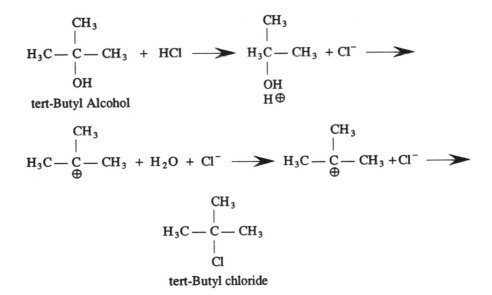

tert-Butyl Alcohol

tert-Butyl chloride

Most 1° alcohols and methanol will react by what is known as an "S_N2" (substitution-2) mechanism. This yields an alkyl halide and water, but there is a rearrangement of the ions formed, and hence the halide is on a different carbon than the -OH group was. This is shown below.

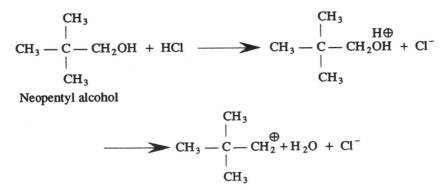

Neopentyl alcohol

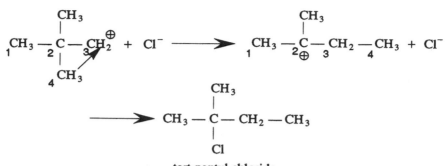

tert-pentyl chloride

Hydrogen bonding affects the physical properties of alcohols. This subject is covered in the inorganic chemistry section. Generally speaking, the more carbons in an alcohol, the less effect hydrogen bonding has, and the alcohol behaves more like a hydrocarbon. Branching, as in most organic compounds, increases melting point, boiling point, and density.

2. Aldehydes and Ketones

Aldehydes have the general formula

$$H - \overset{\overset{\text{O}}{\|}}{C} - R,$$

and ketones have the general formula

$$R - \overset{\overset{\text{O}}{\|}}{C} - R'.$$

Note that they both possess the carbonyl group. $C = O$. It is this carbonyl group that largely determines the chemistry of these compounds, hence they are collectively known as "carbonyl compounds." The names of aldehydes end in "-al" (e.g., propanal, butanal, pentanal, etc.). The names of ketones end in "-one" (e.g., 3-pentanone, 3-methyl-2-butanone, propanone, etc.). These carbonyl compounds react typically by nucleophilic addition. Aldehydes will undergo nucleophilic addition even more readily than ketones, due to electronic and steric factors. One such example of this electrophilic addition is the Grignard synthesis of alcohols from carbonyl compounds.

A Grignard reagent is prepared by mixing an appropriate organic halide with metal magnesium, using dry ether as a solvent, such that

$$RX + Mg \longrightarrow RMgX.$$

The $C - Mg$ bond is highly polar, the carbon being somewhat negative and the magnesium being somewhat positive. Due to differences in electronegativity between carbon and oxygen in the carbonyl group, the carbon has a somewhat positive charge, and the oxygen is somewhat negative. The two carbons will be attracted to one another, as will the magnesium and oxygen atoms, the result being the magnesium salt of an alcohol, which upon the addition of water, becomes the alcohol itself. The Grignard synthesis with formaldehyde (methanal) yields a 1° alcohol, higher

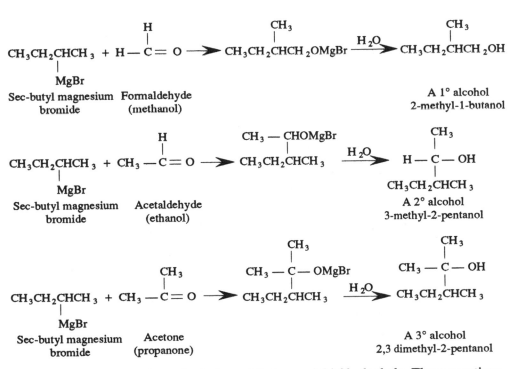

aldehydes yield secondary alcohols, and ketones yield 3° alcohols. These reactions are shown above.

Dry alcohols, in the presence of anhydrous acids, can add to the carbonyl group of aldehydes and yield an acetal. For example:

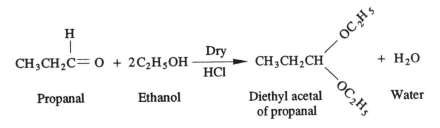

In an alcoholic solution, there is strong evidence that an aldehyde exists in equilibrium with a compound known as a "hemiacetal." A hemiacetal is formed by the addition of the nucleophilic alcohol to the carbonyl group. Hemiacetals are usually too unstable to be isolated. Below shows such a reaction:

$$R' - \overset{\displaystyle H}{\underset{\displaystyle H}{C}} = O + ROH \underset{\longleftarrow}{\overset{H^+}{\rightleftharpoons}} R' - \overset{\displaystyle H}{\underset{\displaystyle OH}{C}} - OR$$

A Hemiacetal

The basis of acetal chemistry is the "carbonium ion," whose resonating structure is as follows:

$$\left[\begin{array}{ccc} & \overset{\displaystyle H}{\underset{\displaystyle \oplus}{|}} & \\ R - C - OR & \longleftrightarrow & R - \overset{\displaystyle H}{\underset{\displaystyle \oplus}{C}} = OR \end{array} \right]$$

In like manner, "ketals" can be made using ketones. (Simple ketals are difficult to prepare by the reaction of ketones and alcohols.) The same is true of "hemiketals". Many aldehydes and ketones are converted to amines by reductive amination, a process that involves reduction in the presence of ammonia. An intermediate compound, an "imine" (RCH=NH), is formed, which is then reduced to an amine. Two examples of such a procedure are shown here.

$$CH_3(CH_2)_5 \overset{\displaystyle H}{\overset{|}{C}} = O \xrightarrow{NH_3,\ Ni} CH_3(CH_2)_5 \overset{\displaystyle H}{\overset{|}{C}} = NH \xrightarrow{H_2,\ Ni} CH_3(CH_2)_5CH_2NH_2$$

n-Heptanal   An Imine   1-aminoheptane a 1° Amine

$$CH_3(CH_2)_2 \overset{\displaystyle O}{\overset{\|}{C}} CH_3 \xrightarrow{NH_3,\ Ni} CH_3(CH_2)_2 \underset{\underset{H}{N}}{\overset{\|}{C}} CH_3 \xrightarrow{H_2,\ Ni} CH_3(CH_2)_2 \underset{NH_2}{\overset{|}{C}H} CH_3$$

2-Pentanone   An Imine   2-aminopentane a 2° Amine

Imines can exist in tautomerization with "enamines." Compounds whose structures differ markedly in arrangement of atoms, but which exist in equilibrium, are called "tautomers." "Tautmerism" is the term that describes this equilibrium. An enamine contains two double-bonded carbons, single bonded to an amine group. An imine contains a carbon double-bonded to a nitrogen. An example of such tautomerism is shown below.

$$-\underset{H}{\overset{|}{C}} - \overset{|}{C}\ \ O + R_2NH \rightleftharpoons -\underset{H}{\overset{|}{C}} - \underset{OH}{\overset{|}{C}} - \overset{R}{\overset{|}{N}} - R \rightleftharpoons -\overset{|}{C} = \overset{|}{C} - \overset{R}{\overset{|}{N}} - R$$

A 2° Imine   An Enamine

Two moles of an aldehyde or two moles of a ketone can combine with one another in the presence of a dilute base. This is known as an "aldol condensation," and the product, having the combined properties of a carbonyl (-al) and al alcohol (-ol), is called an "aldo." An aldo can react with a weak base to form a carbonyl compound that was larger than the starting material. A couple of examples follow.

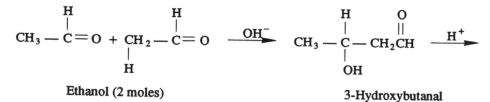

$$CH_3 - \overset{\displaystyle H}{\overset{|}{C}} = O + \underset{\underset{H}{|}}{\overset{\displaystyle H}{\overset{|}{C}}} H_2 - \overset{\displaystyle H}{\overset{|}{C}} = O \xrightarrow{OH^-} CH_3 - \underset{OH}{\overset{\displaystyle H}{\overset{|}{C}}} - CH_2 \overset{\displaystyle O}{\overset{\|}{C}} H \xrightarrow{H^+}$$

Ethanol (2 moles)   3-Hydroxybutanal

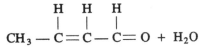

2-Butenal

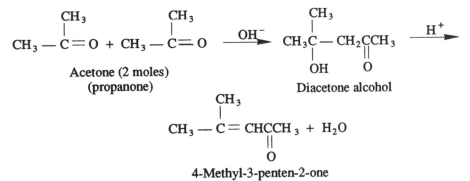

Acetone (2 moles)
(propanone)

Diacetone alcohol

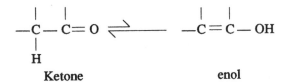

4-Methyl-3-penten-2-one

An enol is a compound that contains two doubly bonded carbons (-ene), and an alcohol group (-ol). These will exist in tautomerism with ketones, with the ketone structure the most stable, and hence the most preferred. An example follows:

$$-\overset{|}{\underset{H}{C}}-\overset{|}{C}=O \rightleftharpoons -\overset{|}{C}=\overset{|}{C}-OH$$

Ketone   enol

Aldehydes will reduce Tollen's reagent. Methyl ketones are oxidized by hypohalite. An aldol condensation will not occur if an aldehyde or ketone lack an alpha hydrogen. This is a hydrogen on the carbon next to the carbonyl group. It is the acidity of this hydrogen that causes the reaction to proceed. These are chemical tests that can be used to identify the various types of carbonyl compounds present.

A compound that not only has a carbonyl group, but also a carbon-carbon double bond, has properties that are characteristic of both functional groups. In α, β-unsaturated (containing carbon-carbon multiple bonds) carbonyl compounds, the carbon-carbon double bond and the carbon-oxygen double bond are separated by just one carbon-carbon single bond.

$$\overset{\beta}{\underset{|}{C}}=\overset{\alpha}{\underset{|}{C}}-\overset{|}{C}=O$$

Because of this conjugation, these compounds have properties of both the double carbon-carbon bond and the carbonyl group, and some special properties as well. The presence of the carbonyl group lowers the reactivity of the carbon-carbon double bond toward electrophilic addition, and also controls the orientation of the addition.

A – 133

3. Carboxylic Acids

Carboxylic acids possess both a carbonyl and a hydroxide group,

$$R - \overset{\displaystyle O}{\underset{\displaystyle |}{C}} - OH .$$

It is the -OH group that undergoes change in nearly every reaction, but does so in a way that is only due to the effect of the C=O. These acids are named ending with "-oic acid" (e.g., methanoic acid, ethanoic acid, propanoic acid, etc.), common names also abound, such as formic acid, acetic acid, etc.

When an alcohol is mixed with a carboxylic acid, the result is an ester. The type of alcohol used determines the degree of the reaction, methanol being the most reactive; methanol > 1° > 2° > 3°. Below is such a reaction:

$$CH_3C - OH + CH_3OH \longrightarrow CH_3C - OCH_3$$

Lithium aluminum hydride can decarboxylate an acid to an alcohol. A typical reaction might be the conversion of pentanoic acid to pentanol:

$$CH_3(CH_2)_3COOH \xrightarrow{\text{LiAlH}_4} CH_3(CH_2)_3CH_2OH .$$

Hydrogen bonding, as would be expected, plays an important role in intermolecular forces in carboxylic acids. Intramolecular forces include the inductive effect of substituents on the acid chain. The inductive effect is an effect that is caused by a highly electronegative or electropositive group or atom located on the chain. This effect is felt throughout the whole molecule, although it does decrease with the distance from the group. Electron withdrawing groups (e.g., NO_2, halogens, etc.) increase the stability of the carboxylate ion, and hence strengthen the acid. Again the farther away this group is from the carboxylic acid functional group, the less effect it will have. An electron releasing group, such as an amine, will destabilize the ion, and weaken the acid; again, the distance rule also comes into play. The carboxylate ion is further stabilized by resonance. It really has one and one-half bonds from the carbon to each oxygen:

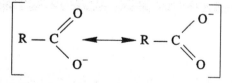

4. Common Acid Derivatives

Fats, glycerides, and esters can be "saponificated," literally, made into soap, by using excess base of a known concentration. This is analogous to an acid-base titration. The amount of base needed to convert these materials into soap is known as the "saponification equivalent," and can be used to determine their equivalent weight. The saponification reaction of an ester and a base are shown below:

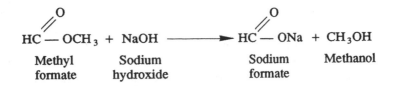

Methyl formate + Sodium hydroxide → Sodium formate + Methanol

Amides have the general formula

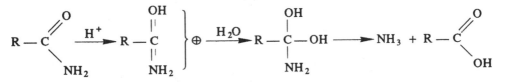

and can be hydrolized to carboxylic acids. Under acid conditions, hydrolysis involves attack by the hydroxide ion on the amide itself:

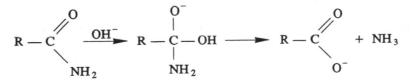

As a rule, carboxylic acid derivatives are more reactive toward nucleophilic substitution than are their non-carboxylic acid counterparts. In other words, for nucleophilic substitution, an acid chloride (R-COOCl) is more reactive than a comparable alkyl chloride (R-Cl), an amide (R-COONH$_2$) is more reactive than an amine (R-NH$_2$), and an ester (R-COOR´) is more reactive than an ether (R-OR´). Steric effects must also be taken into account for these reactions. These are the repulsive forces of the positive hydrogen nuclei on the carbons. Carbon atoms are free to swing about single bonds, and it is the repulsion of hydrogens from one carbon atom to another that keep them from getting too close.

5. Ethers

Ethers are of the general formula R-O-R´, and are named by the two groups attached to the oxygen, followed by the word "ether." (e.g., di-ethyl ether, methyl ethyl ether, butyl propyl ether, ect.) Ethers are generally very unreactive since the oxygen bond is quite stable. However, under extreme conditions (high temperatures and concentrations) ethers can be cleaved acids to yield alkyl halides and alcohols. HI is the most reactive acid used for this, followed by HBr, followed by HCl. Cleavage includes a nucleophilic attack by a halid ion on a protonated ether. (Since the ether accepts this proton, it is weakly basic.) The result is the displacement of the weakly basic alcohol molecule. This is shown below:

$$H_3 - O - CH_3 + HI \rightleftharpoons CH_3 - \underset{\underset{H\oplus}{|}}{O} - CH_3 + I^-$$

$$\xrightarrow{S_N1 \text{ or } S_N2} CH_3I + CH_3OH$$

6. Phenols

Phenols are of the general formula of an -OH group attached directly to an aromatic ring. Intermolecular hydrogen bonding causes the boiling and melting points of phenols to be high. Phenols are much stronger acids than are alcohols, which is due to the influence of the aromatic ring. Electron attracting groups, such as halides, or NO_2, increase the acidity of phenols, while electron releasing groups, such as CH_3, decrease acidity.

XIV. Amines

An amine has the general formula

$$RNH_2, \ R_2NH, \text{ and } R_3N,$$

for primary (1°), secondary (2°), and tertiary (3°) amines. Amines are quite basic, strong enough to turn litmus blue. Amines are named by naming the group to which they are attached, and adding the word "-amine" at the end. Stereoisomers of amines do exist, but the energy barrier between two possible arrangements about the nitrogen atom is so low that optical isomers are rapidly interconverted before they can be isolated. Primary and secondary amines can be converted to amides by reaction with acyl chlorides; tertiary amines will not react in this manner. This is shown below.

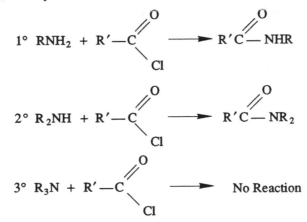

Amines can also react by alkylation. By this method, a primary amine can react with an alkyl halide to yield a secondary amine and a hydrogen halide. This process can be repeated with the secondary to form a tertiary amine. Finally the tertiary amine can be converted by the same reaction to a quaternary ammonium salt. This can be summarized as

$$RNH_2 \xrightarrow{\text{HX}} R_2NH \xrightarrow{\text{HX}} R_3N \xrightarrow{\text{HX}} R_4N^+ + X^-.$$

Quaternary ammonia salts have four organic groups covalently bonded to a nitrogen atom, and the positive charge of this ion is balanced by some negative ion. When the salt of a 1°, 2°, or 3° amine is treated with hydroxide ion, the nitrogen gives a hydrogen ion and the free amine is liberated. The quaternary ammonium ion has no proton to give up, and hence it is not affected by hydroxide ion. The quaternary ammonium salt does react with silver oxide to form a quarternary ammonium hydroxide and a precipitate of silver halide:

$$R_4N^+ + X + Ag_2O \longrightarrow P_4N^+OH^- + AgX.$$

Amines are more basic than water, and less basic than hydroxide ions. Aliphatic amines have k_b's that are from 10^{-3} to 10^{-4}, stronger than that of ammonia. Aromatic amines have lower k_b's, 10^{-9} or less, far lower than ammonia. Electron releasing groups on the aromatic rings, such as CH_3, stabilize the cation, and increase the basicity. Electron withdrawing groups. on the aromatic ring, such as COOH, halogens, etc. destabilize the cation, and decrease the basicity.

XV.   Hydrocarbons

1. Saturated

Compounds made up only of hydrogen and carbon are appropriately called "hydrocarbons." If they contain all of the hydrogen atoms the carbons will allow (i.e., no double or triple bonds), they are referred to as "saturated." These saturated hydrocarbons are known properly as "alkanes." The alkanes are named according to the number of carbon atoms present. Methane has one carbon, ethane has two, propane has three, butane has four, and from there Greek roots are used (e.g., pentane for five, hexane, for six, heptane for seven, etc.). If the alkane is in a ring structure, rather than a chain, the name is prefixed with "cyclo-." The higher the molecular weight of an alkane, and the more branching and side chains, the higher the melting and boiling points are.

Alkanes are generally very unreactive; however, they can be combusted to carbon dioxide and water (this is covered in the inorganic section), and they can be halogenated by a free radical reaction. A free radical chain reaction consist of

1)   A chain initiating step,

2)   Chain propagating steps, and

3)   Chain terminating steps.

The overall reaction for the halogenation of alkanes can be shown as

$$RH + X_2 \longrightarrow RX + HX.$$

The ease of the halogenation decreases as the molecular weight of the halogen increases. In decreasing order of reactivity,

$$F > Cl > Br > I.$$

If ethane is mixed with chlorine gas, nothing will occur. However, if the mixture is heated, or light is shone upon it, the halogenation reaction will begin, and continue even after the source of heat or light has been removed. The light or heat is involved in the chain initiating step. In this instance it is the dissociation of a chlorine molecule into two chlorine radicals:

$$Cl_2 + light/heat \longrightarrow 2Cl\cdot.$$

The chlorine radicals can then react with the ethane to form hydrogen chloride and an ethane radical:

$$Cl\cdot + CH_3CH_3 \longrightarrow HCl + CH_3CH_2\cdot.$$

The ethane radical can react with a molecule of chlorine to produce chloroethane and a chlorine radical:

$$CH_3CH_2\cdot\ Cl_2 \longrightarrow CH_3CH_2Cl + Cl\cdot.$$

The last two reactions presented are chain propagation steps. So long as they continue, so does the chain reaction. Chain terminating steps halt the reaction. One such example is a chlorine radical reacting with an ethane radical. The result is still chloroethane, but no free radicals are generated to continue the reaction:

$$CH_3CH_2{}^\bullet + Cl^\bullet \longrightarrow CH_3CH_2Cl.$$

Other chain terminating reactions will not only stop the chain reaction, but will not produce any product. Such an example would be two chlorine radicals coming together to form a chlorine molecule. The chain is broken, but no product is formed:

$$Cl^\bullet + Cl^\bullet \longrightarrow Cl_2{}^\bullet.$$

Sometimes a chain terminating step will actually produce a contaminant. Such an example would be two ethane radicals coming together to produce a molecule of butane;

$$2CH_3CH_2{}^\bullet \longrightarrow CH_3CH_2CH_2CH_3.$$

Inhibitors can also absorb these radicals and terminate the chain reaction. An example of such would be oxygen. Were oxygen to get into the system, it would react with the ethane radical as shown:

$$CH_3CH_2{}^\bullet + O_2 \longrightarrow CH_3CH_2 - O - O^\bullet.$$

The radical formed can do very little to continue the chain reaction.

Small cyclic hydrocarbons undergo a great deal of bond strain. Cyclopropane has bond angles of 60°, and cyclobutane 90°. Both rapidly undergo ring opening reactions to relieve this strain. Cyclopentane has carbon bond angles of 108°, and hence is fairly stable. Ring structures over five carbons are quite stable, as the three-dimensional shape of the molecule allows the ring to pucker, and all carbon bonds are at the 109.5° angle of the tetrahedron bond.

2. Unsaturated

Alkenes possess a double carbon-carbon bond, and as a result possess two less hydrogens than their alkane counterparts. The general formula for an alkene is C_nH_{2n}, where n is the number of hydrogen bonds. Geometric isomers can exist for alkenes, due to their double bonds. If two identical functional groups are on the same side of the double-bond, the molecule is referred to as a "cis" isomer. If they are on different sides, it is a "trans" isomer. If there are more than two different groups around the double bond, a different system is used. The atoms on each carbon are ranked according to their atomic weight. if the two higher weight atoms are on the same side of the double bond, it is the Z isomer, and if they are on the opposite sides, it is the E isomer. The illustrations below give examples of how this nomenclature is used.

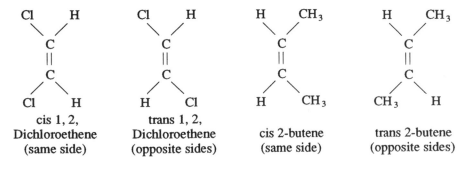

| cis 1, 2, Dichloroethene (same side) | trans 1, 2, Dichloroethene (opposite sides) | cis 2-butene (same side) | trans 2-butene (opposite sides) |

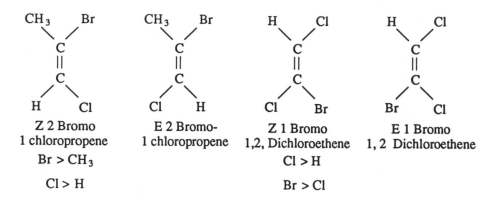

Z 2 Bromo
1 chloropropene

Br > CH₃

Cl > H

E 2 Bromo-
1 chloropropene

Z 1 Bromo
1,2, Dichloroethene

Cl > H

Br > Cl

E 1 Bromo
1, 2 Dichloroethene

Alkenes are named using the same system for alkanes, but the "-ane" ending is replaced with "-ene." (e.g., ethene, propene, hexane, etc.) The boiling and melting points of alkenes are relatively similar to their alkane counterparts.

A notable difference between alkanes and alkenes involves the reaction of alkenes with hydrogen halides and with water. The addition of hydrogen halides destroys the double bond, and the hydrogen and halide will add to the two carbons that were once double bonded. The hydrogen will go to the carbon atom attached to the fewest carbons. This is shown in the following example.

$$CH_3 - CH = CH_2 + HI \longrightarrow CH_3 - CH - CH_2 \quad \textbf{NOT} \quad CH_3 - C - CH_2$$

Propene   2-iodopropane   1-iodopropane

Alkenes will also react with water in the presence of acids to form alcohols. The water is split into H and OH groups, and these groups add at the site of the double bond, destroying it. Again, if there is a difference among the two carbons, the hydrogen will add to the one attached to the fewest other carbons. This is shown below.

$$CH_3 - C = CH_2 + H_2O \xrightarrow{H^+} CH_3 - CH - CH_2 \quad \textbf{NOT} \quad CH_3 - CH - CH_2$$

Propene   2-propanol   1-propanol

3. Aromatic

Aromatic compounds are benzene and compounds that resemble benzene in behavior. Benzene is a molecule of cyclohexane with three double bonds, but it behaves so chemically different than the properties that would be predicted for "cyclohexatriene" that it has not only been given its own name, but its own category, the aromatics. Benzene is a very flat molecule with all of the hydrogens and carbons lying in the same plane. It is also very symmetrical, with every bond angle equalling 120°. The electrons in the carbon atoms are shared in several bonds in benzene; it is this delocalization of these electrons that make the molecule so stable. Cyclohexadene, (one double-bond) and cyclohexadiene (two double bonds) differ as predicted from cyclohexane. Benzene, with the three double bonds, differs greatly from its predicted properties due to this delocalization of electrons and resonance stability. Below is a diagram of the resonance hybrid to represent benzene, and the symbol that is commonly used to represent benzene, or an aromatic ring.

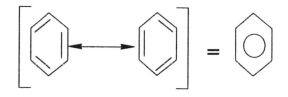

XVI. Molecular Structure of Organic Compounds

1. σ AND π BONDS

All double and single bonds consist of what is called a sigma (σ) bond. In addition to the sigma bond, double bonds also possess a pi (π) bond. Triple bonds are made of two pi's and a sigma. For example, in a molecule of ethene, $CH_2=CH_2$, there is only one pi bond and five sigmas (don't forget to count the single bonds going to the hydrogens!). Hybrid orbitals and the determination of molecular structure are covered in the inorganic section. The delocalization of electrons in resonating structures are covered in the above section on aromatics.

2. Multiple Bonding

The length of a carbon-carbon single bond is 0.153 nm, a double carbon-carbon bond is 0.134 nm, and a carbon-hydrogen bond is 0.110 nm. Generally speaking, the shorter a bond is, the less energetic it is, hence the *easier it is to break*. Atoms are free to rotate about single bonds, hence we speak of rotational energy and rotational isomers. The more single bonds (carbon-carbon) a molecule has, the more rotational configurations are possible. Double and triple bonds do not permit this type of rotation. For each double bond in a compound, a number of rotational isomers are eliminated. but the possibility of cis, trans, Z, and E isomers are introduced.

3. Stereochemistry

General isomerism is covered in the inorganic chemistry section. Isomers that differ only in the way their atoms are oriented in space are called "stereoisomers." Stereoisomers that are mirror images of one another are called "enantiomers." If they are not mirror images of one another, they are called "diastereomers." The rotational isomers mentioned in the section above are also known as "conformational isomers." Many organic substances possess the ability to rotate a beam of polarized light when in solution. This phenomenon is known as "optical activity." The symbols + and - are used to denote materials that rotate polarized light to the right and to the left, respectively. Enantiomers of the same substance will have identical physical properties, and can only be distinguished by measuring the direction they bend polarized light. Their biological properties, however, can be markedly different. A "racemic mixture" is one that contains equal amounts of both enantiomers, and hence, does not bend polarized light at all. A racemic mixture is designed by ±. Optically active molecules can be classified as R or S. Holding the central atom of such a molecule, with the lowest molecular weight group pointing away, the sequence of the remaining three groups is noted. If they decrease in molecular weight in a clockwise direction, the isomer is denoted R, and if they decrease in a counter-clockwise direction, they are denoted as S. It must be stressed that one cannot tell simply from the R or S nomenclature whether or not they are + or -.

XVII. Separations and Purification

1. Extraction

A material can be transferred from one solvent to another that it is even more soluble in. The only requirement is that the two solvents used must not be miscible in one another. It is standard procedure for the solvent used in the extraction to be volatile so it can be evaporated, and the solute recovered. In practice the extraction solvent is added to the solution and the container agitated. The container is then left to stand so the two phases of the immiscible solvents are completely separated. The phase contining the solute is then drained off. In practice, it is more efficient to perform several extractions using a small amount of solvent each time, rather than one extraction using a large amount of solvent.

2. Chromatography

Chromatography uses the differences in polarity and molecular weight to separate compounds. Any chromatography system consists of a mobile phase, such as a gas, which moves with the sample(s), and a stationary phase, which is a type of support that the samples and mobile phase move along.

In gas chromatography, the stationary phase is a liquid that is absorbed onto an inert solid. The two most commonly used solid "support phases" are crushed firebrick, and a similar material, kieselguhr. A variety of liquid stationary phases are used depending upon the nature of the material that is being analyzed. Materials such as squalene, n-hexadecane, polyethylene gylcol, and silicone oils and gums are commonly used. The choice depends upon the polarities of the substances being separated, and the maximum temperature that is being used for the separation.

The mobile phase is a gas that will not react with the stationary phase nor the materials being separated. Helium, nitrogen, and hydrogen are commonly used, with helium by far being the most common. The stationary phases are packed into a column, usually stainless steel or glass, connected to a source of gas, and placed into an oven. A small amount of sample (liquid or gas) is injected into one end of the column in the direction of the gas flow. Depending upon the oven temperature and differences in polarity between the stationary phase and the materials being separated, the different components of the sample will have a "retention time," which is the time it takes them to pass through the column under the given conditions.

The two most common detectors at the end of these columns to detect the species as they leave the column are the thermal conductivity detector (TCD, or "hot-wire") and the flame ionization detector (FID). The TCD consists of a simple platinum glow-wire which emits infrared radiation (heat) that passes through the end of the column. A detector on the other side of the column records any changes in the

I.R. radiation passing through the column. Since the most commonly used carrier gas, helium, has such a vastly different thermoconductivity than anything else (except possibly hydrogen), any material that passes through the column other than the carrier gas will absorb the I.R. radiation differently, which will be registered by the detector, and converted into an electronic signal.

An FID consists of a hydrogen-oxygen flame. This flame is hot enough to volatilize and ionize most substances. An ion detector above the flame registers the ions thus produced. This type of detector is about two orders of magnitude more sensitive than the TCD. It has the disadvantage that it cannot detect water, since water is a product of the hydrogen-oxygen flame.

Thin-layer chromatography is technically simpler, but no less useful of a technique. The stationary phase is coated in a thin layer onto some rigid sheet, which is usually a glass plate. Common stationary phases include starch, silica gel, cellulose, and aluminum oxide. These are applied by dissolving them in a solvent (usually water) to make a slurry, and spreading a thin film on a glass plate, which is then baked in an oven to remove the solvent. The sample, in liquid form, or dissolved in the mobile phase, is spotted at the bottom of the plate by a pipet and allowed to dry. The plate is then placed in a jar with a small amount of the mobile phase in the bottom. The jar is covered so the vapor pressure of the mobile phase is in equilibrium with the liquid.

Mobile phases are usually organic solvents such as benzene, chloroform, ethanol, butanol, acetone, and many more, as well as combinations of these. Differences between the polarity of the mobile phase and the substances of analysis cause the mobile phase to carry these to different heights on the plate. Once the solvent has run all of the way to the top of the plate, the plate is removed from the chamber and dried. Unless the materials of interest are naturally colored, they must be stained before they can be detected. Commonly used stains include ninhydrin, pH indicators, iodine, and potassium permanganate. Sometimes, ultraviolet light is used if the material fluoresces.

3. Distillation

A mixture of two substances, only one of which is volatile, can be separated by distillation. A very common example is the removal of salt from water. The water will boil and vaporize at a far lower termperature than the salt. The water boils away, leaving the salt behind. The water vapor can then be condensed, and pure water recovered. This is known as simple distillation, and is a technique that has been used for centuries.

If both components in a mixture have boiling points that are fairly close, such as benzene (b.p. = 85°) and pentane (b.p. = 36°), simple distillation will not do a good job of separating them. In the above example, analysis would show that after distillation, there would be a slight enrichment of pentane (the lower boiling material) in the distillate (that which was vaporized and recondensed), and a slight enrichment of benzene, (the higher boiling component) in the residue (the material that remains in the distillation flask), but not enough of an enrichment in either to be of any use. In such instances "fractional distillation" is used. A column filled with glass beads, or any other inert material offering a larger surface area, is placed between the distillation apparatus and the condenser. The less volatile vapors (in this case, benzene) will condense onto the surface of the beads, and fall back into the distillation flask, directly below. In such a manner a good separation of the above mixture can be realized.

4. Recrystallization

Recrystallization is commonly used to prepare high purity chemicals. The original batch of impure crystals are dissolved in a "good" solvent, one that they are readily soluble in. A second "bad" solvent is then added. The material of interest is insoluble in this, but the impurities present are. The substance of interest will then precipitate out of the solution. Another method involves the use of only one solvent. The impure crystals are dissolved in a minimum amount of hot solvent. The solution is then cooled. If the solubility of the material of interest is minimal in the cold solvent, but the impurities are still soluble, the material of interest will crystallize out. To do this, one must know the solubilities of the material of interest, as well as the impurities that are present, at a variety of temperatures in the solvent used.

XVIII. Use of Spectroscopy in Structural Identification

1. Infrared Spectroscopy

Infrared radiation consists of wavelengths from 0.78 to 1,000 um. The wavelengths from 2.5 to 15 um are most commonly used in spectroscopy, which corresponds to a frequency range of 1.2×10^{14} to 2.0×10^{13} Hz. In spectroscopy, wavenumber (δ), which is the reciprocal of the wavelength in centimeters, is commonly used. This would correspond to a wavenumber of 4,000 to 670 cm^{-1}. Recall from organic chemistry that atoms attached to a carbon atom can rotate about a single bond. The bonds can also vibrate and stretch. The energy needed to excite these molecules in such a manner is present in the I.R. band. For example, I.R. radiation from 3700 to 3100 cm^{-1} (2.7 to 3.2 um) will be absorbed by O-H and N-H bonds and cause them to vibrate. Hence, if I.R. radiation is sent through an organic sample, and the detector finds that the above region has been absorbed by the sample, then it is likely that the sample contains O-H and/or N-H bonds. In a like manner, triple bonds absorb between 2700 to 1850 cm^{-1}, and carbon-carbon double bonds absorb between 1950 to 1550 cm^{-1}. Tables are available that list various organic functional groups, and the I.R. regions that they absorb. Books are available that show the I.R. spectrum of known compounds.

2. NMR Spectroscopy

A strong magnetic field can cause the nuclei of certain atoms to be split into two or more quantisized energy levels. Absorption of electromagnetic energy in the range of 0.1 to 100 MHz (3,000 to 3 m) will cause transitions in the magnetically induced energy levels of the nuclei. This process is known as nuclear magnetic resonance (NMR). The sample is placed in a strong magnetic field and spun. Radiofrequencies of the above wavelengths are then introduced to the sample. The nucleus in question must have a quantum spin of $1/_2$ to be of any use with this procedure. Ordinary hydrogen, H^1, is the most commonly observed nuclei in this procedure; however, none of the solvent used to dissolve the sample can contain H^1. Hence, ordinary water would not be suitable, nor would ordinary pentane. (However, heavy water, or deuterated pentane, where the ordinary H^1 has been replaced by H^2 (deuterium) will work find, provided the sample is soluble in one of them.)

C^{13} is the next most commonly used atom for NMR studies. However, its natural abundance is only 1% (H^1 is 99.99% abundant in nature, by comparison), which limits its use. Other nuclei used in NMR are F^{19} and P^{31}. The frequency of the radiation absorbed (usually calibrated in ppm) gives a clue to the surroundings of the atom of interest (e.g., for a proton, whether it is on a methyl or ethyl group, or a benzene ring.) The number of splits in the absorpotion peak indicate how many other atoms of the same element are adjacent. The ratio of area of the peaks in the total NMR spectra show the relative ratios of the different types of atoms present.

GRE

BIOLOGY TEST

MODEL TEST I

THE GRADUATE RECORD EXAMINATION

BIOLOGY TEST

ANSWER SHEET

1. Ⓐ Ⓑ Ⓒ Ⓓ Ⓔ
2. Ⓐ Ⓑ Ⓒ Ⓓ Ⓔ
3. Ⓐ Ⓑ Ⓒ Ⓓ Ⓔ
4. Ⓐ Ⓑ Ⓒ Ⓓ Ⓔ
5. Ⓐ Ⓑ Ⓒ Ⓓ Ⓔ
6. Ⓐ Ⓑ Ⓒ Ⓓ Ⓔ
7. Ⓐ Ⓑ Ⓒ Ⓓ Ⓔ
8. Ⓐ Ⓑ Ⓒ Ⓓ Ⓔ
9. Ⓐ Ⓑ Ⓒ Ⓓ Ⓔ
10. Ⓐ Ⓑ Ⓒ Ⓓ Ⓔ
11. Ⓐ Ⓑ Ⓒ Ⓓ Ⓔ
12. Ⓐ Ⓑ Ⓒ Ⓓ Ⓔ
13. Ⓐ Ⓑ Ⓒ Ⓓ Ⓔ
14. Ⓐ Ⓑ Ⓒ Ⓓ Ⓔ
15. Ⓐ Ⓑ Ⓒ Ⓓ Ⓔ
16. Ⓐ Ⓑ Ⓒ Ⓓ Ⓔ
17. Ⓐ Ⓑ Ⓒ Ⓓ Ⓔ
18. Ⓐ Ⓑ Ⓒ Ⓓ Ⓔ
19. Ⓐ Ⓑ Ⓒ Ⓓ Ⓔ
20. Ⓐ Ⓑ Ⓒ Ⓓ Ⓔ
21. Ⓐ Ⓑ Ⓒ Ⓓ Ⓔ
22. Ⓐ Ⓑ Ⓒ Ⓓ Ⓔ
23. Ⓐ Ⓑ Ⓒ Ⓓ Ⓔ
24. Ⓐ Ⓑ Ⓒ Ⓓ Ⓔ

25. Ⓐ Ⓑ Ⓒ Ⓓ Ⓔ
26. Ⓐ Ⓑ Ⓒ Ⓓ Ⓔ
27. Ⓐ Ⓑ Ⓒ Ⓓ Ⓔ
28. Ⓐ Ⓑ Ⓒ Ⓓ Ⓔ
29. Ⓐ Ⓑ Ⓒ Ⓓ Ⓔ
30. Ⓐ Ⓑ Ⓒ Ⓓ Ⓔ
31. Ⓐ Ⓑ Ⓒ Ⓓ Ⓔ
32. Ⓐ Ⓑ Ⓒ Ⓓ Ⓔ
33. Ⓐ Ⓑ Ⓒ Ⓓ Ⓔ
34. Ⓐ Ⓑ Ⓒ Ⓓ Ⓔ
35. Ⓐ Ⓑ Ⓒ Ⓓ Ⓔ
36. Ⓐ Ⓑ Ⓒ Ⓓ Ⓔ
37. Ⓐ Ⓑ Ⓒ Ⓓ Ⓔ
38. Ⓐ Ⓑ Ⓒ Ⓓ Ⓔ
39. Ⓐ Ⓑ Ⓒ Ⓓ Ⓔ
40. Ⓐ Ⓑ Ⓒ Ⓓ Ⓔ
41. Ⓐ Ⓑ Ⓒ Ⓓ Ⓔ
42. Ⓐ Ⓑ Ⓒ Ⓓ Ⓔ
43. Ⓐ Ⓑ Ⓒ Ⓓ Ⓔ
44. Ⓐ Ⓑ Ⓒ Ⓓ Ⓔ
45. Ⓐ Ⓑ Ⓒ Ⓓ Ⓔ
46. Ⓐ Ⓑ Ⓒ Ⓓ Ⓔ
47. Ⓐ Ⓑ Ⓒ Ⓓ Ⓔ
48. Ⓐ Ⓑ Ⓒ Ⓓ Ⓔ

49. Ⓐ Ⓑ Ⓒ Ⓓ Ⓔ
50. Ⓐ Ⓑ Ⓒ Ⓓ Ⓔ
51. Ⓐ Ⓑ Ⓒ Ⓓ Ⓔ
52. Ⓐ Ⓑ Ⓒ Ⓓ Ⓔ
53. Ⓐ Ⓑ Ⓒ Ⓓ Ⓔ
54. Ⓐ Ⓑ Ⓒ Ⓓ Ⓔ
55. Ⓐ Ⓑ Ⓒ Ⓓ Ⓔ
56. Ⓐ Ⓑ Ⓒ Ⓓ Ⓔ
57. Ⓐ Ⓑ Ⓒ Ⓓ Ⓔ
58. Ⓐ Ⓑ Ⓒ Ⓓ Ⓔ
59. Ⓐ Ⓑ Ⓒ Ⓓ Ⓔ
60. Ⓐ Ⓑ Ⓒ Ⓓ Ⓔ
61. Ⓐ Ⓑ Ⓒ Ⓓ Ⓔ
62. Ⓐ Ⓑ Ⓒ Ⓓ Ⓔ
63. Ⓐ Ⓑ Ⓒ Ⓓ Ⓔ
64. Ⓐ Ⓑ Ⓒ Ⓓ Ⓔ
65. Ⓐ Ⓑ Ⓒ Ⓓ Ⓔ
66. Ⓐ Ⓑ Ⓒ Ⓓ Ⓔ
67. Ⓐ Ⓑ Ⓒ Ⓓ Ⓔ
68. Ⓐ Ⓑ Ⓒ Ⓓ Ⓔ
69. Ⓐ Ⓑ Ⓒ Ⓓ Ⓔ
70. Ⓐ Ⓑ Ⓒ Ⓓ Ⓔ
71. Ⓐ Ⓑ Ⓒ Ⓓ Ⓔ
72. Ⓐ Ⓑ Ⓒ Ⓓ Ⓔ

73. Ⓐ Ⓑ Ⓒ Ⓓ Ⓔ	97. Ⓐ Ⓑ Ⓒ Ⓓ Ⓔ	121. Ⓐ Ⓑ Ⓒ Ⓓ Ⓔ
74. Ⓐ Ⓑ Ⓒ Ⓓ Ⓔ	98. Ⓐ Ⓑ Ⓒ Ⓓ Ⓔ	122. Ⓐ Ⓑ Ⓒ Ⓓ Ⓔ
75. Ⓐ Ⓑ Ⓒ Ⓓ Ⓔ	99. Ⓐ Ⓑ Ⓒ Ⓓ Ⓔ	123. Ⓐ Ⓑ Ⓒ Ⓓ Ⓔ
76. Ⓐ Ⓑ Ⓒ Ⓓ Ⓔ	100. Ⓐ Ⓑ Ⓒ Ⓓ Ⓔ	124. Ⓐ Ⓑ Ⓒ Ⓓ Ⓔ
77. Ⓐ Ⓑ Ⓒ Ⓓ Ⓔ	101. Ⓐ Ⓑ Ⓒ Ⓓ Ⓔ	125. Ⓐ Ⓑ Ⓒ Ⓓ Ⓔ
78. Ⓐ Ⓑ Ⓒ Ⓓ Ⓔ	102. Ⓐ Ⓑ Ⓒ Ⓓ Ⓔ	126. Ⓐ Ⓑ Ⓒ Ⓓ Ⓔ
79. Ⓐ Ⓑ Ⓒ Ⓓ Ⓔ	103. Ⓐ Ⓑ Ⓒ Ⓓ Ⓔ	127. Ⓐ Ⓑ Ⓒ Ⓓ Ⓔ
80. Ⓐ Ⓑ Ⓒ Ⓓ Ⓔ	104. Ⓐ Ⓑ Ⓒ Ⓓ Ⓔ	128. Ⓐ Ⓑ Ⓒ Ⓓ Ⓔ
81. Ⓐ Ⓑ Ⓒ Ⓓ Ⓔ	105. Ⓐ Ⓑ Ⓒ Ⓓ Ⓔ	129. Ⓐ Ⓑ Ⓒ Ⓓ Ⓔ
82. Ⓐ Ⓑ Ⓒ Ⓓ Ⓔ	106. Ⓐ Ⓑ Ⓒ Ⓓ Ⓔ	130. Ⓐ Ⓑ Ⓒ Ⓓ Ⓔ
83. Ⓐ Ⓑ Ⓒ Ⓓ Ⓔ	107. Ⓐ Ⓑ Ⓒ Ⓓ Ⓔ	131. Ⓐ Ⓑ Ⓒ Ⓓ Ⓔ
84. Ⓐ Ⓑ Ⓒ Ⓓ Ⓔ	108. Ⓐ Ⓑ Ⓒ Ⓓ Ⓔ	132. Ⓐ Ⓑ Ⓒ Ⓓ Ⓔ
85. Ⓐ Ⓑ Ⓒ Ⓓ Ⓔ	109. Ⓐ Ⓑ Ⓒ Ⓓ Ⓔ	133. Ⓐ Ⓑ Ⓒ Ⓓ Ⓔ
86. Ⓐ Ⓑ Ⓒ Ⓓ Ⓔ	110. Ⓐ Ⓑ Ⓒ Ⓓ Ⓔ	134. Ⓐ Ⓑ Ⓒ Ⓓ Ⓔ
87. Ⓐ Ⓑ Ⓒ Ⓓ Ⓔ	111. Ⓐ Ⓑ Ⓒ Ⓓ Ⓔ	135. Ⓐ Ⓑ Ⓒ Ⓓ Ⓔ
88. Ⓐ Ⓑ Ⓒ Ⓓ Ⓔ	112. Ⓐ Ⓑ Ⓒ Ⓓ Ⓔ	136. Ⓐ Ⓑ Ⓒ Ⓓ Ⓔ
89. Ⓐ Ⓑ Ⓒ Ⓓ Ⓔ	113. Ⓐ Ⓑ Ⓒ Ⓓ Ⓔ	137. Ⓐ Ⓑ Ⓒ Ⓓ Ⓔ
90. Ⓐ Ⓑ Ⓒ Ⓓ Ⓔ	114. Ⓐ Ⓑ Ⓒ Ⓓ Ⓔ	138. Ⓐ Ⓑ Ⓒ Ⓓ Ⓔ
91. Ⓐ Ⓑ Ⓒ Ⓓ Ⓔ	115. Ⓐ Ⓑ Ⓒ Ⓓ Ⓔ	139. Ⓐ Ⓑ Ⓒ Ⓓ Ⓔ
92. Ⓐ Ⓑ Ⓒ Ⓓ Ⓔ	116. Ⓐ Ⓑ Ⓒ Ⓓ Ⓔ	140. Ⓐ Ⓑ Ⓒ Ⓓ Ⓔ
93. Ⓐ Ⓑ Ⓒ Ⓓ Ⓔ	117. Ⓐ Ⓑ Ⓒ Ⓓ Ⓔ	141. Ⓐ Ⓑ Ⓒ Ⓓ Ⓔ
94. Ⓐ Ⓑ Ⓒ Ⓓ Ⓔ	118. Ⓐ Ⓑ Ⓒ Ⓓ Ⓔ	142. Ⓐ Ⓑ Ⓒ Ⓓ Ⓔ
95. Ⓐ Ⓑ Ⓒ Ⓓ Ⓔ	119. Ⓐ Ⓑ Ⓒ Ⓓ Ⓔ	143. Ⓐ Ⓑ Ⓒ Ⓓ Ⓔ
96. Ⓐ Ⓑ Ⓒ Ⓓ Ⓔ	120. Ⓐ Ⓑ Ⓒ Ⓓ Ⓔ	144. Ⓐ Ⓑ Ⓒ Ⓓ Ⓔ

145. Ⓐ Ⓑ Ⓒ Ⓓ Ⓔ
146. Ⓐ Ⓑ Ⓒ Ⓓ Ⓔ
147. Ⓐ Ⓑ Ⓒ Ⓓ Ⓔ
148. Ⓐ Ⓑ Ⓒ Ⓓ Ⓔ
149. Ⓐ Ⓑ Ⓒ Ⓓ Ⓔ
150. Ⓐ Ⓑ Ⓒ Ⓓ Ⓔ
151. Ⓐ Ⓑ Ⓒ Ⓓ Ⓔ
152. Ⓐ Ⓑ Ⓒ Ⓓ Ⓔ
153. Ⓐ Ⓑ Ⓒ Ⓓ Ⓔ
154. Ⓐ Ⓑ Ⓒ Ⓓ Ⓔ
155. Ⓐ Ⓑ Ⓒ Ⓓ Ⓔ
156. Ⓐ Ⓑ Ⓒ Ⓓ Ⓔ
157. Ⓐ Ⓑ Ⓒ Ⓓ Ⓔ
158. Ⓐ Ⓑ Ⓒ Ⓓ Ⓔ
159. Ⓐ Ⓑ Ⓒ Ⓓ Ⓔ
160. Ⓐ Ⓑ Ⓒ Ⓓ Ⓔ
161. Ⓐ Ⓑ Ⓒ Ⓓ Ⓔ
162. Ⓐ Ⓑ Ⓒ Ⓓ Ⓔ
163. Ⓐ Ⓑ Ⓒ Ⓓ Ⓔ
164. Ⓐ Ⓑ Ⓒ Ⓓ Ⓔ
165. Ⓐ Ⓑ Ⓒ Ⓓ Ⓔ
166. Ⓐ Ⓑ Ⓒ Ⓓ Ⓔ

167. Ⓐ Ⓑ Ⓒ Ⓓ Ⓔ
168. Ⓐ Ⓑ Ⓒ Ⓓ Ⓔ
169. Ⓐ Ⓑ Ⓒ Ⓓ Ⓔ
170. Ⓐ Ⓑ Ⓒ Ⓓ Ⓔ
171. Ⓐ Ⓑ Ⓒ Ⓓ Ⓔ
172. Ⓐ Ⓑ Ⓒ Ⓓ Ⓔ
173. Ⓐ Ⓑ Ⓒ Ⓓ Ⓔ
174. Ⓐ Ⓑ Ⓒ Ⓓ Ⓔ
175. Ⓐ Ⓑ Ⓒ Ⓓ Ⓔ
176. Ⓐ Ⓑ Ⓒ Ⓓ Ⓔ
177. Ⓐ Ⓑ Ⓒ Ⓓ Ⓔ
178. Ⓐ Ⓑ Ⓒ Ⓓ Ⓔ
179. Ⓐ Ⓑ Ⓒ Ⓓ Ⓔ
180. Ⓐ Ⓑ Ⓒ Ⓓ Ⓔ
181. Ⓐ Ⓑ Ⓒ Ⓓ Ⓔ
182. Ⓐ Ⓑ Ⓒ Ⓓ Ⓔ
183. Ⓐ Ⓑ Ⓒ Ⓓ Ⓔ
184. Ⓐ Ⓑ Ⓒ Ⓓ Ⓔ
185. Ⓐ Ⓑ Ⓒ Ⓓ Ⓔ
186. Ⓐ Ⓑ Ⓒ Ⓓ Ⓔ
187. Ⓐ Ⓑ Ⓒ Ⓓ Ⓔ
188. Ⓐ Ⓑ Ⓒ Ⓓ Ⓔ

189. Ⓐ Ⓑ Ⓒ Ⓓ Ⓔ
190. Ⓐ Ⓑ Ⓒ Ⓓ Ⓔ
191. Ⓐ Ⓑ Ⓒ Ⓓ Ⓔ
192. Ⓐ Ⓑ Ⓒ Ⓓ Ⓔ
193. Ⓐ Ⓑ Ⓒ Ⓓ Ⓔ
194. Ⓐ Ⓑ Ⓒ Ⓓ Ⓔ
195. Ⓐ Ⓑ Ⓒ Ⓓ Ⓔ
196. Ⓐ Ⓑ Ⓒ Ⓓ Ⓔ
197. Ⓐ Ⓑ Ⓒ Ⓓ Ⓔ
198. Ⓐ Ⓑ Ⓒ Ⓓ Ⓔ
199. Ⓐ Ⓑ Ⓒ Ⓓ Ⓔ
200. Ⓐ Ⓑ Ⓒ Ⓓ Ⓔ
201. Ⓐ Ⓑ Ⓒ Ⓓ Ⓔ
202. Ⓐ Ⓑ Ⓒ Ⓓ Ⓔ
203. Ⓐ Ⓑ Ⓒ Ⓓ Ⓔ
204. Ⓐ Ⓑ Ⓒ Ⓓ Ⓔ
205. Ⓐ Ⓑ Ⓒ Ⓓ Ⓔ
206. Ⓐ Ⓑ Ⓒ Ⓓ Ⓔ
207. Ⓐ Ⓑ Ⓒ Ⓓ Ⓔ
208. Ⓐ Ⓑ Ⓒ Ⓓ Ⓔ
209. Ⓐ Ⓑ Ⓒ Ⓓ Ⓔ
210. Ⓐ Ⓑ Ⓒ Ⓓ Ⓔ

THE GRE BIOLOGY TEST

MODEL TEST I

Time: 170 Minutes
        210 Questions

DIRECTIONS: *Choose the best answer for each question and mark the letter of your selection on the corresponding answer sheet.*

1. The posterior lobe of the pituitary gland in humans releases

    (A) TSH and FSH.

    (B) ACTH and LH.

    (C) oxytocin and vasopression.

    (D) FSH and LH.

    (E) prolactin and growth hormone.

2. If we wished to obtain a 3-dimensional image of a very minute structure, we would use

    (A) scanning-electron microscopy.

    (B) fluorescence microscopy.

    (C) transmission-electron microscopy.

    (D) ultraviolet microscopy.

    (E) light microscopy.

3. Penicillin inhibits bacterial proliferation by

    (A) blocking ribosomal function.

(B) blocking the glycolytic pathway.

(C) stopping the electron transport chain.

(D) blocking cell wall synthesis.

(E) interrupting the active sites of vital enzymes.

4. The golgi apparatus primarily functions in

(A) packaging protein for secretion.

(B) synthesizing protein for secretion.

(C) packaging protein for hydrolysis.

(D) synthesizing protein for hydrolysis.

(E) all of the above

5. Cellulose is a natural polymer composed of the monomer

(A) glucagon.

(B) amino acids.

(C) glucose.

(D) amides.

(E) lipids and amino acids.

6. The most recent theories of the origin of life include all of the following elements in the primitive atmosphere except

(A) free oxygen.

(B) hydrogen.

(C) methane.

(D) ammonia.

(E) carbon dioxide.

7. Albinism is a recessive trait. In a certain community of 200 people, 18 persons are albinos. How many people are normal homozygotes?

(A) 182

(B) 164

(C) 100

(D) 98

(E) 84

8. Which of the following is/are characteristic(s) of the phylum Chordata?

(A) a vertebral column

(D) a notochord

(B) a dorsal hollow nerve cord

(E) B, C, and D

(C) gill slits

9. A harmless animal that imitates the appearance of another species that is dangerous to the predator is an example of

(A) Müllerian mimicry.

(D) mutualism.

(B) Batesian mimicry.

(E) altruism.

(C) cryptic appearance.

10. How many ATPs are derived from one molecule of pyruvate via the Krebs cycle and the electron transport system?

(A) 12

(D) 18

(B) 14

(E) 20

(C) 15

11. The extracellular fibers found in all connective tissues are composed mainly of

(A) collagen.

(D) glycans.

(B) calcium.

(E) both A and C

(C) elastin.

12. Chloroplasts, the basic unit cells of photosynthesis, contain/are composed of all of the following except

(A) thylakoids.

(D) starch granules.

(B) stroma.

(E) lamellae.

(C) cristae.

13. During the anaphase stage of mitosis,

    (A) chromosomes begin to contract and coil.

    (B) chromosomes take up a central position.

    (C) astral rays are formed.

    (D) sister chromatids move to opposite poles.

    (E) the cell undergoes cytokinesis.

14. To find the genetic order of three bacterial genes we do not need to know the

    (A) number of wild type cells.

    (B) frequency of recombin- ation.

    (C) dominance and recessive- ness of the alleles.

    (D) number of double cross- over events.

    (E) phenotypes of the cells.

15. Nitrogeneous wastes are excreted by different species of animals in all of the following forms except

    (A) creatinine.

    (B) uracil.

    (C) ammonia.

    (D) urea.

    (E) none of the above

16. Migration in birds is guided, in part, by

    (A) celestial navigation.

    (B) temperature changes.

    (C) olfactory clues.

    (D) territorial aggression.

    (E) all of the above

17. Muscle fatigue is due, in part, to the accumulation of

    (A) lactic acid.

    (B) citric acid.

    (C) pyruvic acid.

    (D) ACTH.

    (E) ATP.

18. The pituitary regulates all of the following except the

    (A) thyroid.   (D) testes.

    (B) adrenal cortex.   (E) adrenal medulla.

    (C) ovaries.

19. The sequence of differentiative events that leads to the formation of mature sperm cells is

    (A) primary spermatocytes → secondary spermatocytes → spermatids → spermatogonia → sperm.

    (B) spermatids → spermatogonia → primary spermatocytes → secondary spermatocytes → sperm.

    (C) spermatogonia → spermatids → secondary spermatocytes → primary spermatocytes → sperm.

    (D) spermatogonia → primary spermatocytes → secondary spermatocytes → spermatids → sperm.

    (E) secondary spermatocytes → primary spermatocytes → spermatogonia → spermatids → sperm.

20. A plant with no meristematic tissue will be unable to

    (A) photosynthesize.   (D) produce fruits.

    (B) transport water.   (E) respire.

    (C) transport nutrients.

21. Which of the following has an open circulatory system?

    (A) earthworms   (D) sandworms

    (B) fish   (E) none of the above

    (C) clams

22. Protozoans can reproduce in a number of ways; they are however incapable of

(A) sporulation.

(B) binary fission.

(C) sexual reproduction.

(D) viviparity.

(E) budding.

23. The axons of motor neurons are located in the spinal cord in

(A) the ventral horn.

(B) the dorsal horn.

(C) the ventral root ganglia.

(D) the dorsal root ganglia.

(E) both A and B

24. All of the following enzymes are involved in the digestion of food
except

(A) pepsin.

(B) trypsin.

(C) maltase.

(D) amylase.

(E) ligase.

25. Which is not a currently accepted example of an evolutionary
event?

(A) An individual in a population synthesizes a new variation of
an enzyme.

(B) Giraffes develop longer necks as they strain to reach leaves
at the tops of trees.

(C) Individuals leave their dwelling and establish a new feeding
and breeding ground.

(D) An outside group joins an established population.

(E) Antelopes that can run extremely fast survive to reproductive age.

26. Which of the following factors affect enzymatic activity?

I.   temperature

II. hydronium ion concentration

III. enzyme poisoning

IV. water concentration

(A) I

(D) I, III and IV

(B) II and III

(E) I, II, III and IV

(C) I, II, and III

27. An exocrine gastric product which combines with vitamin B_{12} so that it can be absorbed later in the small intestine is

(A) pepsin.

(D) intrinsic factor.

(B) hydrochloric acid.

(E) trypsin.

(C) mucus.

28. Which of the following is found only in birds?

(A) Carina

(D) Cloaca

(B) Syrinx

(E) A and B

(C) Epidermal scales

29. Which of the following is not a proteolytic enzyme?

(A) Trypsin

(D) Pepsin

(B) Lipase

(E) Chymotrypsin

(C) Carboxypeptidase

30. In the overall reaction for photosynthesis

$$CO_2 + H_2O \xrightarrow{\eta\nu} (CH_2O) + O_2$$

where does the carbohydrate oxygen come from?

(A) CO_2

(B) H_2O

(C) O_2

(D) all of the above can donate the oxygen atom

(E) none of the above can donate the oxygen atom

31. The connective tissue sac enclosing the heart is called the

(A) endothelium.

(B) myocardium.

(C) pericardium.

(D) vena cava.

(E) endocardium.

32. Spores have all of the following characteristics except that

(A) they are haploid.

(B) they are usually uni-
cellular.

(C) they are formed by
mitosis.

(D) they germinate and develop
into a gametophyte.

(E) none of the above

33. If black and short hair are dominant characteristics of guinea
pigs, the expected offspring from a cross between a heterozygous
black and short-haired male and a homozygous white and long-
haired female are

(A) ¼ black, short-haired; ¼ black, long-haired; ¼ white, short-
haired; ¼ white, long-haired.

(B) 3/4 black, short-haired; ¼ white, long-haired.

(C) ½ black, short-haired; ½ white, long-haired.

(D) 9 black, short-haired; 3 black, long-haired; 3 white, short-
haired; 1 white, long-haired.

(E) 5/16 black, short-haired; 3/16 black, long-haired; 5/16
white, short-haired; 3/16 white, long-haired.

34. The main factor that determines the uptake and dissociation of
oxygen and carbon dioxide in the blood is

(A) the partial pressure of
oxygen.

(B) the partial pressure of
carbon dioxide.

(C) the level of carbonic
anhydrase.

(D) all of the above

(E) both A and B

35. Striated (skeletal) muscle fibers exhibit

    (A) few mitochondria.

    (B) alternating A bands and I bands in a transverse pattern.

    (C) only one nucleus.

    (D) no orderly arrangement.

    (E) all of the above

36. Lichens are an example of plant

    (A) parasitism.   (D) altruism.

    (B) commensalism.   (E) socialism.

    (C) mutualism.

37. The corpus luteum can secrete progesterone at high levels without being shut off because

    (A) it is not under regulatory control.

    (B) the pituitary secretes FSH.

    (C) the levels of estrogen have decreased.

    (D) the hypothalamus is actively secreting.

    (E) the placenta secretes chorionic gonadotrophin which is like luteinizing hormone (LH).

38. The stomata of a plant leaf opens when

    (A) it is light.   (D) cellular pH is low.

    (B) it is dark.   (E) both A and C are true

    (C) the CO_2 level in the plant is high.

39. All of the following are secreted by the pancreas except

    (A) chymotrypsin.

(B) trypsin.

(D) carboxypeptidase.

(C) pepsin.

(E) lipase.

40. Which occur(s) during the light reaction of photosynthesis?

(A) production of ATP

(D) A and C

(B) release of oxygen derived from CO_2

(E) all of the above

(C) photophosphorylation

41. Hydrogen ions are not free to lower the blood's pH because they are

(A) removed through the action of carbonic anhydrase.

(B) bound to water.

(C) bound to hemoglobin.

(D) removed by diffusion.

(E) bound to carbon dioxide.

42. An organism can survive with a mutation in a gene whose product is essential for survival only if the mutation is

(A) temperature sensitive.

(D) dominant to the wild type.

(B) in a somatic cell.

(E) in both somatic and germ cells.

(C) in a germ cell.

43. Electron microscopes have a higher resolution than light microscopes because

(A) staining procedures are more efficient when the electron microscope is used.

(B) electrons have shorter wavelengths than light.

(C) the wave properties of light interfere with high resolution whereas electrons have no wave properties.

(D) the magnification of the electron microscope is much higher.

(E) electrons have longer wavelengths than light.

44. All flagella and motile cilia have the following pattern of micro-tubules:

(A) 7 outer and 2 inner.   (D) 9 outer and 2 inner.

(B) 9 outer and 0 inner.   (E) 11 outer and 3 inner.

(C) 8 outer and 2 inner.

45. The last part of the human small intestine before entering the large intestine is the

(A) cecum.   (D) ileum.

(B) jejenum.   (E) pylorus.

(C) duodenum.

46. Unlike plant cells, most animal cells possess

(A) a cell wall.   (D) a nuclear membrane.

(B) centrioles.   (E) mitochondria.

(C) chloroplasts.

47. Which of the following does not perform excretory functions?

(A) kidneys   (D) liver

(B) lungs   (E) lymph nodes

(C) skin

48. Epiphytes are plants which use other plants as bases of attachment. They do not obtain any nourishment from their host. This relationship is an example of

(A) mutualism.

(B) parasitism.

(C) commensalism.

(D) altruism.

(E) isolationism.

49. Colchicine inhibits the formation of microtubules. Which cellular process would be disrupted by this inhibitor?

(A) growth

(B) respiration

(C) cytokinesis

(D) enzyme activity

(E) meiosis

50. An undifferentiated cell can be induced to follow the developmental patterns of a different species by

(A) replacing the original nucleus with one from another (chosen) species.

(B) introducing cytoplasm from another species.

(C) removing the native nucleus.

(D) no known means.

(E) adding mitochondria from another species.

51. The acrosome of a sperm cell contains

(A) mitochondria.

(B) enzymes.

(C) centrioles.

(D) flagella.

(E) hormones.

52. Which of the following is not a function of auxin hormones?

(A) They cause the plant to bend and elongate towards the light.

(B) They direct tissue differentiation in the vascular cambium.

(C) They stimulate development of lateral buds and inhibit growth of the terminal bud.

(D) They control the shedding of leaves, flowers, fruits, and branches.

(E) They determine growth correlations of plant parts.

53. A wood louse becomes restless after it becomes dry. In response, it moves around randomly until it finds a new, moist home. This is an example of

(A) taxis.

(B) kinesis.

(C) biological communication.

(D) operant conditioning.

(E) all of the above

54. Meiotic drive is

(A) preferential segregation of genes during meiosis in a heterozygous individual.

(B) preferential segregation of genes during meiosis in a homozygous individual.

(C) a force that prevents meiosis from occurring.

(D) a force that disrupts the normal segregation processes in meiosis.

(E) a force that causes somatic cells to lyse.

55. The gram stain, which is used to differentiate bacterial cells, is based on

(A) the protein content in the respective bacterial cell wall.

(B) the carbohydrate content in the respective bacterial cell wall.

(C) the lipid content in the respective bacterial cell wall.

(D) the diffusion rate of staining fluid through the bacterial cell wall.

(E) none of the above

56. Which of the following classification groups are in the proper sequence from the largest to the smallest?

(A) phylum, class, order, family

(B) kingdom, family, class, phylum

(C) family, order, genus, species

(D) kingdom, class, species, genus

(E) class, order, species, genus

57. The main point of contrast between marsupial and placental mammals is

(A) marsupials lay eggs while placentals give birth to live young.

(B) placentals lay eggs while marsupials give birth to live young.

(C) marsupial embryos have a shorter development period.

(D) placental embryos have a shorter development period.

(E) marsupials are hermaphroditic.

58. During conjugation, what is transferred from the Hfr bacterium to the F⁻ bacterium?

(A) the sex factor (F factor)

(B) portions of the Hfr chromosome

(C) the sex factor and portions of the Hfr chromosome

(D) nothing is transferred

(E) none of the above

59. Cycads are remarkable among the seed plants in that they form motile, swimming sperm within their pollen tubes. This fact indicates that

(A) they are aquatic plants.

(B) they are less advanced than the conifers.

(C) they are more advanced than the conifers.

(D) they require moisture in order for pollination to occur.

(E) they are subterranean plants.

60. The myelin sheath of many axons is produced by the

(A) node of Ranvier.   (D) astrocytes.

(B) nerve cell body.   (E) axon hillock.

(C) Schwann cell.

61. All of the following occurs as muscles contract except

(A) Z bands come closer.   (D) I bands decrease.

(B) H zones stay the same.   (E) thick and thin filaments
                                    slide past each other.
(C) A bands stay the same.

62. Unlike noncompetitive inhibition, competitive inhibition of enzymes may be overcome

(A) since the enzyme active   (D) A and B
    site is unaltered.
                               (E) all of the above
(B) by increasing the sub-
    strate concentration.

(C) by the production of
    additional enzymes.

63. The permeability of the walls of the distal convoluted collecting tubules of the kidneys to water is regulated by

(A) the amount of water.   (D) the adrenals.

(B) the concentration of salts. (E) the thymus.

(C) vasopressin.

64. The genes which serve as a binding site for RNA polymerase in DNA transcription are called

(A) operator genes.   (D) regulatory genes.

(B) structural genes.   (E) inhibitor genes.

(C) promoter genes.

65. Which of the following pigments absorbs radiant energy and ultimately transfers it to chlorophyll as high energy electrons?

(A) chlorophyll b

(D) all of the above

(B) carotene

(E) none of the above

(C) xanthophyll

66. The ecological unit composed of organisms and their physical environment is known as a/an

(A) niche.

(D) community.

(B) population.

(E) genus.

(C) ecosystem.

67. A releaser produces

(A) a complex series of reactions not always directly related to the stimulus.

(B) aggressive behavior.

(C) an abnormal behavior.

(D) a group of reactions related to a learned experience.

(E) non-aggressive behavior.

68. Due to the dominant fauna in this period, the Mesozoic Era is often referred to as the

(A) "Age of Reptiles."

(D) "Age of Birds."

(B) "Age of Fishes."

(E) "Age of Mammals."

(C) "Age of Amphibians."

69. How do many non-photosynthetic plants such as the mushroom obtain food from their surroundings?

(A) absorption and subsequent extracellular digestion

(B) absorption and subsequent intracellular digestion

(C) extracellular digestion and subsequent absorption

(D) none of the above

(E) all of the above

70. The placenta originates from

(A) embryonic cells.   (D) both A and B

(B) maternal cells.   (E) both B and C

(C) paternal cells.

71. The prostaglandins, which may eventually prove to be effective birth control substances, are classified as

(A) steroids.   (D) lipoproteins.

(B) 20-carbon fatty acids.   (E) proteins.

(C) carbohydrates.

72. A queen bee which has a yellow body color (Yy) is fertilized by a drone which also has a yellow body color. Assuming that yellow body color is dominant and knowing that the drone, being haploid, possesses only the gene for yellow body color (Y), how is it possible that a brown-bodied female (yy) is produced? Assume the two bees are isolated in a vial. Choose the best explanation.

(A) A mutation has occurred during development.

(B) Body color has been environmentally influenced, following fertilization.

(C) The queen bee has used stored sperm from previous inseminations, resulting in this offspring.

(D) Male bees are not haploid so the above experiment is based on incorrect assumptions.

(E) none of the above

73. Which of the following is not correct concerning regulatory enzymes?

(A) They usually have two binding sites: one for the substrate and one for the "regulator."

(B) They are often found in key metabolic pathways, such as the Krebs cycle.

(C) The end-product of a pathway is never a regulatory enzyme.

(D) Any product half-way down a pathway can inhibit regulatory enzymes.

(E) These enzymes are temperature sensitive.

74. The primitive gut formed during gastrulation is called the

(A) blastopore.

(D) archenteron.

(B) gastrocoel.

(E) ventral pore.

(C) blastocoel.

75. Which researcher(s) did not study cellular function or morphology?

(A) van Leeuwenhoek

(D) Virchow

(B) Schleiden and Schwann

(E) Watson and Crick

(C) Robert Brown

76. In the early 1900's, many fruit growers made a practice of ripening fruits by keeping them in a room with a kerosene stove. Why did the fruits ripen?

(A) Due to heat from the stove.

(B) Due to ethylene released by the stove.

(C) Due to florigen released by the stove.

(D) Due to cytokinins released by the stove.

(E) Due to gibberellins released by the stove.

77. Usually a recessive sex-linked trait

(A) is expressed more often in males than in females.

(B) is expressed more often in females than in males.

(C) is expressed to the same extent in both males and females.

(D) is only expressed in males.

(E) is only expressed in females.

78. The "10 percent rule" in ecology

(A) refers to the precentage of similar species that can coexist in one ecosystem.

(B) refers to the average death total of all mammals before maturity.

(C) is the percent of animals not affected by DDT.

(D) refers to the level of energy production in a given trophic level that is used for production by the next higher level.

(E) refers to the average birth rate in a climax community.

79. Which of the following occurs during the dark reaction of the photosynthetic process of an angiosperm?

(A) Production of carbohydrates

(B) Production of ATP

(C) Oxidative phosphorylation

(D) A and B

(E) all of the above

80. Gene duplication takes place during

(A) interphase.   (D) anaphase.

(B) prophase.   (E) telophase.

(C) metaphase.

81. The bones which are the most frequently found in human anatomical fossils are

(A) femurs.   (D) radii.

(B) hip bones.   (E) ulna.

(C) teeth.

82. When ^{15}N is added to the environment of bacteria, 50% of the nitrogen in the DNA of the first new generation of bacteria is ^{15}N. If this generation were isolated from ^{15}N and only exposed to ^{14}N, what isotopic makeup of nitrogen would you expect to find in the second generation of bacteria? (Assume that there is no ^{15}N in the bacterial environment).

(A) All the second generation bacteria have 50% of their DNA's nitogen in the ^{15}N form.

(B) Half of the second generation bacteria have 100% of their DNA's nitrogen in the ^{15}N form, and half have no ^{15}N in their DNA.

(C) Half of the second generation bacteria have 50% of their DNA's nitrogen in the ^{15}N form, and half have no ^{15}N in their DNA.

(D) One-quarter of the second generation bacteria have 100% of their DNA's nitrogen in the ^{15}N form, and three-quarters have no ^{15}N in their DNA.

(E) One-quarter of the second generation bacteria have 50% of their DNA's nitrogen in the ^{15}N form, and three-quarters have no ^{15}N in their DNA.

83. Choose the statement that best describes the climax stage of an ecological succession.

(A) It is usually populated only by plants.

(B) It is usually populated only by animals.

(C) It represents the initial phases of evolution.

(D) It changes rapidly from season to season.

(E) It remains until there are severe changes in the environment.

84. Which of the following serves as a hydrostatic mechanism in a paramecium?

(A) Oral groove

(B) Food vacuole

(C) Cilia

(D) Contractile vacuole

(E) A and D

85. Parasites transmitted by mosquitoes are usually carried by their

(A) lungs.

(B) salivary glands.

(C) probosises.

(D) intestines.

(E) feet.

86. Translocation is a type of chromosomal mutation where

(A) a segment of the chromosome is missing.

(B) a portion of the chromosome is represented twice.

(C) a segment of one chromosome is transferred to another non-homologous chromosome.

(D) a segment is removed and reinserted.

(E) a segment is removed and destroyed.

87. The denaturing of a protein by heat or radiation is caused by the destruction of which structures?

(A) secondary and tertiary

(B) secondary and quarternary

(C) secondary, tertiary, and quarternary

(D) tertiary and quaternary

(E) none of the above

88. Unlike collenchyma and sclerenchyma tissues, parenchyma tissue does not function in

(A) support.

(B) gas exchange.

(C) nutrient exchange.

(D) B and C

(E) A and B

89. A cell is placed in a solution of dye. After a while, the intracellular concentration of dye becomes much greater than the extracellular concentration. Upon addition of a metabolic inhibitor to the solution, the dye equilibrates across the cell membrane until the intra- and extra-cellular concentrations are equal. A possible role for this metabolic inhibitor might be to

(A) inhibit protein synthesis.   (D) inhibit ATP production.

(B) delay chromosomal replica-   (E) accelerate meiotic
     tion.   processes.

(C) accelerate aerobic
    respiration.

90. A prophase chromosome does not contain which of the following?

(A) centromere   (D) DNA

(B) centrosome   (E) all of the above

(C) chromatid

91. What is meant by the "counter-current system" with regard to the respiratory mechanism of gilled-fish?

(A) CO_2 and O_2 flow in opposite directions to each other.

(B) The water inside the gills flows in the opposite direction to that of the water flow outside the gill.

(C) The fish is capable of putting its gills on the outside.

(D) Blood inside the gills of a fish flows in the opposite direction to the water flow over the gill.

(E) Blood flow alternates directions inside the fish.

92. Aggressive behavior in animals

(A) consists mostly of encounters between members of different species (excluding predation).

(B) occurs most frequently in contests over food.

(C) usually consists of nonviolent displays which avoid serious injury.

(D) occur equally in both sexes.

(E) tends to occur at only one time of the year.

93. The climax organism growing above the tree line on a mountain would be the same as the climax organism found in the

(A) taiga.

(D) desert.

(B) tundra.

(E) temperate regions.

(C) tropical forest.

Questions 94-96 refer to the typical flower of an angiosperm.

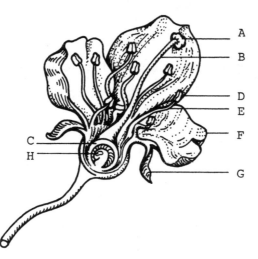

94. The profile of structure D is a(an)

(A) stigma.

(D) filament.

(B) pedicel.

(E) anther.

(C) calyx.

95. The function of this flower structure is to secrete a moist, sticky substance to which pollen grains can adhere.

(A) A

(D) D

(B) B

(E) F

(C) C

96. Which one of these structures may be absent if the flower is pollinated by the wind?

(A) A   (D) F

(B) C   (E) H

(C) D

97. An individual whose genitalia and internal ducts are male but whose testes are underdeveloped and do not produce sperms is most likely a victim of

(A) Down Syndrome.   (D) Edwards Syndrome.

(B) Turner Syndrome.   (E) Klinefelter Syndrome.

(C) Patau Syndrome.

98. The undifferentiated duct in the female embryo is the

(A) Müllerian duct.   (D) Ductless gland.

(B) Endocrine gland.   (E) Wollfian duct.

(C) Exocrine duct.

99. The formation of pyrimidine dimers particularly between two thymine residues is effected by

(A) Acradine Orange.   (D) Infrared Radiation.

(B) UV Radiation.   (E) Cosmic rays.

(C) X-Rays.

100. Which of the following is a realistic chart illustrating the relationship between enzymatic activity and pH?

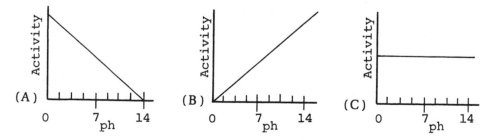

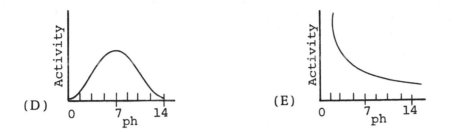

(D) 0 7 14
      ph

(E) 0 7 14
      ph

101. The stage during development in which there is a hollow ball of cells is called

(A) the blastula.

(D) the gastrula.

(B) the morula.

(E) ovulation.

(C) the isolecithal stage.

102. The structural framework of the apo-enzyme or the transcriptional enzyme is made up of

(A) β sub units

(D) α subunits.

(B) β' sub units

(E) α_2 subunits.

(C) σ subunits.

103. Separation of the first two cells of a cleaving human embryo would presumably result in the production of

(A) fraternal twins.

(D) monozygotic twins

(B) zygotic twins

(E) dizygotic twins

(C) paternal twins

104. Carnivores and rodents exhibit the best _____ in land vertebrates.

(A) gustatory sense

(D) none of the above is correct

(B) auditory sense

(E) all of the above is correct

(C) olfactory sense

105. The percentage of individuals that when carrying a given gene in proper combination for its expression, actually express the gene's phenotype is known as

(A) interference.

(D) coincidence.

(B) expressivity.

(E) none of the above is correct

(C) penetrance.

106. The chemicals which work like hormones between individuals are known as

(A) pheromones.

(D) chemotransmitters.

(B) bombykols.

(E) effectors.

(C) hemotoxins.

107. In Skinner's operant conditioning an(a)

(A) unconditioned stimulus → unconditioned response.

(B) unconditioned response → reinforcement.

(C) unconditioned stimulus → reinforcement.

(D) conditioned stimulus → response.

(E) conditioned response → reinforcement.

108. Damage to the right temporal lobe of the cerebral hemisphere results in

(A) poor performance on IQ test.

(B) poor performance on perceptual tests.

(C) no sensation on the right side of the body.

(D) lack of voluntary movement in the right arm, hand, leg or foot.

(E) lack of movement in the body parts.

109. Which of these is not characteristic of procaryotes?

(A) polysaccharide cell wall

(D) centrioles and asters

(B) flagella or cilia

(E) sexual recombination

(C) cell membrane

110. Electron microscopy differs from light microscopy in the

(A) staining technique used on material.

(B) preparation of material.

(C) material studied.

(D) resolution of microscopes.

(E) all of the above

111. The enzyme which adds the final nucleotide to seal the Okazaki fragments is

(A) Polymerase I.

(D) Polymerase II.

(B) Ligase.

(E) Polymerase III*.

(C) Gyrase

112. The energy molecule used to attach the 50 S ribosomal subunit to the 30 S subunit is

(A) GTP.

(D) NTP.

(B) ATP.

(E) ADP.

(C) NMP.

113. The genes of a plasmid are moved from one bacterium to another by

(A) F^+ episomes.

(D) M^+ episomes.

(B) F^- episomes.

(E) M^- episomes.

(C) Pili.

114. In the human excretory system blood components such as glucose and amino acids are returned to the blood from the filtrate by

(A) facililated diffusion.

(D) osmosis.

(B) pinocytosis.

(E) passive transport.

(C) simple diffusion.

115. Mesosomes are characteristic of prokaryotic

(A) ribosomes.

(D) nucleiod.

(B) cell wall.

(E) flagella.

(C) cell membrane.

116. In using the gram staining technique to separate and classify bacteria, a mordant in the form of an iodine solution is applied to

(A) identify gram - negative bacteria.

(B) identify gram - positive bacteria.

(C) counterstain gram - positive bacteria.

(D) Both A and B are correct.

(E) None of the above are correct.

117. To which of the following taxonomic levels do the names Rana and Canis belong?

(A) Species

(D) Family

(B) Genus

(E) Class

(C) Order

118. Which of the following is characteristic of homotherms?

(A) They maintain a body temperature that is in keeping with the external circumstance.

(B) They maintain a constant internal body temperature.

(C) Their skin temperature is usually cooler than the temperature of their other organs.

(D) A and C only are correct.

(E) B and C only are correct.

DIRECTIONS: *For each group of questions below, match the numbered word, phrase, or sentence to the most closely related lettered heading and mark the letter of your selection on the corresponding answer sheet. A lettered heading may be chosen as the answer once, more than once, or not at all for the question in each group.*

Questions 119-121

(A) Protochordata
(B) Hemichordata
(C) Urochordata
(D) Cephalochordata
(E) Vertebrata

119. Tunicates or sea-squirts

120. Lancelets, most notably, the amphioxus

121. Marine animals, often called, acorn worms

Questions 122-126

(A) Interphase
(B) Prophase
(C) Metaphase
(D) Anaphase
(E) Telophase

122. This stage occurs when the chromatin threads begin to condense.

123. This stage occurs when chromosomes appear as vague, dispersed thread-like structures.

124. This stage occurs after the chromosomes have lined up along the equatorial plane of the cell.

125. This stage occurs before the daughter chromosomes (or separated chromatids) have moved to opposite sides of the cell.

126. Cytokinesis occurs after this stage.

Questions 127-130

(A) Acetylcholine
(B) Cytokinin
(C) Progesterone
(D) Ecdysone
(E) Actin

127. Neural Transmission

128. Human Reproduction

129. Muscle Contraction

130. Insect Metamorphosis

Questions 131-135

(A) Cotyledons
(B) Rhizoids
(C) Sporophyte
(D) Xylem
(E) Palisade layer

131. Simple filaments performing water absorption

132. Chiefly responsible for photosynthesis

133. Reduced in size, cannot exist independently, especially in primitive plants

134. Functions mainly in nutrient absorption from endosperm in monocots

135. Transports nutrients up to leaves

Questions 136-138

(A) Holoenzyme
(B) Apoenzyme
(C) Metalloenzyme
(D) Coenzyme
(E) Prosthetic group

136. An organic, non-protein co-factor

137. A coenzyme tightly bound to the apoenzyme

138. A cofactor joined to an apoenzyme

Questions 139-142

(A) Medulla
(B) Midbrain
(C) Cerebellum
(D) Cerebrum
(E) Thalamus

139. Immediately connected to the spinal cord

140. Relay center for sensory impulses

141. Regulates and coordinates muscle contraction

142. Contains grey matter

Questions 143-146

(A) DNA
(B) Ribosomal RNA
(C) Transfer RNA
(D) Messenger RNA
(E) RNA polymerase

143. Involved in semi-conservative replication

144. Mediates transcription of DNA.

145. Is the anticodon which carries amino acids.

146. Transports codons to the ribosomes.

Questions 147-151

(A) Gastrin
(B) Duodenum
(C) Lipase
(D) Colon
(E) Oral cavity

147. Digestion begins at this site.

148.　Produced by food distending the stomach walls.

149.　Structure where villi begin to appear.

150.　Is found in man, where very large amounts of bacteria exist.

151.　Is released by pancreas.

Questions 152–155

(A)　Capillaries
(B)　Birds
(C)　Sino-atrial node
(D)　Atrio-ventricular node.
(E)　Reptiles

152.　Initiates heartbeat in mammals.

153.　Possess a characteristic four-chambered heart.

154.　Most oxygen exchange takes place at this site.

155.　Particularly active during fever, causes increased heart rate.

Questions 156–160

(A)　Pheromones
(B)　Display
(C)　Courtship
(D)　Müllerian mimicry
(E)　Adaptive behavior

156.　Female moths attract males over very large areas using air currents.

157. Black-headed gulls remove conspicuous objects and broken egg-shells from nests.

158. Two black-headed gulls meet; one gives the agonistic "upright" posture.

159. Two species of inedible insects evolve to resemble each other.

160. A Eyrebird sings his favorite song.

Questions 161-165

(A) Cortex
(B) Adventitious roots
(C) Growth zone
(D) Lenticel
(E) Cutin

161. After main roots have been severed, this structure can clearly be seen.

162. Meristematic cells composed of small, thin-walled cells with large nuclei.

163. Composed of parenchyma cells in both stem and root.

164. Helps stems to breathe.

165. Accounts for retarding amount of water loss due to evaporation.

Questions 166-169

(A) Echinoderms

(B) Coelenterates
(C) Arthropods
(D) Annelids
(E) Flatworms

166. Two part double ventral nerve cord which is segmented and which originates in the esophagus.

167. Netlike system composed of separate neurons that cross each other but do not touch.

168. Nervous system consists of specialized neuroectoderm which forms shallow grooves along the surface of the arms.

169. Advanced planarians which have a longitudinal, bilaterally symmetrical body plan comprised of two nerve cords and whose neurons are concentrated in the head region.

Questions 170-174

(A) Adenyl cyclase
(B) Cyclic AMP (cAMP)
(C) Epinephrine
(D) Steroid hormone
(E) Adrenocorticotrophic hormone (ACTH)

170. Mediates the activation of an enzyme which causes an increase in blood glucose level through the enzymatic breakdown of glycogen.

171. Is known as the second messenger.

172. Activated by receptor-hormonal complex and incorporated into the membrane of a liver cell.

173. Is lipid soluble and has no apparent membrane receptors.

174. Stimulates the adrenal cortex to secrete steroids.

Questions 175-179

(A) Tautomeric Shift
(B) Ionizing Radiation
(C) Non-Ionizing Radiation
(D) Mutagenicity
(E) Carcinogenicity

175. Oncogenes

176. Point by point mutations result when the purines and pyrimidines are altered.

177. Spontaneous abortions, stillbirths and congenital malformations have increased due to the high incidence of noxious compounds in the environment.

178. Keto-enol pairs are substituted for thymine and guanine and amino-imino pairs are substituted for cytosine and adenine during this mutation.

179. Electrons absorb radiant energy and are raised to a higher energy level. This can lead to the formation of dimers.

Questions 180-184

(A) Succession
(B) Variation
(C) Aggression
(D) Cooperation
(E) Symbiosis

180. A ciliate protozoan containing mutualistic green algae has undergone fission producing two daughter cells.

181. The spreading of the wings of a female goshawk.

182. A queen honeybee surrounded by the smaller workers.

183. The invasion of the Okefenokee Swamp by vegetation.

184. Penguins forming large populations of sexually reproducing individuals.

Questions 185-189

(A) Adductors
(B) Summation
(C) Tetanus
(D) Abductors
(E) Extensors

185. Opposing muscles that straighten the limb out.

186. Muscles that pull a limb away from the median body line.

187. A great degree of contraction which is caused by electric shocks being applied to a muscle or groups of muscles.

188. Muscle maintained in state of contraction due to no period of relaxation.

189. Muscles which pull a limb toward the median body line.

Questions 190-194

(A) Rattlesnakes
(B) Bees

43

(C) Bats
(D) Fish
(E) Spiders

190. Touch receptors are commonly located in the hair.

191. Sensory endings are located in the lateral line.

192. Heat sensors are located in the depressions near the eye.

193. Are sensitive to polarized light.

194. Are sensitive to high-frequency sounds.

Questions 195-199

(A) Turner's Syndrome
(B) Klienfelter's Syndrome
(C) Patau's Syndrome
(D) Müllerian duct
(E) Wolffian duct

195. Is involved in the development of the oviduct, uterus and vagina.

196. Is a genetic disorder which is found in females causing them to
      have one X chromosome only.

197. Is found in both males and females but promotes mainly the devel-
      opment of male characteristics.

198. Is a genetic disorder which affects males primarily, causing them
      to have 2 X and 1 Y chromosomes.

199. Causes severe mental retardation in affected humans, is characterized by hairlip cleft palate, and polydactyly and is designated
as chromosome 13.

Questions 200-201 concerns the flow diagram below of enzymatic activity.

[A] $\xrightarrow{\text{reaction 1}}$ [B] $\xrightarrow{\text{reaction 2}}$ [C] $\xrightarrow{\text{reaction 3}}$ [D] $\xrightarrow{\text{reaction 4}}$ [E]

(A) A only   (D) B only

(B) C only   (E) E only

(C) B, C and D

200. The substrate would most likely be found at

201. Metabolic intermediate(s) would be at this stage of the reaction.

Questions 202-205 concern the terms below.

(A) Adaptive radiation   (D) Phylogeny

(B) Speciation   (E) Balanced polymorphism

(C) Hybridization

202. Process of evolution from a single ancestral species of a variety of
forms that occupy several different habitats.

203. The adaptive change of the genetic pool of fruit flies, even in response to such environmental changes as the alternations of the
seasons.

204. A mutation which is caused when geographic and/or genetic isolation sets up reproductive barriers.

205.   The combination of the best characters of each of two original, related species into a single descendant thereby creating a new type of species better able to survive.

DIRECTIONS: *The following groups of questions are based on laboratory or experimental situations. Choose the best answer for each question and mark the letter of your selection on the corresponding answer sheet.*

Questions 206-207 concern the curves which show the effect of competition between two species of Paramecium. The solid curves show the growth of population volume of each species alone in a controlled environment with a fixed supply of food. The dotted curve shows the change in population volume of the same species when in competition with each other under similar conditions.

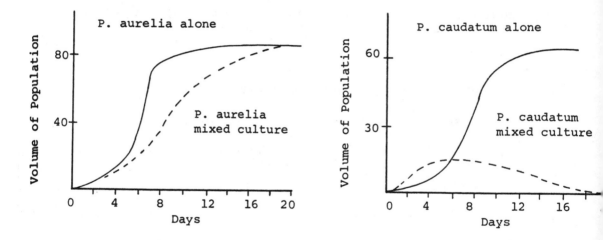

206.   It was found that when grown in the mixed culture

(A)   P aurelia grew normally.

(B)   P caudatum was inhibited.

(C)   there was no noticeable effect on any of the species.

(D)   the two species rapidly evolved in divergent directions.

(E)   both (B) and (D) are correct.

207. The effect when the two cultures were mixed was most likely due to

    (A) mutalism.

    (B) cooperation.

    (C) parasitism.

    (D) competition.

    (E) eutrophication.

Questions 208-210

Genetic damage in certain organisms was traced to mutagenic screening systems applied. The chart below shows a reflection of the types of genetic damage detected.

Screening System		Type of Damage Detected						
		Chromosome Aberrations				Mutations		
Category	Organisms	Deletions & Duplications	Dominant Lethality	Trans-locations	Nondis-junction	Forward or reverse or both	Multiple Specific Locus	Induced Recombination
Bacterial	E. Coli					●		
Fungal	Yeasts	●	●		●	●	●	●
Plant	Vica fibre	●		●	●	●		
Insect	Bombyx	●	●	●	●	●	●	●
Mammals	Chinese Hamster	●		●	●	●		
	Mouse	●		●	●	●		
	Humans	●	●	●	●	●	●	

208. Which of the following statements is true regarding the data obtained from these tests?

    (A) E.coli suffered no particular mutation.

    (B) Fungal organisms especially yeasts produced spores under these circumstances.

    (C) Insects were more adapted to screening systems and therefore suffered little or no damage.

    (D) Vica fibre and mouse cells had similar damages.

(E) Only vica faba and humans did not suffer forward or reverse mutations.

209. From this data we can infer that

(A) screening systems can detect cancer.

(B) screening systems are more helpful than hazardous to somatic cells.

(C) the nature of the screening tests and the outcome are helpful in understanding what causes chromosome aberrations and mutations in some organisms.

(D) None of the above can be inferred.

(E) All of the above are correct.

210. The most frequent type of chromosomal aberrations were

I. Dominant lethality

II. Translocations and forward or reverse mutations

III. Deletions and duplications

IV. Nondisjunction

V. Forward or reverse or both mutations

(A) II only   (D) V only

(B) II and IV   (E) III and IV

(C) II and III

THE GRADUATE RECORD EXAMINATION BIOLOGY TEST

MODEL TEST I

ANSWERS

1.	C	20.	D	39.	C
2.	A	21.	C	40.	D
3.	D	22.	D	41.	C
4.	A	23.	A	42.	A
5.	C	24.	E	43.	B
6.	A	25.	B	44.	D
7.	D	26.	C	45.	D
8.	E	27.	D	46.	B
9.	B	28.	E	47.	E
10.	C	29.	B	48.	C
11.	E	30.	A	49.	E
12.	C	31.	C	50.	A
13.	D	32.	C	51.	B
14.	C	33.	A	52.	C
15.	B	34.	E	53.	B
16.	A	35.	B	54.	D
17.	A	36.	C	55.	C
18.	E	37.	E	56.	A
19.	D	38.	A	57.	C

58.	B	89.	D	120.	D
59.	B	90.	B	121.	B
60.	C	91.	D	122.	B
61.	B	92.	C	123.	A
62.	E	93.	B	124.	D
63.	C	94.	E	125.	C
64.	C	95.	A	126.	E
65.	D	96.	D	127.	A
66.	C	97.	E	128.	C
67.	A	98.	A	129.	E
68.	A	99.	B	130.	D
69.	C	100.	D	131.	B
70.	D	101.	A	132.	E
71.	B	102.	E	133.	C
72.	C	103.	D	134.	A
73.	C	104.	C	135.	D
74.	D	105.	C	136.	D
75.	E	106.	A	137.	E
76.	B	107.	C	138.	A
77.	A	108.	A	139.	A
78.	D	109.	D	140.	E
79.	A	110.	E	141.	C
80.	A	111.	B	142.	D
81.	C	112.	A	143.	A
82.	C	113.	C	144.	E
83.	E	114.	A	145.	C
84.	D	115.	C	146.	D
85.	B	116.	B	147.	E
86.	C	117.	B	148.	A
87.	A	118.	E	149.	B
88.	A	119.	C	150.	D

151.	C	171.	B	191.	D
152.	C	172.	A	192.	A
153.	B	173.	D	193.	B
154.	A	174.	E	194.	C
155.	C	175.	E	195.	D
156.	A	176.	B	196.	A
157.	E	177.	D	197.	E
158.	B	178.	A	198.	B
159.	D	179.	C	199.	C
160.	C	180.	E	200.	A
161.	B	181.	C	201.	C
162.	C	182.	D	202.	A
163.	A	183.	A	203.	E
164.	D	184.	B	204.	B
165.	E	185.	E	205.	C
166.	D	186.	D	206.	E
167.	B	187.	B	207.	D
168.	A	188.	C	208.	D
169.	E	189.	A	209.	C
170.	C	190.	E	210.	E

THE GRE BIOLOGY TEST

MODEL TEST I

DETAILED EXPLANATIONS
OF ANSWERS

1. (C)

The pituitary gland, also known as the hypophysis, lies in a pocket in the skull known as the sella turcica. It is connected to the brain by the infundibular stalk. The pituitary is a compound organ made up of anterior, posterior, and intermediate lobes. The anterior lobe is made up of glandular tissue which produces at least six different protein hormones: TSH, ACTH, FSH, LH, prolactin, and growth hormone. The posterior lobe which is true neural tissue releases two hormones: oxytocin and vasopressin.

2. (A)

Scanning-electron microscopy provides a 3-dimensional, high-resolution image of cells and tissues. In scanning-electron microscopy, the surface of the tissue is studied. Such microscopes have a resolution of 25 to 75 angstroms.

3. (D)

Penicillin is effective only against actively growing bacteria because it blocks the synthesis of cell walls.

Penicillin prevents the incorporation of N-acetyl muramic acid into the structure that comprises the bacteria's cell wall. If cell wall formation is complete, penicillin has no effect. Thus, only actively growing cells are killed by this antibiotic.

4. (A)
The golgi apparatus is an organelle that is responsible only for the packaging of protein for secretion.

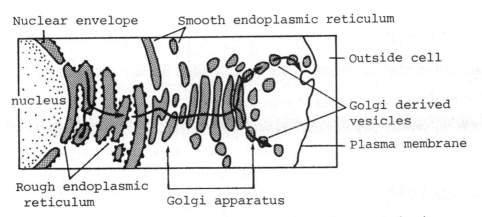

Schematic representation of the secretion of a protein in a typical animal cell. The solid arrow represents the probable route of secreted proteins.

5. (C)
Glucose is the monomer that makes up both cellulose and starch.

6. (A)
The primitive atmosphere had essentially no free oxygen. All oxygen was in the form of water and oxides.

7. (D)
First, we must find the percentage of albinos in the population:

$$\frac{18}{200} \times 100 = 9\%.$$

We know that albinism is the homozygous recessive trait (aa), so to find the frequency of the recessive allele, a, we take the square root of 9% (0.09) = $\sqrt{.09}$ = 0.30. We also know that the frequency of the recessive allele (a) added to the frequency of the dominant allele (A) must equal 1. Thus, the frequency of the non-albino allele = 1 - 0.30 = 0.70. Squaring the frequency of the non-albino allele will give us the frequency of homozygous non-albinos (AA).

$$(0.70)^2 = 0.49 \text{ or } 49\%.$$

Then we multiply 0.49 × 200 and see that there are 98 homozygous dominant non-albinos (AA).

8. (E)
All chordates display the following three characteristics: 1) the central nervous system, which is a tube containing a single continuous cavity situated on the dorsal side of the body, 2) the presence of clefts in the wall of the throat region, usually referred to as gill slits, and 3) the presence of a notochord, a rod which lies dorsal to the intestine and extends anterior to posterior to serve as skeletal support.

These characteristics need only to be present at some time in the life of the organism for it to be considered a chordate; they don't have to be present in the adult form.

9. (B)
Cryptic appearance, Müllerian mimicry, and Batesian mimicry are ways in which organisms avoid becoming the victims of predators. The cryptic appearance of some organisms enables them to blend into their background, becoming invisible to their potential attackers. Mullerian mimicry involves the evolution of two or more inedible or unpleasant-tasting species to resemble one another. Batesian mimicry, however, involves the resemblance of an unprotected, harmless species to a dangerous species. This makes it difficult for the predator to distinguish between the two forms. Once the predator has tasted the noxious species, it tends to stay away from both species.

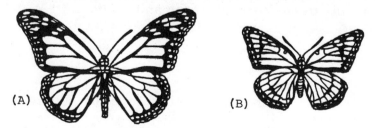

(A)   (B)

An example of Batesian mimicry.
(A) The Monarch butterfly, a distasteful species. (B) The Viceroy a species that mimics the Monarch. Species in the group to which the Viceroy belongs ordinarily have a quite different appearance.

10. (C)

The electron transport system yields 6 ATP and the Krebs cycle yields 9 ATP for a total of 15 ATP.

11. (E)

The connective tissues are defined as the complex of cells and extracellular materials which provide the supporting and connecting framework for all other body tissues. Connective tissues consist of extracellular fibers, amorphous ground substance, and connective tissue cells. The fibers are composed mainly of the proteins collagen and elastin. The ground substance occupies the space between the cells and fibers and contains proteoglycans, glycoproteins, and other molecules secreted from the cells.

12. (C)

Cristae refers to the inner membrane of the mitochondria, while all the others, thylakoids, stroma, starch granules and lamellae are of the chloroplasts.

13. (D)

Anaphase is the stage of mitosis characterized by the separation of sister chromatids from one another and their movement to opposite poles of the spindle. The lengthwise separation of chromatids begins at the centromeres and spreads distally as the respective sister chromatids move apart in opposite directions. When they reach the respective poles, movement stops and telophase begins.

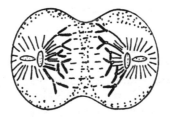

Anaphase

14. (C)

Bacterial genomes are haploid. This means that they have

only one copy of each gene. Thus no recessive traits can be masked by dominant traits since the alleles are in single copies. This fact is very useful to geneticists who wish to follow traits that would be masked in diploid organisms.

15. (B)
Of all the choices, only uracil is not a nitrogeneous, excretory component. It is, however, a nitrogeneous base found in RNA.

16. (A)
Birds utilize celestial clues to guide them during migration. If a caged bird is allowed to face the sun during the migration period, it will fly along its migratory route.

17. (A)
A muscle that has contracted strenuously and repeatedly and has exhausted its stored supply of organic phosphates and glycogen, will accumulate lactic acid. This lactic acid is a product of glycolysis and fermentation. The muscle has incurred what is known as an oxygen debt. When the violent activity is over, the muscle cells consume large quantities of oxygen as they convert lactic acid into pyruvic acid. Pyruvic acid is oxidized via the Krebs cycle and electron-transport process. The cells utilize the energy obtained to resynthesize glycogen from the lactic acid that remains.

18. (E)
The pituitary is the master gland which regulates all of these structures except the adrenal medulla. The adrenal medulla arises from the same source as the nervous system. Therefore, it is regulated by the stimulation of sympathetic preganglionic nerve fibers. Once stimulated, the adrenal medulla can secrete adrenalin (epinephrine) and noradrenalin (norepinephrine) which are responsible for the fight-or-flight response.

19. (D)
Mature sperm cells are the result of two stages of

meiosis. One spermatogonia from the testes produces four
mature sperm cells.

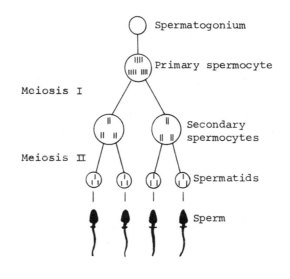

20.   (D)
Fruit production requires active cell division. In plants,
only regions of meristematic tissue are capable of active cell
division. The removal of all of the meristematic tissue from a
plant leaves it incapable of cell division; and therefore,
incapable of producing fruits.

21.   (C)
In an open circulatory system there are sections where
vessels are absent and where blood flows through large open
spaces known as sinuses. All vertebrates have closed
circulatory systems, as do the annelids (earthworms and
their relatives). All arthropods (insects, spiders, crabs,
crayfish, etc.), as well as most mollusks, have open
circulatory systems. Movement of blood through an open
system is not as fast, orderly or efficient as through a
closed system.

22.   (D)
Protozoans are single-celled animals whose cells are often
highly specialized containing many organelles. They can
reproduce both sexually and asexually. However, they
cannot give birth to live progeny (viviparity), in the way
that mammals can.

23. (A)

A cross-section of the spinal cord shows that it is composed of two regions: an inner gray mass composed of nerve cell bodies, and an outer mass of white matter made up of bundles of axons and dendrites. Four protuberances of gray matter are present; these are two anterior processes called ventral horns and two posterior ones called dorsal horns. The spinal cord receives sensory fibers from peripheral receptors at the dorsal root. These sensory fibers pass into the dorsal horns. The spinal cord receives sensory fibers from peripheral receptors at the dorsal root. These sensory fibers pass into the dorsal horns. Axons from the motor neurons leave the spinal cord at the ventral root after arising from the ventral horn.

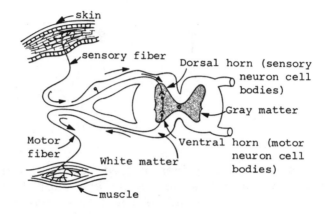

24. (E)

Many enzymes are involved in the breakdown of foods into the simple molecular components needed by the body. Of the enzymes listed, all are involved in digestion except for ligase. Ligase is the enzyme that repairs breakages in the DNA helix.

25. (B)

The process of change in the allele frequencies in the gene pool of a population over time is called evolution. Evolution occurs through the processes of mutation, genetic drift, migration, and natural selection. An individual who synthesizes a new enzyme has had a mutation in one of his genes so that a new, mutant gene product is produced. Genetic drift occurs when individuals move to a new feeding and hunting ground. Migration happens when individuals from one population join another. Natural selection, or survival of the fittest is, in effect, the survival of fast antelopes. The Lamarkian view that evolution occurs through the use and disuse of body parts is illustrated in the example with giraffes. This viewpoint has been overwhelmed by contrary evidence.

26. (C)
Temperature extremes affect enzymatic activities. High temperatures denature the enzyme while low temperatures cause the rate of enzymatic action to slow down. The concentration of hydronium ions, or pH, also has a great effect on enzymatic activity. Finally, enzyme-poisoning may also disrupt the activity of enzymes.

27. (D)
Intrinsic factor is a glycoprotein that combines with vitamin B_{12} to form a complex necessary for the absorption of vitamin B_{12} in the ileum.

28. (E)
Carina and syrinx are found only in birds. The carina is the enlarged sternum; it provides the point of attachment for the powerful flight muscles. The carina is also known as the keel.

The syrinx is located at the lower end of the trachea and is responsible for sound production in birds.

29. (B)
Lipase is not a proteolytic enzyme. It hydrolyzes fats to fatty acids and glycerol. Proteolytic enzymes hydrolyze proteins and very large polypeptides into small peptides and dipeptides.

30. (A)
The molecular oxygen in this reaction comes from the water. Since the only compound left that could donate oxygen is CO_2, it must donate its oxygen to the carbohydrate.

31.   (C)
The pericardium is the membranous sac that encloses the heart. Endothelium is a term that refers to endothelial cells. Endocardium and myocardium refer to the heart muscle. The vena cava is the largest vein in the body.

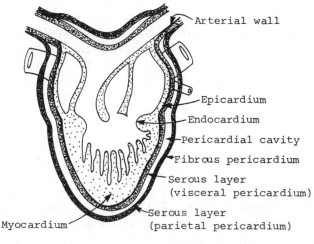

THE SURROUNDING HEART LAYERS

32.   (C)
The only statement that is incorrect is choice (C) which states that spores are formed by mitosis. Spores are actually formed by reduction division in the sporangium of a sporophyte.

33.   (A)
We can represent the traits as follows:

B = black fur	S = short hair
b = white fur	s = long hair

Thus, the parental genotypes are BbSs (heterozygous black, short-haired), and bbss (homozygous white, long-haired). Schematically, the cross proceeds as shown in the diagram.

Parents   BbSs♂ × bbss♀

Gametes ♂	BS	Bs	bS	bs
♀ 2bs	2BbSs	2Bbss	2bbSs	2bbss

black	:	black	:	white	:	white
short	:	long	:	short	:	long
$\frac{1}{4}$	:	$\frac{1}{4}$	:	$\frac{1}{4}$	:	$\frac{1}{4}$

60

34. (E)
The partial pressures of oxygen and carbon dioxide are the main determinants of whether oxygen will be picked up or released by hemoglobin. When there is a large amount of oxygen (a high partial pressure) and a low partial pressure of carbon dioxide, hemoglobin picks up oxygen. This is the case in the alveoli of the lungs. Oxyhemoglobin dissociates oxygen when the partial pressure of carbon dioxide is high and that of oxygen is low. This occurs in the tissues that are serviced by the systemic circulation.

35. (B)
Striated (skeletal) muscle fibers are multinucleated cylindrical cells arranged in parallel bundles. The most striking feature of such fibers is the transverse pattern of alternating dark A and light I bands. Both skeletal and cardiac muscle exhibit such patterns; smooth muscle does not.

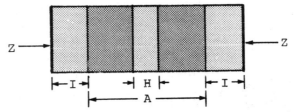

A sarcomere of a myofibril showing bands.

36. (C)
Mutualism is a symbiotic relationship between two organisms in which both organisms benefit. Lichens are composites formed of an alga and a fungus in a close mutualistic relationship. The fungi absorbs water and nutrients and forms most of the supporting structure, while the alga provides nutrients for both organisms via photosynthesis.

37. (E)
The chorion secretes a hormone called chorionic gonadotrophin, which is functionally similar to LH (luteinizing hormone). Its function is to take the place of LH in preserving the corpus luteum because of the inhibition of LH production by a high progesterone level. The corpus luteum is then able to secrete progesterone at high levels without being shut off.

38. (A)
During the day, a plant undergoes photosynthesis, a process which converts CO_2 and H_2O into carbohydrates. The resulting drop in CO_2 concentration decreases the acidity in the cells and promotes the conversion of the insoluble starch into soluble glucose. The high cellular concentration of glucose causes water to enter the guard cells by osmosis, rendering them turgid. The turgor of the guard cells causes them to bend, opening the stomata as a result.

At night, CO_2 is produced. The resulting rise in CO_2 concentration increases the acidity in the cells which inhibits the conversion of starch into sugar. Since starch is insoluble in the cell sap, it does not exert an osmotic pressure. As a result, the guard cells lose their turgor, causing them to collapse and close the stomata.

39. (C)
Pepsin is an enzyme of the stomach. It starts protein digestion in the stomach by splitting the long polypeptides into shorter fragments. These peptides are further digested in the intestine. The pancreatic secretions contain enzymes that finish the digestion of food in the duodenum; the first part of the small intestine.

40. (D)
Both the production of ATP and photophosphorylation occur during the light reaction of photosynthesis.

The light reaction of photosynthesis involves two pathways; cyclic and non-cyclic photophosphorylation. In cyclic photophosphorylation ATP is formed when the energy released by the electrons as they move down the energy gradient combines with ADP and inorganic phosphate. Photophosphorylation is an important aspect of the light reaction which uses light.

41. (C)
Hydrogen ions combine with the ionized form of hemoglobin (Hb-) to form acid hemoglobin (HHb). Acid hemoglobin binds the H^+ ions relatively strongly so that they can be effectively removed from the blood. This keeps the pH constant if there is an excess of hydrogen ions.

42. (A)
Temperature-sensitive mutations are mutant genotypes that are expressed at a restrictive temperature and not expressed at a permissive temperature. Such mutations most likely exist as a result of a faulty protein which is unstable at the restrictive temperature. Hence, genes which are absolutely essential for the life of an organism can carry conditional mutations such as temperature-sensitive mutations. This facilitates the genetic study of mutations on essential gene function.

43. (B)
Resolution is the ability to separate adjacent objects so that they are distinct. Objects lying close to one another cannot be distinguished as separate objects if the distance between them is less than one-half the wavelength of the "light" source being used. Electrons have properties of both particles and waves. Since their wavelength is much smaller than the wavelenth of light waves, the resolving power of a microscope using electrons as its source is higher than a microscope which uses light.

44. (D)
The core of all motile cilia and flagella consist of two central and 9 peripheral pairs of microtubules composed of the proteins alpha and beta tubulin. Nonmotile cilia lack the two central pairs.

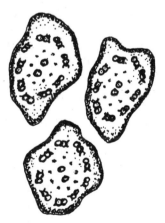

Cross-sections of Motile cilia

45. (D)
The human small intestine is a thin-walled tube about 4

meters in length extending from the pylorus of the stomach to the colon. It consists of three portions which start with the duodenum, then the jejenum, and finally there is the ileum.

46. (B)
Centrioles are present in most types of animal cells. They are present during cell division and seem to have some function in directing the orderly distribution of the genetic material. Most higher plant cells lack centrioles.

47. (E)
Excretion is the removal of metabolic wastes that can become toxic if allowed to accumulate. The kidneys and skin remove salts, urea and other organic compounds. The lungs remove water and carbon dioxide that are waste products of the circulatory system. The liver breaks down bile pigments, red blood cells, and some proteins and drugs. Lymph nodes, on the other hand, are filters for invading bacteria and indigestible particles such as dust and soot. Thus, the lymphatic system is not excretory in the strict sense of the word since it rids the body of foreign, invading substances rather than waste products of its own metabolism.

48. (C)
Commensalism is a type of symbiosis. One species benefits while the other receives neither harm nor benefit in this type of relationship. The plants upon which the epiphytes grow are simply used for support. Commensals are not parasites.

49. (E)
Centrioles are made up of microtubules. Thus, their activity will be blocked by colchicine because they participate in the formation of spindle fibers. If spindle formation is blocked, meiosis (and mitosis) will not proceed beyond metaphase.

50. (A)
The nucleus is the control center of the cell. Experiments

have sucessfully demonstrated that the replacement of one nucleus with that of another species is sufficient for that cell to develop characteristically like the other species. It is therefore concluded that all of the necessary hereditary information is contained in the nucleus.

51. (B)
The acrosome is a membrane-bounded vesicle at the tip of the head of a sperm cell. It contains enzymes which help the sperm to digest its way through the protective jelly coat of the egg. This digestion enables the sperm nucleus to enter the egg cell and fertilize it.

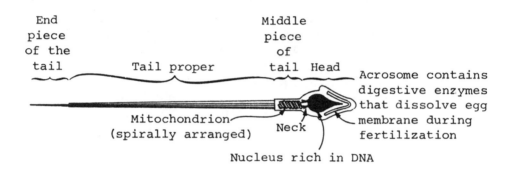

52. (C)
Auxins may be regarded as the most important of the plant hormones, since they have the most marked effects in correlating growth and differentiation to result in the normal pattern of development. The differential distribution of auxin in the stem of the plant as it moves down from the apex (where it is produced) causes the plant to elongate and bend towards light. Auxin from seeds induces the maturation of the fruit. Auxin from the tip of the stem passes down into the vascular cambium below and directs the tissue toward differentiating into secondary phloem and xylem. Auxin also stimulates the differentiation of roots; it has been discovered that by placing a cut stem in a dilute solution of auxins, roots can be easily produced. The auxins, in addition, determine the growth correlations of the several parts of the plant. They inhibit development of the lateral buds and promote growth of the terminal bud. Finally, auxins control the shedding of leaves, flowers, and branches from the parent plant. By inhibiting the formation of the abcission layer between leaf petioles and the stem or branch, the auxins prevent the leaves from being shed.

53. (B)
Kinesis is a simple form of orientation in which the animal does not necessarily direct its body toward the stimulus. The stimulus merely causes the animal to speed up or slow down its movements, resulting in the eventual displacement of the animal toward or away from the stimulus. It is an undirected type of orientation. The wood louse did not direct itself toward the moist spot, but simply encountered it by random movement.

54. (D)
Meiotic drive is a factor that may alter allelic frequencies in a gene pool. Meiotic drive is the term for preferential segregation of genes that may occur in meiosis. For example, if a particular chromosome is continually segregated to the polar body in female gametogenesis, its genes would tend to be excluded from the gene pool since the polar bodies are nonfunctional and will disintegrate. There is significant evidence that, due to physical differences between certain homologous chromosomes, preferential selection of one over the other often occurs of other than random proportions.

55. (C)
Gram-staining is one of the most important differential staining techniques used today to determine differences between bacterial cells. Bacteria may either be gram-positive, staining a violet color, or gram-negative, staining a red color. The difference in staining is based on the varying lipid contents of the cell walls of the bacteria.

56. (A)
The proper sequence of classification is

Kingdom → Phylum → Class → Order → Family → Genus Species → Variety.

57. (C)
The characteristic difference between the marsupials, such as kangaroos, and the placentals, such as the deer, is the time of embryonic development within the uterus. Marsupial embryos undergo a short development period

before they leave the uterus. Embryonic development is completed in an abdominal pouch of the mother where the embryo is attached to a nipple. In contrast to this are the placentals. In this group, the embryo develops completely in the uterus of the mother.

58. (B)
To become an Hfr bacterial cell, the F factor must become integrated into the bacterial chromosome. When an Hfr cell and an F⁻ cell begin conjugation, the F factor portion of the circular Hfr chromosome initiates synthesis of a linear chromosome. This linear chromosome carries a fragment of F factor on both its ends, with one end acting as the origin for the transfer of the chromosome. Since the other end of the chromosome contains the remaining portion of the F factor, and since the whole chromosome rarely gets transferred, the F⁻ cell usually does not receive a complete F factor to become F⁺. Instead, the F⁻ cell receives pieces of the Hfr chromosome.

59. (B)
The presence of swimming sperm is a primitive condition characteristic of lower aquatic plants and early terrestrial plants, such as ferns and mosses (which require water for fertilization). However, swimming sperm is of no advantage to a seed plant, and virtually all have been eliminated through natural selection as the plants become more adapted to terrestrial life. For this reason, the cycad is considered to be a primitive gymnosperm which provides a connecting link between the ferns and lycopods and the more advanced seed plants.

60. (C)
Schwann cells are the myelin-forming cells of the peripheral nervous system. Each Schwann cell forms a single myelin internodal segment around a portion of an axon. Schwann cells may also surround unmyelinated axons, without producing myelin.

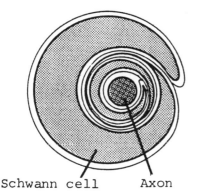

Schwann cell   Axon

Cross-section shows how Schawnn cell wraps around axon to form myelin sheath.

61. (B)
The sliding filament model of muscle contraction states that thick and thin filaments slide past each other while their lengths remain the same. The H zone is comprised of an area of thick filament alone; the overlap with thin filaments is not part of this area. As muscles contract, the Z bands come closer and the H zone decreases.

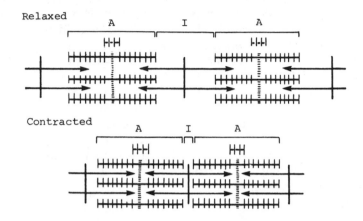

The sliding filament theory of muscle contraction. The thin filaments slide over the thick filaments during contraction. Filaments do not shorten during contraction. The sarcomere itself shortens since adjacent Z-lines are drawn together as actin filaments slide over myosin filaments. Actin filaments on either side of a Z-line have opposite polarity, so that each sarcomere contracts during filament sliding.

62. (E)
Competitive inhibition, unlike non-competitive inhibition, may be overcome. It is possible for this to occur by increasing the substrate and enzyme concentrations, thus allowing for an increase in active sites. It can be overcome when there is no alternation in the enzyme active site.

63. (C)
Vasopressin or antidiuretic hormone (ADH) is produced by a discrete group of hypothalamic neurons whose axons terminate in the posterior pituitary. The release of vasopressin from the pituitary is regulated by specific baroreceptors which, in turn, are stimulated by increased blood volume. When the baroreceptors are stimulated, they inhibit the production of vasopressin.

64. (C)

A controllable unit of transcription is called an operon. An operon consists of a binding site for RNA polymerase (promoter), a binding site for a specific repressor (operator), and one or more structural genes. RNA polymerase is the enzyme responsible for transcription of DNA.

The operator gene is located between the promoter and structural genes on the chromosome. When a repressor binds to the operator, the repressor prevents the movement of RNA polymerase along the DNA molecule, thereby inhibiting transcription of the structural genes. When the repressor is not bound to the operator, transcription is free to occur.

65. (D)

Chlorophyll a is found in all photosynthetic eukaryotic cells and is considered to be essential for photosynthesis of the type carried out by plants. It functions in the capture of light energy by either directly absorbing it or receiving it in the form of high energy electrons from accessory pigments such as chlorophyll b, carotene, and xanthophyll. Carotene and xanthophyll belong to a class of pigments known as carotenoids.

66. (C)

Groups of organisms are characterized by three levels of organization - populations, communities, and ecosystems. A population is a group of organism belonging to the same species which occupy a given area. A community is a unit composed of a group of populations living in a given area. The community and the physical environment considered together make up an ecosystem.

67. (A)

Innate behavior patterns occur in response to certain stimuli detected by an animal's sensory receptors. Of the myriad of stimuli encountered in any situation, an animal responds only to a limited number of stimuli. These stimuli, which elicit specific responses from an animal, are referred to as sign stimuli. The sign stimuli involved in intraspecies communication are called releasers.

68. (A)
The Mesozoic Era began some 230 million years ago and was characterized by a wide variety of reptiles. In fact, the Mesozoic Era is commonly referred to as the "Age of Reptiles"; common reptiles of this era were the primitive lizards: snakes, turtles, alligators, crocodiles, pterosaurs (flying reptiles) and later in the era, the dinosaurs. All of these, and also the mammals which came later, evolved from an important paleozoic group called the stem or root reptiles (cotylosaurs).

69. (C)
Many non-photosynthetic plants, such as mushrooms, cannot produce their own carbohydrates from the air and water. Instead, they must obtain the carbohydrates they require from the surroundings. This is usually supplied by dead plant matter, but plant matter is mainly cellulose and other polysaccharides which must be first digested. Specialized secretory cells in these organisms release the enzymes for digestion into the surrounding organic matter. The products of this extracellular digestion are then directly absorbed through the cells' surfaces; sometimes by specialized absorbing structures.

70. (D)
The placenta is a region where a portion of the embryonic chorion and the maternal uterine wall join. It functions in the exchange of nutrients, wastes, and gases between the mother and the fetus.

71. (B)
The compounds known as prostaglandins have been identified as cyclic, oxygenated, 20-carbon fatty acids. The prostaglandins, which are secreted by the seminal vesicles and other tissues, appear to mediate hormonal action by influencing the formation of cyclic AMP.

72. (C)
A queen bee soon after hatching, mates several times with male drone bees. The queen accumulates enough sperm to last for her lifetime, and stores the sperm internally in a

sperm sac. Thereafter, she uses this sperm to lay fertilized or unfertilized eggs, as many as a thousand a day.

73. (C)
Regulatory enzymes are a key controlling factor in metabolic pathways. If the end product of a pathway is in excess, it inhibits the action of the regulatory enzyme by binding to its regulatory site. The end product shuts off the catalytic activity of the active site by altering the arrangement of the enzyme's polypeptide chains, thus deforming and inactivating the enzyme. This feedback mechanism is known as end product inhibition and is important in preventing the accumulation of unwanted substances.

74. (D)
A gastrula is a two- or three-layered ball of cells; these layers being known as the germ layers. In amphibians and echinoderms, the gastrula is formed by the inward folding of one side of the blastula, partially obliterating the original cavity, or blastocoel. A new cavity is formed which is open to the outside, and is called the archenteron, or primitive gut. The blastopore is the opening from the archenteron to the outside.

75. (E)
In 1953, Watson and Crick proposed the double-helix model of DNA. Their model has been supported by studies by Arthur Kornberg who was able to synthesize DNA in a cell-free environment by combining nucleotides, DNA polymerase, and a DNA primer. Further support for the model came from Meselson and Stahl's experiment which showed the semi-conservative nature of replication.

76. (B)
Ethylene is the gas which controls the ripening process of fruits, and which is synthesized by altering the amino acid methionine. It initiates the ripening process and it therefore must be the gas that was released by the kerosene stove that caused the ripening of the fruits.

77. (A)

If a sex-linked trait is recessive, a male would have a greater chance of expressing the trait. In order for a girl to express that trait, she would have to have two copies of the recessive gene, because its expression would be masked by the presence of a normal X chromosome or one with a dominant allele. Thus both her parents would have to carry the gene. A boy however, need only have one copy of the gene in order to express its trait, because his Y chromosome does not carry any genes that would mask the recessive gene. So only his mother needs to carry the gene. The chances of this happening are much greater than the chance that two people carrying the gene will mate and have a female. Therefore, the trait is more commonly expressed in males.

78. (D)

As energy flows through the various food chains, it is being constantly channeled into three areas. Some of the energy goes into production which is the creation of new tissues by growth and reproduction. Energy is also used for the manufacture of storage products such as fats and carbohydrates. The rest of the energy is lost to the ecosystem by respiration and decomposition. The loss of one energy due to respiration is very high and only a small fraction of energy is transferred successfully from one trophic level to the next.

Each trophic level depends on the preceding level for its energy source. The number of organisms supportable by any given trophic level depends on the efficiency in transforming the energy available in that level to useful energy of the subsequent level. Ecological efficiencies vary widely, but it has been shown that the average ecological efficiency of any one trophic level is about 10%.

79. (A)

The dark reaction of photosynthesis consumes ATP to make carbohydrates.

80. (A)

This phase of the cell cycle is called the resting phase. However, the cell is "resting" only with respect to the visible events of division in later phases. During this phase, however, the nucleus is metabolically very active and chromosomal duplication is occurring.

81. (C)
Of all the fossils discovered which are relevant to the evolution of man, the craniums, jaws, and teeth have been the most numerous and useful. However, teeth have been found in the largest numbers at archaeological sites because their enamel coating gives them great durability.

82. (C)
The first new generation is reproducing on ^{14}N. The progeny will have one-half of its DNA newly synthesized on ^{14}N and half of its DNA already present. It can be pictured as the following:

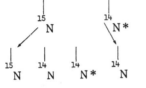

(first generation progeny)
parents

← second generation progeny

The starred nitrogen 14 was that already present. The unstarred was that newly synthesized. From this, it can be inferred that half of the second new DNA generation would contain only ^{14}N strands (starred and unstarred), and the other half would contain 50% ^{14}N and 50% ^{15}N.

83. (E)
If no disruptive factors interfere, most successions eventually reach a stage that is much more stable than those that preceeded it. The community of this stage is called the climax community. It has much less tendency than earlier successional communities to alter its environment in a manner injurious to itself. In fact, its more complex organization, larger organic structure, and more balanced metabolism enable it to control its own physical environment to such an extent that it can be self-perpetuating.

84. (D)
The contractile vacuole is a special excretory organelle which ejects water from the cell. It does not eject nitrogeneous wastes. These organelles are more common in fresh water protozoans than in marine forms because of the hypotonic environment to which they must adapt. In the

fresh water environment, water flows across the cell membrane into the cell and must be removed before the paramecium swells to a fatal size. The job of the contractile vacuole is to remove this excess water.

85. (B)
If a mosquito sucks blood containing a parasite, the blood goes to the stomach where the parasite continues to live. The zygotic stage of the parasite will encyst in the wall of the gut. When the cyst ruptures, the organism migrates to the salivary glands of the mosquito where it is discharged into a new host when the mosquito feeds.

86. (C)
When a segment of one chromosome is transferred to another non-homologous chromosome,the mutation is known as a translocation. A deletion is a mutation in which a segment of the chromosome is missing. In duplication, a portion of the chromosome is represented twice. An inversion results when a segment is removed and reinserted in the same location, but in the reverse direction.

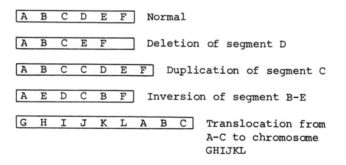

MUTATIONS INVOLVING CHROMOSOME STRUCTURE

87. (A)
The secondary and tertiary structures of polypeptides are destroyed by denaturation.

88. (A)
Parenchyma tissue, which occurs in roots, stems, and

leaves, consists of small cells with a thin cell wall and a thin layer of cyptoplasm surrounding a large vacuole. The cells are loosely packed, resulting in abundant spaces in the tissue for gas and nutrient exchange. Most of the chloroplasts of leaves are found in these cells.

89. (D)
Active transport, initially, enabled the cell to maintain a higher concentration of dye inside the cell. However, if the inhibitor were to interfere with ATP production and, consequently, active transport, normal diffusion processes would take over resulting eventually in equal intra- and extra-cellular concentrations of the dye.

90. (B)
There is no centrosome present in a prophase chromosome.

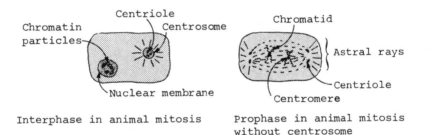

Interphase in animal mitosis   Prophase in animal mitosis without centrosome

91. (D)
In a gilled-fish, the direction of blood flow through a gill is opposite to the direction of water flow over the gill. This "counter-current system" maximizes the amount of gas exchange that can take place. If both the blood flow and water current were in the same direction, the CO_2 and O_2 concentration gradients across the gills would decrease. This would result in a slower rate of diffusion, and the amount of gas exchange would be reduced.

92. (C)
It is necessary to distinguish aggressive behavior between members of the same species from violent predatory behavioral patterns, which are usually directed at members

of different species. Aggressive behavior is usually exhibited between same species males in defending their territory or establishing their status in a social order. The critical factor in aggressive behavior is that most of the fighting consists of display. Display consists of ritualized, highly-exaggerated movement or sound which convey the attack motivation of the contestants. In attempting to appear as formidable as possible, the animal often changes the shape of certain body parts in an attempt to make itself appear as large as possible. The adaptive significance of display is that the two contestants are rarely seriously hurt.

93.　(B)
The environment above the tree line would resemble most closely the climate of the tundra. Because of the tundra's intense coldness, trees are unable to grow and vegetation consists mostly of grasses.

94.　(E)
The parts of a flower are as follows.

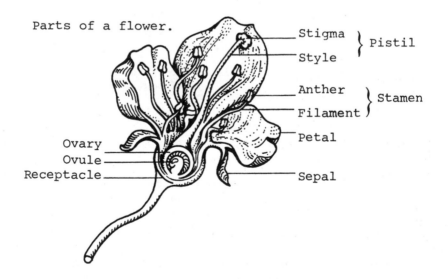

Parts of a flower.

95.　(A)
Structure A is known as the stigma; it is part of the flower's female reproductive organ called the pistil.

96. (D)

Flowers which are pollinated by the wind need not be attractive to birds and insects, and so their petals tend to be less showy, and may even be absent.

97. (E)

Klinefelter syndrome is a human abnormality which causes aberrant sexual development. Individuals with this syndrome have genitalia and internal ducts which are male but underdeveloped testes which do not produce sperms. Some affected individuals may have enlarged breasts and may be mentally retarded. It has been noticed that even though masculine development occurs, feminine development is not suppressed.

98. (A)

The Müllerian duct is the undifferentiated duct which exists in the female embryo. It is connected to the gonad which if it differentiates into the ovary causes the Müllerian system to develop into the female duct system.

99. (B)

Purines and pyrimidines absorb ultraviolet light at a wavelength of around 260 nm. Ultraviolet light is mutagenic and when absorbed by pyrimidines at the point where the addition of water to the ring structure is initiated, dimers are formed, especially between thymine residues. These dimers, it is believed, distort the DNA conformation and inhibit normal replication.

In any population where the species is actively reproducing there will be variations in the species that comprise the population.

100. (D)

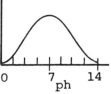

Activity Enzyme

0   7   14
ph

As the pH rises, enzyme activity rises rapidly until an optimum pH of about 7.00 is reached. Then, it begins to drop just as rapidly as it rose.

101. (A)

(A) The developmental stage of cells where the cells resemble a hollow ball is known as the blastula. (B) A morula is a developmental stage found in echinoderms. (C) The isolecithal stage is not a stage of development, but a description of the arrangement of the yolk and albumin in an egg. (D) The gastrula is a 3-layered cell stage formed from the blastula. (E) Ovulation is the release of one or more eggs from the ovaries.

102. (E)

The structural framework of the apoenzyme or transcriptional enzyme is made up of α_2 subunits.

103. (D)

Monozygotic twins are produced when two cells of a cleaving human embryo are separated.

104. (C)

In land vertebrates, carnivores and rodents exhibit the best olfactory sense.

105. (C)

Penetrance of a mutation is defined as the percentage of individuals that show some degree of expression of a mutant genotype.

106. (A)

Pheromones are chemicals which work like hormones but differ from hormones in that they work between individuals.

107. (C)

In Skinner's operant conditioning an unconditioned stimulus →
reinforcement; whereas in classical conditioning an unconditioned
stimulus → unconditioned response.

108. (A)

Damage to the right temporal lobe of the cerebral hemisphere
could result in an individual's poor performance on an IQ test.

109. (D)

Centrioles and asters are not characteristic of procaryotes.

110. (E)

Electron microscopy differs from light microscopy in the staining technique used on material, preparation of material, material
studied and resolution of microscopes.

111. (B)

The DNA enzyme which seals the Okazaki fragments in the lagging strand of a replication fork is ligase.

112. (A)

The energy molecule used to attach the 50S ribosomal subunit
to the 30S subunit is GTP.

113. (C)

A pilus is a slender tube which is formed between two bacteria
of different strains. If a plasmid-containing bacterium and a plas-
mid-free bacterium come into close proximity with each other a pilus is formed. This pilus allows for the transfer of genetic information from one bacterial cell to another.

114. (A)

In the human excretory system large blood components such as
glucose and amino acids are returned to the blood from the filtrate
by facilitated diffusion.

115. (C)
Mesosomes are invaginations of the cell membrane. It has been postulated that they contain respiratory enzymes.

116. (B)
In the gram-staining technique used for staining and separating bacteria into gram-negative and gram-positive bacteria, a mordant in the form of an iodine solution is applied to identify gram– positive bacteria. The mordant binds the dye crystal violet to the cell of gram-positive bacteria and helps resist decolorization. Gram-negative bacteria however does not retain the dye and is decolorized.

117. (B)
Rana and Canis belong to the taxonomic level, genus.

118. (E)
Homotherms are organisms which are characterized according to their ability to maintain a constant internal body temperature and property of their skin temperature, which is usually cooler than the temperature of their other organs.

119. (C) 120. (D) 121. (B)
All chordates display the following three characteristics: first, the central nervous system is a hollow tube containing a single continuous cavity, and is situated on the dorsal side of the body. A second characteristic feature is the presence of clefts in the wall of the throat region, usually called gill slits, originally used perhaps as a food-catching device. The third characteristic is the presence of a notochord, a rod lying dorsal to the intestine, extending from the anterior to posterior end, and serving as skeletal support. This phylom has three subphyla.

In a tunicate, for example, which belongs to the Chordata subphylum, Urochordata, the dorsal hollow nerve cord and the notochord are confined to the tail in the larval stage, and disappear in the adult stage. In the subphylum, Cephalochordata, with its lancelets, all chordate features are retained in the adult. In the Vertebrata, the notochord is partially or wholly replaced by the skull and vertebral column. It should be noted that the dorsal hollow nerve cord, the notochord, and the gill slits need only be present at some time in the life of an organism for it to be considered a chordate.

Finally, it should be noted that the Hemichordata or acorn worms, resemble the Chordata in that they are deuterostomes, have pharyngeal gill slits, and also retain a hollow dorsal nerve cord in the collar.

122. (B)
Prophase begins when the chromatin threads begin to condense and appear as a tangled mass of threads within the nucleus. Each chromosome in prophase is composed of two identical members resulting from duplication in interphase. Each member of the pair is called chromatid. The two chromatids are held together at a dark, constricted area called the centromere. At this point the centromere is a single structure.

123. (A)
During interphase, the chromosomes appear as vague, dispersed thread-like structures, and are referred to as chromatin material.

124. (D)
When the chromosomes have all lined up along the equatorial plane, the dividing cell is in metaphase. The stage after metaphase is anaphase. At metaphase, the centromere divides and the chromatids become completely separate daughter chromosomes. The division of the centromeres occurs simultaneously in all the chromosomes. Anaphase is the phase where the centromeres divide and daughter chromosomes begin to move apart.

125. (C)
Anaphase is marked by the movement of the separated chromatids or daughter chromosomes to opposite poles of the cell. The stage before anaphase is metaphase. It is thought that the chromosomes are pulled as a result of contraction of the spindle fibers in the presence of ATP. The chromosomes

moving toward the poles usually assume a V shape, with the centromere at the apex pointing toward the pole.

126. (E)
    Cytokinesis follows telophase. Telophase itself begins when the daughter chromosomes have reached the poles of the cell. The chromosomes then relax, elongate, and return to the resting condition in which only chromatin threads are visible. A nuclear membrane forms around each new daughter nucleus. This completes karyokinesis and cytokinesis follows. Cytokinesis, also known as cell division and is the actual division of one cell into two daughter cells.

127. (A)
    Acetylcholine is a chemical responsible for transmitting impulses across synaptic junctions. The chemical transmitter passes from the presynaptic axon to the postsynaptic dendrite by simple diffusion across the narrow space, called the synaptic cleft, separating the two neurons involved in the synapse. After the neurotransmitter has exerted its effect on the postsynaptic membrane, it is promptly destroyed by an enzyme called cholinesterase.

128. (C)
    Progesterone acts on the glandular epithelium of the uterus called the endometrium and converts it to an actively secreting tissue. The glands become coiled and filled with glycogen; the blood vessels become more numerous; and various enzymes accumulate in the glands and connective tissue of the endometrium. These changes are ideally suited to provide a favorable environment for implantation of a fertilized ovum. Progesterone also causes the mucus secreted by the cervix (a muscular ring of tissue at the mouth of the uterus which projects into the vagina) to become thick and sticky. This forms a "plug" which may constitute an important blockage against the entry of bacteria from the vagina. This is a protective measure for the fetus should conception occur.

129. (E)
    Actin is a globular shaped molecule which is an essential component of the thin myofilaments which are themselves necessary for the contraction of muscle fibers.

130. (D)
Ecdysone is the hormone that stimulates growth and molting in insects. It is released by the prothoracic gland, has been obtained in pure form, and identified as a steroid.

131. (B)
Rhizoids are usually found in bryophytes such as mosses and liverworts. They are simple filaments of cells or cellular projections performing the function of water absorption. The rhizoids are, however, not very efficient absorbers and in relatively dry areas cannot withdraw adequate materials from the ground.

132. (E)
The cells of the palisade layer, containing abundant amounts of chloroplasts, are chiefly responsible for the photosynthetic functioning of the leaf.

133. (C)
The mature sporophyte is composed of a foot in the archegonium of the gametophyte and a leafless, spindle-like stalk or seta which rises above the gametophyte. The sporophyte is nutritionally dependent on the gametophyte, absorbing water and nutrients from the archegonium via the tissues at the foot.

134. (A)
The cotyledon is a primary leaf structure found in the early angiosperm embryo. In the monocots, the single cotyledon serves primarily to absorb, rather than store as dicots do, the endosperm tissue. The portion of the embryo lying below the point of attachment of the cotyledons is called the hypocotyl and the part above is called the epicotyl. In monocots, the endosperm usually persists even after germination, and the cotyledon continues to absorb the nutrient material for the seedling until it can synthesize its own nutrients.

135. (D)
The xylem of the vascular system found in plants is

involved in the conduction of absorbed water and minerals up the stem to the leaves.

136. (D)
The structure of enzymes differ significantly. Some are composed solely of protein (for example, pepsin). Others consist of two parts, a protein part (also called an apoenzyme), and a non-protein part or co-factor. The co-factor may be either a metal ion or an organic molecule called a coenzyme. Coenzymes usually function as intermediate carriers of the functional groups, atoms, or electrons that are transferred in an overall enzyme transfer reaction.

137. (E)
A coenzyme that is very tightly bound to the apoenzyme is called a prosthetic group.

138. (A)
When an apoenzyme and a cofactor are joined together, they form what is called a holoenzyme. Often, both a metal ion and a coenzyme are required in a holoenzyme. Those apoenzymes needing a metal ion to function are also called metalloenzymes.

139. (A)
The most posterior part of the brain, connected immediately to the spinal cord, is the medulla. Here the central canal of the spinal cord (spinal lumen) enlarges to form a fluid-filled cavity called the fourth ventricle. The medulla has numerous nerve tracts (bundles of nerves) which bring impulses to and from the brain. The medulla also contains a number of clusters of nerve cell bodies, known as nerve centers. These reflex centers help control respiration, heart rate, the dilation and constriction of blood vessels, swallowing and vomiting.

140. (E)
The thalamus of the forebrain serves as a relay center for sensory impulses. Fibers from the spinal cord and parts

of the brain synapse here with other neurons going to the various sensory areas of the cerebrum. The thalamus seems to regulate and coordinate the external signs of emotions. By stimulating the thalamus with an electrode, a sham rage can be elicited in a cat - the hair stands on end, the claws protrude, and the back becomes humped. However, as soon as the stimulation ceases, the rage responses disappear.

141. (C)
Above the medulla is the cerebellum, which is made up of a central part and two hemispheres extending sideways. The size of the cerebellum in different animals is roughly correlated with the amount of their muscular activity. It regulates and coordinates muscle contraction and is relatively large in active animals such as birds. Removal or injury of the cerebellum is accompanied not by paralysis of the muscles but by impairment of muscle coordination. A bird, with its cerebellum surgically removed, is unable to fly and its wings seem to thrash about without coordination.

142. (D)
The cerebrum, consisting of two hemispheres, is the largest and most anterior part of the human brain. In human beings, the cerebral hemispheres grow back over the rest of the brain, hiding it from view. The outer portion of the cerebrum, the cortex, is made up of gray matter which comprises the nerve cell bodies. The gray matter folds greatly, producing many convolutions of the cerebral surface. These convolutions increase the surface area of the gray matter. The inner part of the brain is the white matter which is composed of masses of nerve fibers.

143. (A)
Semiconservative replication is the mechanism of DNA replication; the process by which the two complementary strands of the double helical DNA replicate to form new complementary strands. This model states that the replication of one DNA molecule yields two hybrids, each composed of one parental strand and one newly synthesized strand.

144. (E)
Inside the nucleus, transcription of DNA is mediated by

an enzyme, RNA polymerase. DNA undergoes a localized unfolding in the vicinity of the gene to be transcribed. RNA polymerase then selects precursor ribonucleotides complementary to the DNA template which are then polymerized to give a strand of m-RNA. The newly synthesized single-stranded m-RMA subsequently peels away from the template allowing a new RNA polymerase molecule to attach or the DNA strand to reunite.

145. (C)

Transfer RNA is the smallest type of RNA and is involved in the carrying of amino acids to the ribosomes where translation of mRNA into proteins takes place. There are at least 20 different tRNA's (one specific for each amino acid). Unlike the other two RNA's, mRNA and rRNA, tRNA, forms loops and double-stranded sections. In one of these loops is located the anticodon, which is what distinguishes the different types of tRNA from each other. The function of tRNA is to insert the amino acid specified by the codon on the mRNA into the growing polypeptide chain at the ribosomes, and it is through the complementation of anticodon and codon that the appropriate amino acid is incorporated.

146. (D)

Messenger RNA carries the genetic information coded for in DNA and is responsible for the translation of that information into a polypeptide chain. Each set of 3 bases of the mRNA comprises a codon which directs the incorporation of a specific amino acid into the polypeptide chain. Messenger RNA binds reversibly to the smaller subunit of the ribosome, where protein synthesis is initiated. It can dissociate from the ribosome without jeopardizing the integrity of the ribosome. Messenger RNA is heterogeneous in size due to the different lengths of polypeptide chains which a cell needs to synthesize, therefore, it is the most variable type of RNA.

147. (E)

The human and other animals' digestive system begins at the oral cavity. The teeth break up food by mechanical means, increasing the substrate's surface area available to the action of digestive enzymes. In addition to tasting, the tongue manipulates food and forms it into a semi-spherical ball (bolus) with the aid of saliva.

148. (A)
When food reaches the stomach, the distension of the stomach walls stimulates an increase in the rate of stomach movement and the production of gastrin. The hormone gastrin has several effects, the major of which is the stimulation of gastric juice secretion. It achieves this by stimulating gastric HCl secretion , which in turn produces protease activity . Pepsin, present in the gastric juice, cleaves proteins into many smaller chains . These small peptide chains stimulate receptor cells in the antrum of the stomach to produce more gastrin. Gastrin also has the ability to limitedly stimulate secretion in the intestine and pancreas. Also, gastrin stimulates an increase in gastric motility and helps keep the sphincter of the esophagus tightly closed.

149. (B)
The first part of the small intestine, which is mostly involved in absorption is the duodenum. Small finger-like protrusions,called villi, line the entire small intestine. Villi greatly increase the intestinal surface area and it is through the villi that most of the nutrient absorption takes place.

150. (D)
In an area of the large intestine called the colon, massive numbers of bacteria exist. Their function is not fully understood, but some can synthesize vitamin K which is of great importance to man's blood clotting mechanism.

151. (C)
Lipase, produced by the pancreas, digests the lipids in our foods, with the help of bile.

152. (C)
The initiation of the heartbeat and the beat itself are intrinsic properties and are not dependent upon stimulation from the central nervous system. The initiation of the heartbeat originates from a small strip of specialized muscle in the wall of the right atrium called the sinoatrial (SA) node, which is also known as the pacemaker of the heart. It is the SA node which generates the rhythmic self-excitatory impulse, causing a wave of contraction

across the walls of the atria. This wave of contraction reaches a second mass of nodal muscle called the atrioventricular node, or AV node. The AV node then helps the rest of the heart to undergo contraction.

153. (B)
A four-chambered heart is characteristic of "warm-blooded" animals such as birds and mammals. Since these animals maintain a relatively high constant body temperature, they must have a fairly high metabolic rate. To accomplish this, much oxygen must be continually provided to the body's tissues. A four-chambered heart helps to maximize this oxygen transport by keeping the oxygenated blood completely separate from the deoxygenated blood.

154. (A)
It is in the capillaries that the most important function of circulation occurs; that is, the exchange of nutrients and waste materials between the blood and the tissues.

Capillaries are the blood vessels that connect the arteries with the veins. The capillaries have walls composed of endothelium only one cell thick. It is the thinness of the capillary walls which allows for the diffusion of oxygen and nutrients from the blood into the tissues and for carbon dioxide and nitrogeneous wastes to be picked up from the tissues by the blood.

155. (C)
During a fever or exercise, excess heat is produced which raises the body temperature a few degrees. This extra heat causes the sinoatrial node to increase its rate of stimulation, thus increasing the number of beats per minute. It is thought that increased temperature increases the permeability of the muscle membrane of the SA node to various ions (Na^+ and K^+), thus accelerating the stimulatory process. (It is the change in membrane permeability to Na^+ ions which causes an impulse to arise.) This temperature sensitivity of the SA node explains the increased heart rate accompanying a fever.

156. (A)
Pheromones, specific chemicals involved in social communication, are released by an organism and act upon

another organism at a distance from their point of release and in very specific ways. Pheromones can be classified in two groups: those that possess releaser effects, which elicit immediate behavioral responses, and those that have primer effects, which work by altering the physiology and subsequent behavior of the recipient. Releaser pheromones are used by a variety of animals for different purposes, such as the attraction of mates, individual recognition, or trail or territorial marking. As said before, the females of many moth species secrete pheromones to attract males over large areas.

157. (E)
Behavior that contributes to the immediate survival and reproduction of the animal is adaptive. The eggshell-and-object-removing habit of these gulls is such adaptive behavior. The gulls remove not only broken shells, but also any conspicuous objects placed in the nest during the breeding season. They seem to discard conspicuous objects as a defense mechanism against visual predators. When investigators placed conspicuous objects along with egg in nests, these nests were robbed of eggs (by other gulls) more often than nests having only eggs. Thus object-removing behavior is significant and adaptive in that it reduces the chances of a nest being robbed, thus enhancing the survival of offspring.

158. (B)
Displays, mostly visual, serve as signals to convey their potential behavior. Such displays as observed with the gulls convey an animal's readiness to mate, to attack, or to retreat.

159. (D)
Müllerian mimicry involves the evolution of two or more inedible or unpleasant species to resemble each other. Since each species serves as both model and mimic, greater protection is afforded to them because their repellent qualities are more frequently advertised.

160. (C)
Courtship is an important means of communication between some animals that involves a multitude of

precopulatory behavior paterns which serve in many instances to advertise the presence of sexually receptive individual and inhibit aggression by a prospective mate.

161. (B)
Additional roots that grow from the stem or leaf, or any structure other than the primary root, or one of its branches are termed adventitious roots. Adventitious roots of climbing plants such as the ivy and other vines attach the plant body to a wall or a tree. Adventitious roots will arise from the stems of many plants when the main root system is removed. This accounts for the ease of vegetative propagation of plants that are able to produce adventitious roots.

162. (C)
Meristematic cells comprise the primary growth zones of both roots and plants. The meristem consists of actively dividing cells from which all the other tissues of the root or stem are formed.

163. (A)
Inside the epidermis of both stem and root is the cortex. The cortex of the root is a wide area composed of many areas of large, thin-walled, nearly spherical parenchymal cells. The cortex of the stem is narrower than that of the root. The cortical cells differ from those of the root in that the former is photosynthetic while the latter is not. In addition, there is an outer layer of thick-walled collenchymal cells in the stem, which serves as a supportive tissue. Collenchymal cells are not present in the root.

164. (D)
Lenticels are masses of cells which rupture the epidermis and form swellings. Lenticels represent a continuation of the inner plant tissues with the external environment, and permits a direct diffusion of gases into and out of the stem or twig. Such direct passages are necessary because the cambium forms a complete sheath around the vascular bundles and effectively obstructs the ventilation of the vascular bundles.

165. (E)
Cutin is a waxy organic substance secreted by the epidermal cells of the stems and leaves, but not the roots. Its waterproof property retards water loss to the atmosphere as a result of evaporation.

166. (D)
Annelids are segmented worms whose central nervous system is a pair of longitudinal segmented cords in which the cell bodies of the neurons form masses called ganglia. The ganglia originates in the esophagus of the head. Almost all the cell bodies are located in the ganglia.

167. (B)
It is in the coelenterates that the most primitive true nervous system appears. This is a netlike system which is composed of separate neurons which cross each other without touching. Coelenterates may have two kinds of neurons. These are slow-conducting neurons and fast-conducting neurons.

168. (A)
Echinoderms are radially symmetrical when mature. Their nervous system consists of a nerve ring around the mouth which gives rise to five nerve trunks.

In some groups of echinoderms the five nerve trunks consists of specialized neurectoderm which form shallow grooves along the surface of the arms.

169. (E)
Flat worms are advanced planarians which possess a longitudinal bilaterally symmetrical body plan comprised of two nerve cords. The neurons are concentrated in the head region.

170. (C)
Epinephine, also known as adrenalin, is a hormone produced by the adrenal medulla. It causes specific changes in the body usually in response to stress. One of its

functions is to cause an increase in blood glucose level
through the enzymatic breakdown of glycogen by mediating
the activation of an enzyme.

171. (B)
Cyclic AMP or cAMP is now known as the second
messenger. It was discovered to be a participant in hormone
activity. It is an intracellular second messenger and
receives "messages" from hormones which bring their
message to the target cell membrane. The response from
within the cell is actually started by cAMP.

172. (A)
The enzyme activated by epinephrine in the
receptor-hormonal complex is adenyl cyclase. Adenyl cyclase
once activated converts cytoplasmic ATP to cAMP thus
starting a chain of events which leads to the release of
glucose.

173. (D)
Steroid hormones are lipid soluble and readily pass
through cell membranes into cells. They have no apparent
membrane receptors and if the cell is not a target cell they
diffuse out. If however, the cells are target cells, they are
caught and bound by a cytoplasmic binding protein, thus forming
steroid-binding protein complex.

174. (E)
Adrenocorticotrophic hormone (ACTH) stimulates the
adrenal cortex to secrete steroids.

175. (E)
Oncogenes are potential cancer-causing genes which are
believed to cause malignancy. Carcinogenicity refers to the
research being done currently on oncogenes.

176. (B)
Ionizing radiation is the use of rays from the

electromagnetic spectrum. Rays used are X-rays, gamma rays and cosmic rays. These rays are strong enough to penetrate deeply into tissues where they cause ionization of the molecules along the way. These sources of ionizing radiation are postulated to be mutagenic.

177. (D)
Any foreign chemical which enters the human body by way of the digestive or respiratory system is regarded as a potential mutagen. Mutagens can be residual materials from air and water pollution, food preservatives, artificial sweeteners, pesticides and pharmaceutical drugs which reach somatic cells.

Recent increases in spontaneous abortions, stillbirths, and congenital malformations in populations exposed to noxious compounds are believed to be due to mutagenicity.

178. (A)
A tautomeric shift is a reversible isomerization in a molecule. The nucleotides can exist in several chemical forms which are different only by a single proton shift in the molecule. It was suggested by Watson and Crick that tautomeric shifts could result in base pair changes. These tautomers involve keto-enol pairs for thymine and gaunine, and amino-imino pairs for cytosine and adenine.

179. (C)
Dimers are formed when electrons absorb radiant energy and are raised to a higher energy level. This is an example of non-ionizing radiation.

180. (E)
Symbiosis means living together. It is divided into three categories. The first is commensalism, the second mutualism and the last is parasitism.

The ciliate protozoan containing mutualistic green algae which has undergone fission producing two daughter cells displays symbiosis.

181. (C)

The spreading of the wings of a female goshawk is an expression of aggression. Aggression has been defined as being different from aggressiveness. Aggressiveness is a mood whereas aggression is behavior which arises because of competition. The female goshawk spreads her wings as a direct result of her sensing competition.

182. (D)

Cooperative behavior occurs with and between species. A queen honeybee surrounded by the smaller workers is an example of cooperation. This type of cooperation is of a high level and the resulting effect on the hive is one of efficiency.

183. (A)

During the period of succession communities come and go. Earlier communities by their activities pave the way for successive ones; these successive ones in turn pave the way for others. This is a gradual process which continues until a climax community is formed. This one is reasonably stable.

The invasion of the Okefinokee Swamp by vegetation is an example of succession.

184. (B)

The forming of large populations of sexually reproducing individuals by penguins is typical of variation.

185. (E)

Extensors are opposing muscles that straighten the limb out.

186. (D)

Muscles that pull a limb away from the median body line are called abductors.

187. (B)
When electric shocks of high intensity are applied to muscles with a gradually increasing frequency, there is, at first, time for complete relaxation between contractions. As the frequency increases, however, the muscle does not relax completely before the next shock comes. This causes a great deal of contraction. This principle is called summation.

188. (C)
When a muscle remains in contraction because there is no period for relaxation between contractions, the condition is known as tetanus.

189. (A)
Muscles which pull a limb toward the median body line are adductors.

190. (E)
Bodies of invertebrates are often covered with tactile receptors; it is from these that bristles or hairs arise. Spiders have hairy legs which respond to vibrations set up by the prey captured in its web.

191. (D)
In fish, sensory endings are located in the lateral line. These receptors are known as distance receptors and are sensitive to water movement.

192. (A)
In bees, heat sensors are located in the depressions near the eye.

193. (B)
Rattlesnakes are sensitive to polarized light. Their

sensors are located in the depressions near the eye and are known as heat sensing devices.

194. (C)
   Bats are sensitive to high-frequency sounds.

195. (D)
   In vertebrates, the Müllerian duct is responsible for promoting the development of the oviduct, uterus and vagina. It is an embryonic structure that is present in both male and female fetuses.

196. (A)
   Turner's syndrome is a genetic disorder which is found in females. Females having this disorder have one X chromosome, do not develop ovaries, and remain sexually immature even in adults unless hormones is administered.

197. (E)
   The Wolffian duct is an organ which is found in both male and female fetuses and promotes the development of male characteristics.

198. (B)
   An individual with the XXY constitution is an almost normal male in external appearance except for his underdeveloped gonads. Such an individual may have gynecomastia (a tendency for formation of female-like breasts) which is typical of Klinefelter's syndrome.

199. (C)
   Patan's syndrome, otherwise known as trisomy 13 or chromosome 13, is an abnormality found in infants. Its characteristics are severe mental retardation, deafness, harelip, cleft palate, and polydactyly. It has been revealed

that abnormal developmental events occur as early as five to six weeks of gestation.

200 and 201

reaction $\quad$ reaction $\quad$ reaction $\quad$ reaction

$$[A] \xrightarrow{\quad 1 \quad} [B] \xrightarrow{\quad 2 \quad} [C] \xrightarrow{\quad 3 \quad} [D] \xrightarrow{\quad 4 \quad} [E]$$

200. (A)
The first step in any enzymatic activity is the combination of the enzyme with its complementary substrate thus forming an enzyme-substrate complex. This is therefore reaction 1.

201. (C)
In the reaction shown, [A] is the substrate and [E] is the product; [B], [C] and [D] are metabolic intermediates.

202. (A)
As groups of organisms encounter constant competition for food and living space, they tend to spread out and occupy as many different habitats as possible. This process of evolution from a single ancestral species of a variety of forms that occupy several different habitats is termed adaptive radiation.

203. (E)
Polymorphism means many shapes. This ensures the survival of the population which has maintained enough variation to permit further adaptive changes. Fruit flies and other organisms have shown that their genetic pools change adaptively even in response to such environmental changes as the alternations of the seasons. This is an example of balanced polymorphism.

204.   (B)
Speciation is a direct result of reproductive barriers caused by geographic and/or genetic isolation. When interbreeding between subgroups of a population becomes progressively less frequent and the resulting hybrids become progressively less fertile, the several groups eventually become different species.

205.   (C)
Members of different species are usually not interfertile but occasionally members of two different but closely related species may interbreed to produce a third species by hybridization. This allows for the best characters of each of the two originally related species into a single descendant thereby creating a new type of species better able to survive.

206.   (E)
From the diagrams of the curves it is obvious that P. caudatum's growth activities were inhibited since instead of reproducing it slumped.

It is also seen that the two species rapidly evolved in divergent directions; while P. aurelia continued to grow, P. caudatum's growth fell.

207.   (D)
When the two cultures were mixed, one continued growing and one stopped. This effect was most likely due to competition between species, the most adapted one winning.

208.   (D)
From the chart, it was observed that vica faba and mouse suffered from the same damages in chromosomal aberrations and mutations.

209.   (C)
Chromosomal aberrations and mutations can only be

understood if tests and experiments are carried out such as the screening tests. The outcome of such tests explain the causes and reasons for the aberration or mutation.

210. (E)
   The most frequent type of chromosomal aberrations, not mutations, were deletions and duplications, and nondisjunction.

GRE

BIOLOGY TEST

MODEL TEST II

THE GRADUATE RECORD EXAMINATION

BIOLOGY TEST

ANSWER SHEET

1. Ⓐ Ⓑ Ⓒ Ⓓ Ⓔ
2. Ⓐ Ⓑ Ⓒ Ⓓ Ⓔ
3. Ⓐ Ⓑ Ⓒ Ⓓ Ⓔ
4. Ⓐ Ⓑ Ⓒ Ⓓ Ⓔ
5. Ⓐ Ⓑ Ⓒ Ⓓ Ⓔ
6. Ⓐ Ⓑ Ⓒ Ⓓ Ⓔ
7. Ⓐ Ⓑ Ⓒ Ⓓ Ⓔ
8. Ⓐ Ⓑ Ⓒ Ⓓ Ⓔ
9. Ⓐ Ⓑ Ⓒ Ⓓ Ⓔ
10. Ⓐ Ⓑ Ⓒ Ⓓ Ⓔ
11. Ⓐ Ⓑ Ⓒ Ⓓ Ⓔ
12. Ⓐ Ⓑ Ⓒ Ⓓ Ⓔ
13. Ⓐ Ⓑ Ⓒ Ⓓ Ⓔ
14. Ⓐ Ⓑ Ⓒ Ⓓ Ⓔ
15. Ⓐ Ⓑ Ⓒ Ⓓ Ⓔ
16. Ⓐ Ⓑ Ⓒ Ⓓ Ⓔ
17. Ⓐ Ⓑ Ⓒ Ⓓ Ⓔ
18. Ⓐ Ⓑ Ⓒ Ⓓ Ⓔ
19. Ⓐ Ⓑ Ⓒ Ⓓ Ⓔ
20. Ⓐ Ⓑ Ⓒ Ⓓ Ⓔ
21. Ⓐ Ⓑ Ⓒ Ⓓ Ⓔ
22. Ⓐ Ⓑ Ⓒ Ⓓ Ⓔ
23. Ⓐ Ⓑ Ⓒ Ⓓ Ⓔ
24. Ⓐ Ⓑ Ⓒ Ⓓ Ⓔ

25. Ⓐ Ⓑ Ⓒ Ⓓ Ⓔ
26. Ⓐ Ⓑ Ⓒ Ⓓ Ⓔ
27. Ⓐ Ⓑ Ⓒ Ⓓ Ⓔ
28. Ⓐ Ⓑ Ⓒ Ⓓ Ⓔ
29. Ⓐ Ⓑ Ⓒ Ⓓ Ⓔ
30. Ⓐ Ⓑ Ⓒ Ⓓ Ⓔ
31. Ⓐ Ⓑ Ⓒ Ⓓ Ⓔ
32. Ⓐ Ⓑ Ⓒ Ⓓ Ⓔ
33. Ⓐ Ⓑ Ⓒ Ⓓ Ⓔ
34. Ⓐ Ⓑ Ⓒ Ⓓ Ⓔ
35. Ⓐ Ⓑ Ⓒ Ⓓ Ⓔ
36. Ⓐ Ⓑ Ⓒ Ⓓ Ⓔ
37. Ⓐ Ⓑ Ⓒ Ⓓ Ⓔ
38. Ⓐ Ⓑ Ⓒ Ⓓ Ⓔ
39. Ⓐ Ⓑ Ⓒ Ⓓ Ⓔ
40. Ⓐ Ⓑ Ⓒ Ⓓ Ⓔ
41. Ⓐ Ⓑ Ⓒ Ⓓ Ⓔ
42. Ⓐ Ⓑ Ⓒ Ⓓ Ⓔ
43. Ⓐ Ⓑ Ⓒ Ⓓ Ⓔ
44. Ⓐ Ⓑ Ⓒ Ⓓ Ⓔ
45. Ⓐ Ⓑ Ⓒ Ⓓ Ⓔ
46. Ⓐ Ⓑ Ⓒ Ⓓ Ⓔ
47. Ⓐ Ⓑ Ⓒ Ⓓ Ⓔ
48. Ⓐ Ⓑ Ⓒ Ⓓ Ⓔ

49. Ⓐ Ⓑ Ⓒ Ⓓ Ⓔ
50. Ⓐ Ⓑ Ⓒ Ⓓ Ⓔ
51. Ⓐ Ⓑ Ⓒ Ⓓ Ⓔ
52. Ⓐ Ⓑ Ⓒ Ⓓ Ⓔ
53. Ⓐ Ⓑ Ⓒ Ⓓ Ⓔ
54. Ⓐ Ⓑ Ⓒ Ⓓ Ⓔ
55. Ⓐ Ⓑ Ⓒ Ⓓ Ⓔ
56. Ⓐ Ⓑ Ⓒ Ⓓ Ⓔ
57. Ⓐ Ⓑ Ⓒ Ⓓ Ⓔ
58. Ⓐ Ⓑ Ⓒ Ⓓ Ⓔ
59. Ⓐ Ⓑ Ⓒ Ⓓ Ⓔ
60. Ⓐ Ⓑ Ⓒ Ⓓ Ⓔ
61. Ⓐ Ⓑ Ⓒ Ⓓ Ⓔ
62. Ⓐ Ⓑ Ⓒ Ⓓ Ⓔ
63. Ⓐ Ⓑ Ⓒ Ⓓ Ⓔ
64. Ⓐ Ⓑ Ⓒ Ⓓ Ⓔ
65. Ⓐ Ⓑ Ⓒ Ⓓ Ⓔ
66. Ⓐ Ⓑ Ⓒ Ⓓ Ⓔ
67. Ⓐ Ⓑ Ⓒ Ⓓ Ⓔ
68. Ⓐ Ⓑ Ⓒ Ⓓ Ⓔ
69. Ⓐ Ⓑ Ⓒ Ⓓ Ⓔ
70. Ⓐ Ⓑ Ⓒ Ⓓ Ⓔ
71. Ⓐ Ⓑ Ⓒ Ⓓ Ⓔ
72. Ⓐ Ⓑ Ⓒ Ⓓ Ⓔ

73. Ⓐ Ⓑ Ⓒ Ⓓ Ⓔ
74. Ⓐ Ⓑ Ⓒ Ⓓ Ⓔ
75. Ⓐ Ⓑ Ⓒ Ⓓ Ⓔ
76. Ⓐ Ⓑ Ⓒ Ⓓ Ⓔ
77. Ⓐ Ⓑ Ⓒ Ⓓ Ⓔ
78. Ⓐ Ⓑ Ⓒ Ⓓ Ⓔ
79. Ⓐ Ⓑ Ⓒ Ⓓ Ⓔ
80. Ⓐ Ⓑ Ⓒ Ⓓ Ⓔ
81. Ⓐ Ⓑ Ⓒ Ⓓ Ⓔ
82. Ⓐ Ⓑ Ⓒ Ⓓ Ⓔ
83. Ⓐ Ⓑ Ⓒ Ⓓ Ⓔ
84. Ⓐ Ⓑ Ⓒ Ⓓ Ⓔ
85. Ⓐ Ⓑ Ⓒ Ⓓ Ⓔ
86. Ⓐ Ⓑ Ⓒ Ⓓ Ⓔ
87. Ⓐ Ⓑ Ⓒ Ⓓ Ⓔ
88. Ⓐ Ⓑ Ⓒ Ⓓ Ⓔ
89. Ⓐ Ⓑ Ⓒ Ⓓ Ⓔ
90. Ⓐ Ⓑ Ⓒ Ⓓ Ⓔ
91. Ⓐ Ⓑ Ⓒ Ⓓ Ⓔ
92. Ⓐ Ⓑ Ⓒ Ⓓ Ⓔ
93. Ⓐ Ⓑ Ⓒ Ⓓ Ⓔ
94. Ⓐ Ⓑ Ⓒ Ⓓ Ⓔ
95. Ⓐ Ⓑ Ⓒ Ⓓ Ⓔ
96. Ⓐ Ⓑ Ⓒ Ⓓ Ⓔ

97. Ⓐ Ⓑ Ⓒ Ⓓ Ⓔ
98. Ⓐ Ⓑ Ⓒ Ⓓ Ⓔ
99. Ⓐ Ⓑ Ⓒ Ⓓ Ⓔ
100. Ⓐ Ⓑ Ⓒ Ⓓ Ⓔ
101. Ⓐ Ⓑ Ⓒ Ⓓ Ⓔ
102. Ⓐ Ⓑ Ⓒ Ⓓ Ⓔ
103. Ⓐ Ⓑ Ⓒ Ⓓ Ⓔ
104. Ⓐ Ⓑ Ⓒ Ⓓ Ⓔ
105. Ⓐ Ⓑ Ⓒ Ⓓ Ⓔ
106. Ⓐ Ⓑ Ⓒ Ⓓ Ⓔ
107. Ⓐ Ⓑ Ⓒ Ⓓ Ⓔ
108. Ⓐ Ⓑ Ⓒ Ⓓ Ⓔ
109. Ⓐ Ⓑ Ⓒ Ⓓ Ⓔ
110. Ⓐ Ⓑ Ⓒ Ⓓ Ⓔ
111. Ⓐ Ⓑ Ⓒ Ⓓ Ⓔ
112. Ⓐ Ⓑ Ⓒ Ⓓ Ⓔ
113. Ⓐ Ⓑ Ⓒ Ⓓ Ⓔ
114. Ⓐ Ⓑ Ⓒ Ⓓ Ⓔ
115. Ⓐ Ⓑ Ⓒ Ⓓ Ⓔ
116. Ⓐ Ⓑ Ⓒ Ⓓ Ⓔ
117. Ⓐ Ⓑ Ⓒ Ⓓ Ⓔ
118. Ⓐ Ⓑ Ⓒ Ⓓ Ⓔ
119. Ⓐ Ⓑ Ⓒ Ⓓ Ⓔ
120. Ⓐ Ⓑ Ⓒ Ⓓ Ⓔ

121. Ⓐ Ⓑ Ⓒ Ⓓ Ⓔ
122. Ⓐ Ⓑ Ⓒ Ⓓ Ⓔ
123. Ⓐ Ⓑ Ⓒ Ⓓ Ⓔ
124. Ⓐ Ⓑ Ⓒ Ⓓ Ⓔ
125. Ⓐ Ⓑ Ⓒ Ⓓ Ⓔ
126. Ⓐ Ⓑ Ⓒ Ⓓ Ⓔ
127. Ⓐ Ⓑ Ⓒ Ⓓ Ⓔ
128. Ⓐ Ⓑ Ⓒ Ⓓ Ⓔ
129. Ⓐ Ⓑ Ⓒ Ⓓ Ⓔ
130. Ⓐ Ⓑ Ⓒ Ⓓ Ⓔ
131. Ⓐ Ⓑ Ⓒ Ⓓ Ⓔ
132. Ⓐ Ⓑ Ⓒ Ⓓ Ⓔ
133. Ⓐ Ⓑ Ⓒ Ⓓ Ⓔ
134. Ⓐ Ⓑ Ⓒ Ⓓ Ⓔ
135. Ⓐ Ⓑ Ⓒ Ⓓ Ⓔ
136. Ⓐ Ⓑ Ⓒ Ⓓ Ⓔ
137. Ⓐ Ⓑ Ⓒ Ⓓ Ⓔ
138. Ⓐ Ⓑ Ⓒ Ⓓ Ⓔ
139. Ⓐ Ⓑ Ⓒ Ⓓ Ⓔ
140. Ⓐ Ⓑ Ⓒ Ⓓ Ⓔ
141. Ⓐ Ⓑ Ⓒ Ⓓ Ⓔ
142. Ⓐ Ⓑ Ⓒ Ⓓ Ⓔ
143. Ⓐ Ⓑ Ⓒ Ⓓ Ⓔ
144. Ⓐ Ⓑ Ⓒ Ⓓ Ⓔ

145. Ⓐ Ⓑ Ⓒ Ⓓ Ⓔ	167. Ⓐ Ⓑ Ⓒ Ⓓ Ⓔ	189. Ⓐ Ⓑ Ⓒ Ⓓ Ⓔ
146. Ⓐ Ⓑ Ⓒ Ⓓ Ⓔ	168. Ⓐ Ⓑ Ⓒ Ⓓ Ⓔ	190. Ⓐ Ⓑ Ⓒ Ⓓ Ⓔ
147. Ⓐ Ⓑ Ⓒ Ⓓ Ⓔ	169. Ⓐ Ⓑ Ⓒ Ⓓ Ⓔ	191. Ⓐ Ⓑ Ⓒ Ⓓ Ⓔ
148. Ⓐ Ⓑ Ⓒ Ⓓ Ⓔ	170. Ⓐ Ⓑ Ⓒ Ⓓ Ⓔ	192. Ⓐ Ⓑ Ⓒ Ⓓ Ⓔ
149. Ⓐ Ⓑ Ⓒ Ⓓ Ⓔ	171. Ⓐ Ⓑ Ⓒ Ⓓ Ⓔ	193. Ⓐ Ⓑ Ⓒ Ⓓ Ⓔ
150. Ⓐ Ⓑ Ⓒ Ⓓ Ⓔ	172. Ⓐ Ⓑ Ⓒ Ⓓ Ⓔ	194. Ⓐ Ⓑ Ⓒ Ⓓ Ⓔ
151. Ⓐ Ⓑ Ⓒ Ⓓ Ⓔ	173. Ⓐ Ⓑ Ⓒ Ⓓ Ⓔ	195. Ⓐ Ⓑ Ⓒ Ⓓ Ⓔ
152. Ⓐ Ⓑ Ⓒ Ⓓ Ⓔ	174. Ⓐ Ⓑ Ⓒ Ⓓ Ⓔ	196. Ⓐ Ⓑ Ⓒ Ⓓ Ⓔ
153. Ⓐ Ⓑ Ⓒ Ⓓ Ⓔ	175. Ⓐ Ⓑ Ⓒ Ⓓ Ⓔ	197. Ⓐ Ⓑ Ⓒ Ⓓ Ⓔ
154. Ⓐ Ⓑ Ⓒ Ⓓ Ⓔ	176. Ⓐ Ⓑ Ⓒ Ⓓ Ⓔ	198. Ⓐ Ⓑ Ⓒ Ⓓ Ⓔ
155. Ⓐ Ⓑ Ⓒ Ⓓ Ⓔ	177. Ⓐ Ⓑ Ⓒ Ⓓ Ⓔ	199. Ⓐ Ⓑ Ⓒ Ⓓ Ⓔ
156. Ⓐ Ⓑ Ⓒ Ⓓ Ⓔ	178. Ⓐ Ⓑ Ⓒ Ⓓ Ⓔ	200. Ⓐ Ⓑ Ⓒ Ⓓ Ⓔ
157. Ⓐ Ⓑ Ⓒ Ⓓ Ⓔ	179. Ⓐ Ⓑ Ⓒ Ⓓ Ⓔ	201. Ⓐ Ⓑ Ⓒ Ⓓ Ⓔ
158. Ⓐ Ⓑ Ⓒ Ⓓ Ⓔ	180. Ⓐ Ⓑ Ⓒ Ⓓ Ⓔ	202. Ⓐ Ⓑ Ⓒ Ⓓ Ⓔ
159. Ⓐ Ⓑ Ⓒ Ⓓ Ⓔ	181. Ⓐ Ⓑ Ⓒ Ⓓ Ⓔ	203. Ⓐ Ⓑ Ⓒ Ⓓ Ⓔ
160. Ⓐ Ⓑ Ⓒ Ⓓ Ⓔ	182. Ⓐ Ⓑ Ⓒ Ⓓ Ⓔ	204. Ⓐ Ⓑ Ⓒ Ⓓ Ⓔ
161. Ⓐ Ⓑ Ⓒ Ⓓ Ⓔ	183. Ⓐ Ⓑ Ⓒ Ⓓ Ⓔ	205. Ⓐ Ⓑ Ⓒ Ⓓ Ⓔ
162. Ⓐ Ⓑ Ⓒ Ⓓ Ⓔ	184. Ⓐ Ⓑ Ⓒ Ⓓ Ⓔ	206. Ⓐ Ⓑ Ⓒ Ⓓ Ⓔ
163. Ⓐ Ⓑ Ⓒ Ⓓ Ⓔ	185. Ⓐ Ⓑ Ⓒ Ⓓ Ⓔ	207. Ⓐ Ⓑ Ⓒ Ⓓ Ⓔ
164. Ⓐ Ⓑ Ⓒ Ⓓ Ⓔ	186. Ⓐ Ⓑ Ⓒ Ⓓ Ⓔ	208. Ⓐ Ⓑ Ⓒ Ⓓ Ⓔ
165. Ⓐ Ⓑ Ⓒ Ⓓ Ⓔ	187. Ⓐ Ⓑ Ⓒ Ⓓ Ⓔ	209. Ⓐ Ⓑ Ⓒ Ⓓ Ⓔ
166. Ⓐ Ⓑ Ⓒ Ⓓ Ⓔ	188. Ⓐ Ⓑ Ⓒ Ⓓ Ⓔ	210. Ⓐ Ⓑ Ⓒ Ⓓ Ⓔ

THE GRE BIOLOGY TEST

MODEL TEST II

Time: 170 Minutes
        210 Questions

DIRECTIONS: *Choose the best answer for each question and mark the letter of your selection on the corresponding answer sheet.*

1. Two phenotypic markers are found to be inherited according to Mendel's Law of Independent Assortment. What can you say about the corresponding gene loci?

    (A) Both loci correspond to the phenotypic wild type.

    (B) Both genes are dominant.

    (C) Both genes are recessive.

    (D) The two loci are on different chromosomes.

    (E) The two loci are on the same chromosome.

2. Which of the following are not autotrophs?

    (A) herbivores   (D) photosynthetic bacteria

    (B) flowering plants   (E) A and C

    (C) carnivores

3. Electron transport and oxidative phosphorylation are thought to

be coupled by a chemiosmotic mechanism. This hypothesis involves:

(A) the use of energy derived from electron transport to maintain an equilibrium of hydrogen and hydroxyl ions on either side of a membrane

(B) the use of energy derived from electron transport to form both an electrochemical and pH gradient across a membrane by inducing proton translocation

(C) a transient wave of depolarization, opening voltage-gated ion channels in the plasma membrane

(D) electron transport and phosphorylation in the thylakoid membrane of a chloroplast

(E) phosphorylation and isomerization of glucose to a substrate for triose phosphates

4. Basic drives such as hunger, thirst, sex, and rage, as well as internal environmental parameters of blood pressure, heart rate, and body temperature, have all been linked to the functioning of

(A) the basal ganglia   (D) the hypothalamus

(B) the adrenal gland   (E) the corpus callosum

(C) the pineal gland

5. Nicotinamide adenine dinucleotide (NAD) and its reduced form, NADH are primarily involved in

(A) cAMP mediated feedback regulation of glucose levels

(B) photosynthetic electron transport

(C) transporting of protons and electrons and mediation of redox reactions

(D) production of deoxyribonucleotides by reduction of corresponding ribonucleotides

(E) the reduction of folate (F) to tetrahydrofolate (FH_4)

6. Protein synthesis involves the following structures and/or components:

(A) rough endoplasmic reticulum (RER)

(B) ribosomes   (D) transfer RNA (tRNA)

(C) messenger RNA (mRNA)   (E) all of the above

7. Which of the following structures requires the use of an electron
    microscope for   visualization?

    (A) A typical bacterial cell.   (D) all of the above

    (B) Coated pits and coated   (E) (B) and (C), but not (A)
         vesicles.

    (C) The 9 + 2 arrangement of
         microtubules typical of cilia
         and flagella.

8. Neither of the alleles for stem length in a certain plant show dom-
    inance and the heterozygote is of intermediate length. Which of
    the following corresponds to the progeny resulting from the cross
    of a long-stemmed individual with a short-stemmed individual?

    (A) 50 long and 50 short-stemmed individuals

    (B) 100 long-stemmed individuals

    (C) 25 long, 25 short and 50 individuals of intermediate length

    (D) 100 individuals of intermediate length

    (E) 100 short-stemmed individuals

9. Amino acids complex with transfer RNA through

    (A) an aminoacyl link with the cytosine residue at the 5' end of
         the chain

    (B) a phosphodiester bond

    (C) ionic and van der Waals attractions involving the phosphate
         and hydroxyl moieties of the nucleotide chain

    (D) hydrophobic interactions

    (E) an aminoacyl linkage with a 3' terminal adenosine residue

10. Respiration consists of four distinct stages. The process

illustrated below is known as

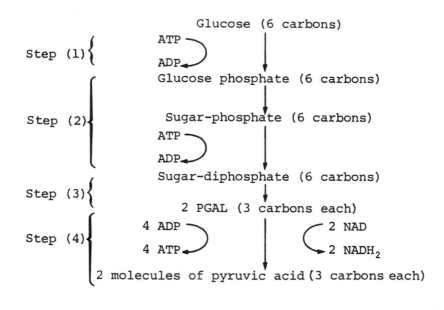

(A) the Krebs cycle   (D) gluconeogenesis

(B) the tricarboxylic acid   (E) the citric acid cycle
     cycle

(C) glycolysis

11. The Hardy-Weinberg Law postulates that if a population exists
     under suitably stable conditions, known as "equilibrium", then
     allelic frequencies and genotypic ratios within the population will
     remain constant. Which of the following is not a requirement for
     the Hardy-Weinberg equilibrium?

     (A) The population must be sufficiently large to insure that gen-
          etic drift would not be a significant factor.

     (B) The frequency of forward mutation and backward mutation
          of a given allele must be approximately equal.

     (C) Reproduction must be random.

     (D) The total number of individuals in the population must remain
          constant.

     (E) There must be no immigration or emigration.

12. Which of the following general statements about the concept of
     biological succession is correct?

(A) The total number of species represented increases until a steady state is reached (with possible slight decline before the final distribution is reached).

(B) The height and masses of plants in the community increase and lead to more clearly delineated vertical strata.

(C) The total biomass in the community increases, as does the amount of non-biologic organic matter.

(D) The food chains within the community become more complex, and interspecies relations become more highly specialized.

(E) All of the above

13. Vascular plants generally absorb nitrogen in the form of

(A) amino acids

(B) molecular nitrogen

(C) nitrose and nitrite compounds

(D) ammonium or nitrate ions

(E) all of the above

14. In Escherichia coli, the mating type known as Hfr is characterized as showing a "high frequency of recombination" in conjugation experiments. This behavior is due to

(A) the presence of a plasmid "sex factor" in the cytoplasm

(B) the presence of viral DNA in the bacterial chromosome

(C) the integration of the plasmid into the bacterial chromosome

(D) the integration of the plasmid into the bacterial chromosome and the subsequent excision of the plasmid along with some chromosal DNA prior to conjugation

(E) a mutation in the episomal DNA

15. If 2 species of herbaceous dicots discovered to have a similiar appearance (but have been found in different continents) are placed in distinct taxonomic subcategories, one may conjecture that

(A) the species were described and classified by different investigators

(B) similar ecological niches elicited similar mutations

(C) this is probably an example of the "founder effect"

(D) similar selection pressures preserved corresponding pheno-
typical adaptations and suppressed the others

(E) the phenotypic similarities are due to chance similarities between the random mutations of each species

16. The following list describes the functions of a particular hormone:

   I. stimulates glycogen formation and storage
  II. inhibits formation of new glucose
 III. stimulates synthesis of protein and fat

Which of the following hormones is primarily responsible for the
functions described in the list above?

(A) epinephrine and /or norepinephrine

(B) glucocorticoids

(C) insulin

(D) glucagon

(E) ACTH

17. The volume of air entering or leaving the lungs during a single
normal breath is called the

(A) vital capacity   (D) residual volume

(B) tidal volume   (E) inspiratory or expiratory
                            reserve volume
(C) breathing capacity

18. Which of the following is arranged in proper order, starting with
the most inclusive and ending with the most specific?

(A) Kingdom, phylum (division), order, class, genus, species

(B) Kingdom, order, phylum (division), genus, class, species

(C) Kingdom, class, order, phylum (division), genus, species

(D) Kingdom, phylum (division), class, order, genus, species

(E) Kingdom, phylum (division), class, genus, species, order

19. Double stranded DNA, when exposed to high temperature, changes in pH, or denaturing agents such as urea or formamide, may be induced to significantly increase its absorption of ultraviolet radiation at 260 nm. This effect is due to

(A) an unwinding of the helix into unpaired, random coils

(B) tautomerization of guanine and cytosine bases

(C) disruption of the tertiary structure of the chain

(D) the formation of thymine dimers

(E) a change in the chirality, or "handedness", of the coils in the superhelix

20. Which of the following statements apply to prokaryotes?

(A) They lack a cell wall.   (D) all of the above

(B) Their ribosomes consist of   (E) B and C, but not A
a 50S subunit and a 30S
subunit.

(C) They lack a proper
nuclear envelope.

21. A plausible directional flow for membrane components in a eukaryotic cell is

(A) golgi apparatus → rough ER → smooth ER → nuclear envelope

(B) nuclear envelope → rough ER → smooth ER → golgi apparatus
→ secretory vesicles → plasma membrane

(C) nuclear envelope → mitochondria → rough ER → smooth ER →
secretory vesicles → plasma membrane

(D) plasma membrane → secretory vesicles → rough ER → smooth
ER → nuclear envelope

(E) plasma membrane → smooth ER → rough ER → secretory vesicles → nuclear envelope

22. Prokaryotic (bacterial) cells differ from eukaryotic cells insofar as

   (A) bacterial cells contain only a single circular chromosome composed exclusively of nucleic acid, while eukaryotes have several chromosomes composed of protein-nucleic acid complexes

   (B) bacterial cells - even those that are photosynthetic - lack chloroplasts

   (C) bacterial flagella lack a 9.2 microtubule structure

   (D) functions performed by mitochondria in eukaryotes are carried out in prokaryotes by structures in the plasma membrane

   (E) all of the above

23. The conformation of polypeptides in solution is maintained in large part by hydrophobic interactions. These interactions are characterized by

   (A) a slight negative heat of reaction

   (B) a large negative heat of reaction

   (C) specific interactions between nearby side chains

   (D) a slight positive heat of reaction, whose unfavorable effect is outweighed by a large increase in entropy

   (E) formation of internal hydrogen bonds and coulomb interactions

24. Which of the following statements apply to adult human erythrocytes?

   (A) They are produced in white bone marrow.

   (B) Their normal survival time is approximately 120 days.

   (C) They lack nuclei, mitochondria, golgi apparatus and other subcellular structures.

   (D) B and C, but not A

   (E) A, B and C

25. Which of the following structures would you expect to find in epithelial tissue, such as the absorptive tissue lining the intestine?

(A) tight junctions

(B) gap junctions

(C) desmosomes

(D) all of the above

(E) A and C, but not B

26. The compound dynein is

(A) an opoid (opiate-like) peptide

(B) a major protein component of myofibrils

(C) a basic protein associated with DNA in nucleosomes

(D) the polypeptide which forms the "arms" by which one microtubule doublet moves along another

(E) a secretion of the pineal gland

27. An axoneme is

(A) an elongate, distal neuronal process

(B) a type of protozoan

(C) a stinging cell found in certain coelenterates

(D) the 9 + 2 arrangement of microtubules, linkers, spokes and other components found in eukaryotic cilia and flagella

(E) a type of algae

28. The basic contractile unit in striated muscle is

(A) the muscle fiber

(B) the myofibril

(C) the sarcomere

(D) the myofilament

(E) the thin filaments

29. The primary advantage of phase-contrast and interference microscopy is that they provide

(A) particularly sharp resolution of surface contours

(B) resolution only slightly inferior to that of an electron (transmission) microscope, at a fraction of the cost

(C) the ability to observe living specimens

(D) a "full color" image

(E) the ability to detect intracellular regions of highly organized structures by applying polarized light

30. The auxins, such as indoleacetic acid, are plant hormones that

(A) stimulate cell division

(B) stimulate a nonspecific, nondirectional elongation in cells in all parts of the plant

(C) control the ripening of fruit through hydrolysis of storage materials and subsequent release of sugars

(D) produce the metabolic inhibitions involved in dormant stages of plant life cycles

(E) regulate the elongation of cells in growing regions of the plant, such as roots and stems

31. Protoplasts are

(A) undifferentiated plant cells

(B) precursors to spermatids

(C) (bacterial) cells whose walls have been selectively removed

(D) the precursor cells from which all types of blood cells are produced

(E) found at the unicellular stage in the life cycle of Dictyostelium discoideum

32. Bacteria may be classified into physiological groups according to the range of temperatures which will permit their growth. The type most suited for cold conditions are the

(A) Mesophiles

(B) Psychrophiles   (D) Thermophobes

(C) Thermophiles   (E) Poikilotherms

33. Osmoregulation in freshwater bony fishes is accomplished by

(A) constant drinking, relatively impermeable skin and scales and active excretion of salts by specialized cells in the gills

(B) rarely drinking, highly permeable regions of skin and production of copious amounts of dilute urine

(C) rarely drinking, impermeable skin and scales, active absorption of salts through specialized cells in the gills, and production of copious, dilute urine

(D) constant drinking, impermeable skin and scales, active excretion of salts through specialized cells in the gills, and production of relatively little isotonic urine

(E) virtually continual drinking, excretion of excess salt through specialized cells in the rectum, impermeable skin and scales and production of copious dilute urine

34. A cell undergoing mitosis is observed to be devoid of a nuclear envelope and to have its double stranded chromosomes arranged so that each is attached by its centromere to spindle microtubules at the spindle equator. The cell is

(A) in early anaphase   (D) in late anaphase

(B) in telophase   (E) in metaphase

(C) in late prophase

35. Xylem is one type of vascular tissue found in plants. Which of the following best describes its functions in flowering plants?

(A) Its exclusive function is transport of water and dissolved substances upward in the plant.

(B) It functions exclusively to carry water and solutes in both directions throughout the plant.

(C) It carries water and solutes in both directions through the plant and gives mechanical support to aerial parts of the plant.

(D)　It transports organic materials, particularly sugars and amino acids, in both directions.

(E)　It transports carbohydrates, amino acids and other organic materials from the leaves to other parts of the plant and lends mechanical support.

36.　A cross section of tissues of a dicot stem differs from those of a dicot root in that

(A)　the phloem and xylem form concentric rings in the stem, whereas the two sets of tissues alternate in the (young) root

(B)　dicot stems usually have pith; dicot roots tend not to have pith

(C)　dicot roots lack vascular cambium

(D)　all of the above

(E)　A and B, but not C.

37.　The model of antigenic stimulation known as clonal selection states that

(A)　antigens react with a lymphocyte of the proper specificity, stimulating proliferation of a clone of antigen-specific lymphocytes

(B)　interaction with an antigen confers the appropriate specificity on a lymphocyte

(C)　antigens are generally able to react with a large variety of lymphocytes

(D)　antigens bind only antibodies produced by a lymphocyte of the correct specificity

(E)　binding of antigen prevents the formation of clones of incorrect specificity

38.　In the Jacob-Monod operon theory, an inducer is

(A)　a low molecular weight compound which binds to the promoter enhancing polymerase binding

(B)　a molecule, often the translated product of the operon, which binds to the repressor and inactivates it

(C)   a protein which binds to the promoter, displacing the pre-
      viously bound repressor

(D)   a small molecule which binds to the regulator gene, prevent-
      ing transcription of the repressor

(E)   a peptide which binds to the operator

39.   Which of the following statements correctly describe /s interferons?

(A)   They are small ($M_r \sim$ 20,000) antiviral glycoproteins.

(B)   They are nontoxic to animal cells in large doses.

(C)   They are nonantigenic.

(D)   When a virally infected cell lyses, interferon is released with
      viral particles.

(E)   all of the above

40.   The cell walls of gram positive bacteria contain

(A)   a peptidoglycan structure composed of perhaps 40 cross-
      linked layers of NAM-NAG

(B)   teichoic acids

(C)   an outer covering of lipopolysaccharides

(D)   all of the above

(E)   A and B, but not C

41.   Membrane-associated proteins are studied by obtaining SDS-
      polyacrylamide gel patterns from "ghosts" and membrane frag-
      ments isolated under different conditions. A group of proteins
      shows the following characteristics:

   I. They may be extracted from the membranes by altering the
       ionic strength or pH of the medium.

   II. They are thoroughly digested by treating leaky ghosts with
       proteases, but remain unaffected if sealed ghosts or intact
       red cells are treated with the same enzymes.

What can you conclude about these proteins?

119

(A) They are peripheral proteins which are associated with the cytoplasmic face of the membrane.

(B) They are integral proteins.

(C) They are peripheral proteins associated with the outer face of the membrane.

(D) They are relatively high molecular weight basic proteins.

(E) They are relatively low molecular weight acidic proteins.

42. The basic dihybrid ratio, 9:3:3:1, is frequently encountered in Mendelian genetics. In a dihybrid cross which follows two traits that obeyed the law of Independent Assortment and showed effects of dominance, this ratio would characterize

(A) the phenotypic ratio obtained in the F_1 generation

(B) the phenotypic ratio obtained in the F_2 generation

(C) the phenotypic ratio obtained in the P_4 generation

(D) the genotypic ratio obtained in the F_2 generation

(E) the phenotypic ratio obtained in the second final generation if the two genes involved are complementary.

43. Which of the following are forms of trisomy of the sex chromosomes?

(A) Down's syndrome

(B) Kleinfelter's syndrome

(C) Turner's syndrome

(D) B and C, but not A

(E) A, B, and C

44. A famous experiment by Hershey and Chase may be summarized as follows:

Bacteriophage was grown on a medium containing radioactive sulfur (^{35}S) and radioactive phosphorus (^{32}P). This radioactive bacteriophage was then introduced to normal, nonradioactive bacteria. Some time later, before the phage was able to reproduce, both the phage and the bacteria were put into a blender. The empty "ghosts" of the bacteriophage were dislodged from the bacterial surface and were separated by centrifugation. It was discovered that all of the radioactive sulfur was located in the empty bacteriophage, while all the radioactive phosphorus was found inside the infected bacteria. What did this experiment indicate?

(A) That DNA is the transforming principle and carrier of heredity in microbial genetics.

(B) That the phages were not inactivated by the radioactivity they absorbed.

(C) That bacteria may be transformed by viral infection.

(D) That bacteriophage was capable of injecting both proteins and nucleic acid into bacteria.

(E) That the transformed bacteria produced viral specific proteins.

45. Which of the following tissues in vertebrates come from mesoderm?

(A) Respiratory pathways and lungs

(B) The pancreas and liver

(C) Blood and bone

(D) Neural tissue

(E) The epidermis and epidermal structures

46. Passing from the nucleus, through the cytoplasm and into the internal matrix of a mitochondrion involves passing through how many membranes? (Without using pores in nuclear envelope)

(A) 1

(B) 2

(C) 3

(D) 4

(E) 5

47. Consider the following model Mendelian pattern. Pure breeding red flowers are crossed with pure breeding white flowers. The progeny of this cross are all red. Subsequently they are used in two further crosses:

I. The first filial generation above (F_1) is crossed with itself to form a second filial generation.

II. The F_1 plants are crossed with true-breeding white plants (i.e. a test cross).

The results of the 2 crosses are:

(A) The F_2 generation is 3:1 (red:white) flowers while the test cross results in a 1:1 ratio of red and white plants.

(B) The F_2 generation is 3:1 (red:white) and the test cross gives only red flowers.

(C) The F_2 generation is 3:1 (red:white) as are the test cross progeny.

(D) The F_2 generation is 3:1 (white:red) as are test cross progeny.

(E) The F_2 generation is uniformly red and the test cross gives a 1:1 ratio of the types crossed.

48. This experiment illustrates

(A) Mendel's second law (Independent Assortment)

(B) Mendel's first law (Segregation)

(C) the concept of linkage

(D) all of the above

(E) A and B, but not C

49. Now consider a slightly more involved Mendelian pattern. A parental cross is made whereby true-breeding red, tall plants are crossed with true-breeding white, short plants. The resulting F_1 generation of plants are all tall and red. As in the monohybrid cross, the F_1 plants are crossed with each other to obtain the F_2 generation. Out of every 16 plants in the second filial generation there are

(A) 12 tall red plants, 2 short red, 1 short white, 1 tall white

(B) 12 tall red plants, 2 short white, 1 tall white, 1 short red

(C) 9 short red, 3 tall white, 3 tall red, 1 short white

(D) 9 tall red, 3 short red, 3 tall white, 1 short white

(E) 9 tall white, 3 short red, 3 short white, 1 tall red

50. Glycerinated muscle fibers may be stimulated to contract; in the early 1950's, H. Huxley observed changes in the banding pattern

of striated muscle during such contraction. He observed which of the following length changes?

(A) The H zone decreased, the I band decreased while the A band remained constant in length.

(B) The H zone decreased and the A band decreased while the I band remained constant in length.

(C) The A band and H zone both remained constant in length, while the I band decreased in length.

(D) The A band, H zone and I band all decreased in length but the Z lines remained in a fixed position.

(E) The A band, H zone, and I band all remained constant in length, while the Z lines moved closer together.

51. Hyperaldosteronism is a hormonal condition where too much aldosterone is secreted into the bloodstream. It's effects include

(A) an increase in blood pressure

(B) excessive retention of sodium

(C) depletion of potassium

(D) an elevation of the ratio of $NaHCO_3$ to H_2CO_3 in the blood and consequently a tendency towards alkalosis

(E) all of the above

52. Insulin deficiency is associated with

(A) an increase in the blood levels of long chain fatty acids

(B) an accumulation of ketones which may lead to acidosis and dehydration

(C) a decrease in the blood level of long chain fatty acids

(D) A and B

(E) A and C

53. Consider the sense strand in a length of genomic DNA. Which of the following represents the correct order of structures travelling in the direction taken by a functioning polymerase?

(A) 3'-terminus-antileader-TAC-structural gene-antitrailer-terminator-5' terminus

(B) 5'-terminus-antileader-TAC-structural gene-antitrailer-terminator-3' terminator

(C) 5'-terminus-leader-AUG-gene-trailer-3' terminus

(D) 5'-terminus-leader-TAC-structural gene-trailer-terminator-3' terminus

(E) 5'-terminus-promoter-leader-AUG-gene-antitrailer-terminator-5' terminus

54. The so-called Pribnow box (TATAATG consensus sequence) is described best as

(A) a prokaryotic terminator, a signal for polymerase and transcriptase to dissociate

(B) a eukaryotic promoter sequence

(C) a promoter found in E. coli, an upstream binding site for RNA polymerase

(D) the leader terminus in eukaryotes responsible for signalling the addition of a 7-methyl-guanosine cap structure to the transcription product

(E) a structure found at intron-exon boundaries

55. Which of the following are not aspects of promoter function?

(A) Binding site for RNA polymerase.

(B) Unwinding of DNA into single stranded regions accessible to polymerase.

(C) Establishing the correct sense and antisense strands.

(D) Base pairing with rRNA and stabilizing mRNA-ribosomal complex in the proper geometry.

(E) Specifying the correct direction for transcription to proceed.

56. Which of the following does not "belong with" the others?

(A) Satellite DNA

(B) palindromic sequences

(D) constituitive heterochromatin

(C) highly repetitive DNA

(E) rapidly reassociating DNA

57. An individual is characterized as being heterozygous with respect to ABC locus; specifically he is I^A/I^B. He

(A) can accept transfusions from any donor but cannot produce any of the antibodies associated with blood typing

(B) can accept transfusions from type O donors only, but cannot produce any of the antibodies involved

(C) can accept transfusions from type O donors only and can produce anti-A and anti-B

(D) can accept transfusions from type A or type O donors and can produce anti-B

(E) can accept transfusions from type B or type O donors and can produce anti-A

58. Two lengths of DNA along the same chromosome encode very sim- ilar products. They produce enzymes identical in catalytic prop- erties and substrate selectivity which are very similar in electro- phoretic mobility and other physical properties. These genes are

(A) isozymes

(D) epistatic

(B) isologous

(E) pleiotropic

(C) allelic variants

59. An individual is diagnosed as possessing a sickle celled trait. This individual's hemoglobin is correctly described as

(A) having 2 α chains and 2 S chains in each tetramer

(B) possessing 2 α chains, one S chain and one β chain

(C) having half of the hemoglobin being composed of 2 α chains and 2 β chains and the other half being composed of 2 α and 2 S chains

(D) having each tetramer being composed of 22 β chains, 1 α and 1 S chain

(E) half being composed of 2 β chains and 2 S chains, while the other half is found to be composed of 2 β chains and 2 α chains

60. Numerous antibiotics work by interfering with protein synthesis. An example is cycloheximide, which

(A) inhibits the peptidyl transferase activity of 60 S ribosomal subunits in eukaryotes

(B) binds to the 30 S subunit and inhibits binding of aminoacyl-tRNAs (prokaryotes)

(C) causes premature chain termination by acting as an analogue of aminoacyl-tRNA (prokaryotes and eukaryotes)

(D) inhibits translocation by binding to 50 S ribosomal subunit (prokaryotes)

(E) inhibits initiation and causes misreading of mRNA in prokaryotes

61. In the inheritance of Down's syndrome, a balanced translocation carrier is best characterized as

(A) an individual whose gametes are such that their progeny falls into 3 types: 1/3 will demonstrate Down's syndrome, 1/3 will be normal, 1/3 will be balanced translocation carriers (assuming that the spouse has a normal karyotype)

(B) a person whose parent produced a gamete which had undergone a translocation event such that virtually the whole of chromosome #21 had been translocated to chromosome #13

(C) an individual developed from a zygote which was virtually haploid for chromosome #21

(D) all of the above

(E) A and B, but not C

62. Which of the following statements correctly describes the organization and contents of eukaryotic chromatin?

(A) Histones contain large amounts of the basic amino acids lysine and arginine, particularly near the N-termini.

(B) Histones and DNA are present in approximately equal proportions (by weight) in mammalian chromatin.

(C) The ratio of proteins to DNA is much higher in eukaryotes than prokaryotes.

(D) The structure found consists primarily of nucleosomes connected by stretches of linker DNA.

(E) all of the above

63. Continuing with the subject matter and format of the previous question which of the following statements is/are true?

(A) Nucleosome core particles are composed of approximately 145 DNA base pairs wrapped around a histone octamer.

(B) In the chromatosome, approximately two full turns of the left-handed superhelix (ca. 165 bp) wrap around the core, and histone H1 is located at sites where DNA enters and leaves the core.

(C) The term mononucleosome refers to a chromatosome and associated linker DNA, the total DNA content being approximately 200 base pairs.

(D) At physiological salt concentrations the nucleosomes form a solenoidal structure consisting of approximately six nucleosomes per turn.

(E) all of the above

64. In E. coli, the enzyme primase functions in DNA replication by

(A) catalysing the energy-dependent unwinding of the duplex using 2 ATPs per base pair separated

(B) stabilizing the single stranded DNA thus formed

(C) catalysing polymerization of ribonucleoside 5'-triphosphates to form 3'-5' phosphodiester bonds with a concomitant release of diphosphate

(D) catalysing addition of deoxyribonucleoside to 3' ends of the chains formed in (C)

(E) catalysing stepwise excision of ribonucleoside residues from the 5' end and concurrent polymerization of 5'-deoxyribonucleoside triphosphates to form 3'-5' phosphodiester bonds with release of inorganic diphosphate

65. Dinitrophenylated Bovine serum albumin (DNP-BSA) is injected into a rabbit. The expected course of events would include

   (A) a rise in the bloodstream concentration of immunoglobulin M (IgM) specific for the haptenic determinant

   (B) an increase in anti-DNP of a different class, IgG, beginning approximately 10 days after administration of immunogen

   (C) a generalized increase in immunoglobulins of all classes approximately 2 weeks after administration

   (D) all of the above

   (E) A and B, but not C

66. Which of the following statements incorrectly describe(s) the functions and/or characteristics of nephrons in the human kidney?

   (A) Active transport of Na^+ ions.

   (B) Passive transport of Na^+ ions.

   (C) The walls of ascending limb of loop of Henle must be water tight.

   (D) The urine in the distal convoluted tubule is less concentrated than the initial filtrate.

   (E) none of the above (i.e. they are all correct)

67. Which of the below correctly lists the compounds in order of decreasing free energy? (Under intracellular conditions)

   a – phosphoenol pyruvate
   b – fructose-1,6-diphosphate
   c – glucose-6 phosphate
   d – pyruvate
   e – 1,3 diphosphoglycerate

   (A) c-b-e-a-d   (D) b-c-e-a-d

   (B) c-b-a-e-d   (E) none of the above

   (C) c-b-e-d-a

68. The following Lineweaver-Burke plot represents the kinetics of an enzyme-mediated reaction and illustrates the course of the reaction without any inhibitor present and in the presence of increasing concentrations of inhibitor.

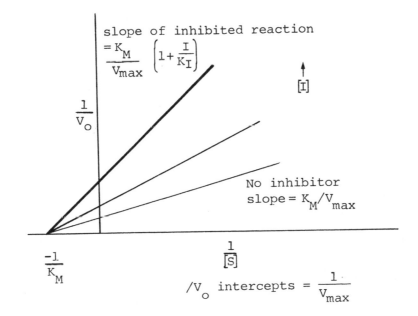

These idealized kinetics reflect which of the following mechanisms of inhibition?

(A) Competitive Inhibition

(D) Irreversible Inhibition

(B) Steric Inhibition

(E) none of the above

(C) Noncompetitive Inhibition

69. A certain bacterial infection is characterized by extreme virulence due to toxins released by the bacteria. The toxin cannot be inactivated by heating or by decreasing the pH. It may be concluded that

(A) the bacteria is probably gram positive and is elaborating an exotoxin

(B) the bacteria is probably gram negative and is producing an endotoxin

(C) the bacteria is probably gram positive, the virulence is due to an endotoxin

(D) the symptoms are caused by a gram negative bacteria and its exotoxin

(E) none of the above

70. A short, leafy plant is treated with gibberellins. What are the most prominent effects you would expect to see?

(A) The plant would sprout more foliage but would not grow taller.

(B) The plant would grow taller, and develop larger leaves

(C) The plant would grow taller and develop more leaves.

(D) The plant would grow taller without any accompanying increase in foliage.

(E) none of the above

71. During the human menstrual cycle, peak levels of estrogen and lutenizing hormone are associated with

(A) the flow phase

(B) the early part of the follicular phase

(C) the latter part of the follicular phase

(D) the early part of the luteal phase

(E) the latter part of the luteal phase

72. In a typical action potential

(A) the initial change is an increase in the Na^+ conductivity

(B) the membrane potential never quite becomes as positive as the Nernstian Na^+ potential

(C) there is an increase in K^+ conductance which occurs when the Na^+ pump resumes work

(D) all of the above

(E) A and B, but not C

73. Language, defined as communication through commonly understood symbols and sounds, is found to be employed in rudimentary form by

(A) bees

(B) chimpanzees

(C) marine animals

(D) B and C, but not A

(E) all of the above

74. Which of the following are examples of primer pheromones?

   (A) queen substance ("royal jelly")

   (B) alarm substances

   (C) trail substances

   (D) sex attractant released by female silkworm moth

   (E) death substances

75. Which of the following statements correctly describe(s) characteristics of ecological communities?

   (A) There is more productivity from the carnivores of a community than from its herbivores.

   (B) Total biomass tends to increase at successive trophic levels.

   (C) It is common for there to be five or more levels in a food chain.

   (D) Secondary consumers tend to be individually smaller than primary consumers.

   (E) none of the above

76. Which of the following correctly describes reflex arcs in vertebrates?

   (A) A minimum of three cells may be involved in a given pathway.

   (B) There is no more than one sensory neuron in a given pathway.

   (C) The cell body of a sensory neuron is always outside the spinal cord in the dorsal root ganglion.

   (D) Axons of motor neurons always enter the spinal cord dorsally, whereas axons of motor neurons leave the spinal cord ventrally.

   (E) all of the above

77. Which of the following statements correctly describes photosynthesis in green plants?

   (A) Cyclic photophosphorylation involves both Photosystem I and Photosystem II.

   (B) Cyclic photophosphorylation involves only Photosystem II.

(C) Photosystem I is directly involved in the oxidation of water.

(D) In cyclic photophosphorylation, both ATP and reduced electron carrier (NADPH) are produced.

(E) none of the above

78. With reference to question 77, it can be concluded that

(A) in chloroplasts, ATP synthesis is coupled with photoelectron
transfer through the translocation of protons from the inside
to the outside of the grana and stroma

(B) although each photosystem can transfer only a single electron
at a time, the stoichiometry of cyclic photophosphorylation
probably involves 2 einsteins (and 2 faradays) for each mole
of ATP produced

(C) oxygen is evolved in both the cyclic and noncyclic modes

(D) all of the above

(E) none of the above

79. Which of the following statements best describes transport across
cell membranes?

(A) Those transport proteins involved in active transport maintain a permanently fixed asymmetry with respect to the 2 sides
of the membrane, this maintains the directional sense of the
transport.

(B) All active transport processes are coupled to the hydrolysis
of phosphate bonds.

(C) The flow of protons plays a role in prokaryotic transport
processes analogous to that of Na^+ ions in eukaryotes.

(D) Mediated transport may be differentiated experimentally from
nonmediated transport by the presence or absence of changes
in the direction of transport if the concentration gradient is
reversed.

(E) Active transport may be differentiated from passive transport
such as mediated diffusion by saturation effects on the initial
velocity of transport.

80. Large, herbivorous mammals generally

(A) have large territories but small home ranges

(B) have large home ranges with no defended territory

(C) have small defended territories and extended home ranges

(D) have large habitats but small home ranges

(E) have large niches but small home ranges

81. Walking along a fence behind which stands several houses, you find yourself being barked at loudly by a German Shepherd. When the animal realizes that the fence will not allow him to get at you, he bites at a stick with exaggerated ferocity. This last behavior is an example of

(A) habituation

(B) intentional movements

(C) redirected activity

(D) all of the above

(E) A and C, but not B

82. Flowering plants generally meet their need for fixed nitrogen by absorbing

(A) amino acids and nucleotides

(B) ammonium ions

(C) nitrates

(D) nitrites

(E) diazonium salts

83. The organism chlamydomonas is characterized by

(A) possession of a macronucleus as well as a micronucleus

(B) a life cycle generally based on asexual reproduction but occasionally on sexual reproduction under conditions of low ambient nitrogen

(C) a life cycle in which the prevalent multicellular, motile form is under conditions such as low ambient nitrogen, replaced by a unicellular, amoeboid form

(D) sexual reproduction which alternates between isogamy and anisogamy

(E)   none of the above

84.   5'-3' cyclic adenosine monophosphate (e.g. cAMP) has been found
      to play many roles as a "second messenger" and regulator. In
      which of the following cases, if any, has cAMP not been shown to
      be involved?

      (A)   Phosphorylation of a protein kinase as part of a reaction
            cascade triggered by the binding of epinephrine to membrane
            receptors which leads to the hydrolysis of glycogen.

      (B)   Binding to CAP (Catabolite Gene Activator Protein),
            thereby stimulating the transcription of a number of induc-
            ible catabolic operons (such as the Lac operon in E. coli).

      (C)   The aggregation of amoeboid cells into a pseudoplasmodium.

      (D)   The stimulation of lipolysis by the phosphorylation of a
            protein kinase which in turn activates a lipase.

      (E)   none of the above

85.   A tissue found in vascular plants is composed largely of cells
      which are dead by the time of functional maturity. These cells
      are characterized by thick and heavily lignified secondary walls
      which confer mechanical support to the plant body. These walls
      are frequently so thick that the intracellular lumen is virtually
      nonexistent. The tissue described (a fundamental tissue) is
      called

      (A)   Parenchyma   (D)   Endodermis

      (B)   Collenchyma   (E)   Periderm

      (C)   Sclerenchyma

86.   Consider a hypothetical cell in a young woody plant. Proceeding
      inwardly from an intercellular space to the interior of the cell,
      which list correctly gives the order in which the given structures
      would be encountered?

      (A)   middle lamella-primary cell wall-plasma membrane-secondary
            cell wall

      (B)   middle lamella-secondary cell wall-primary cell wall-plasma
            membrane.

(C) plasma membrane-secondary cell wall-primary cell wall-middle lamella

(D) middle lamella-primary cell wall-secondary cell wall-plasma membrane

(E) middle lamella-secondary cell wall-plasma membrane-primary cell wall

87. The balance of forces involved in the filtration of materials across capillary walls in human beings is correctly described by which of the following?

(A) At the arteriole end of a capillary bed the osmotic pressure outweighs the hydrostatic pressure, while the converse is true at the venule end.

(B) Materials may move by diffusion through the membranes of the endothelial cells lining the capillary, through the cytoplasm and out the cell membrane on the other side.

(C) The blood in the capillaries has a higher hydrostatic and osmotic pressure than the tissue fluid.

(D) All of the above

(E) B and C, but not A

88. In the transmission electron microscope (TEM)

(A) structures coated with metals appear dark on the flourescent screen as well as on the prints (positives) made by a photo-plate

(B) structures coated with metals appear light on the screen but dark on the prints

(C) structures coated with metals appear light on the screen and light on the positives

(D) metal coated structures show up dark on the screen but light on the positives

(E) metal coated structures appear dark or light on the screen depending on which metallic stain is employed

89. Which of the following cytochemical fixing agents are appropriate for in situ detection of proteins?

(A) OsO$_4$

(D) PAS (Periodic acid–Schiff reagent)

(B)

$$\underset{\text{HCH}}{\overset{\text{O}}{\|}} \quad \text{and/or}$$

(E) Feulgen stain

$$\overset{\text{O}}{\underset{\text{HCCH}_2\text{CH}_2\text{CH}_2\text{CH}}{\|}} \qquad \overset{\text{O}}{\|}$$

(C) Schiff reagent:

90. Rat liver cells are homogenized in .25M sucrose and subsequently centrifuged for 10 minutes at 600 g. Which of the following would you expect to find in the pellet formed?

(A) nuclei and whole cells

(D) synaptosomes

(B) mitochondria and lysosomes

(E) free ribosomes and soluble material

(C) microsomes

91. Monoclonal antibodies of a desired specificity are to be produced. Activated B-lymphocytes are removed from the spleen of an immunized mouse. Which of the following reagents or cell types might be employed in the overall procedure leading to the isolation of the desired antibodies?

(A) aminopterin

(D) all of the above

(B) mutant tumor B-lympho-cytes (not activated)

(E) B and C, but not A

(C) polyethylene glycol or sindai virus

92. An investigator wishes to determine the intracellular location of a protein P via immunoflorescence. Purified P is injected into a

rabbit and anti-P is obtained from rabbit blood serum. The investigator is on a limited budget and therefore must use the indirect method of immunoflorescence. Which of the following steps is/are required in the indirect method and not in the direct method?

(A) Injection of rabbit immunoglobulins into a goat.

(B) Incubation of the cells to be studied with nonflorescent anti-P.

(C) Brief fixation of cells to be studied in formaldehyde, and extraction with acetone.

(D) all of the above

(E) A and B, but not C

93. The following proteins or protein subunits have all been isolated from various filamentous or tubular systems and play some contractile and/or cytoskeletal role. Of these which has been shown to possess endogenous ATPase activity?

(A) Heavy meromyosin

(D) Microtubule associated proteins

(B) α-Tubulin

(E) Actin

(C) β-Tubulin

94. Consider the life cycles characteristic of primitive plants and of common multicellular plants. Which of the following statements correctly describe(s) the various types of lifestyles?

(A) In most multicellular plants, haploid and diploid stages can occur as independent, free living forms.

(B) In a multicellular plant gametes may be produced by mitosis.

(C) In certain primitive plants, the zygote is the only diploid stage in the entire life cycle.

(D) In multicellular plants spores may be produced meiotically.

(E) all of the above

95. Which of the following correctly describes polyoma virus?

(A) It is a DNA tumor virus.

(B) When a permissive cell is infected, a transformation event may occur in one out of every 10^4 to 10^5 cases.

(C) An RNA-DNA hybrid is formed from viral RNA and host deoxyribonucleotides.

(D) It contains a reverse transcriptase.

(E) none of the above

96. In eukaryotes, a region of nucleic acid that specifies a polypeptide sequence is known as

(A) an intron

(B) a cistron

(C) a structural gene

(D) all of the above

(E) B and C, but not A

97. Consider the use of tetrad analysis to follow the inheritance of two markers in Chlamydomonas. Which statement/s correctly describe/s the interpretation of the data generated by this type of experiment?

(A) If the number of parental ditypes (PD) equals the number of nonparental ditypes (NPD), the allele pairs are linked.

(B) If the number of PD greatly exceeds the number of NDP, then the two markers are linked.

(C) Two unlinked markers may produce a tetratype if a crossover event occurs during the first meiotic prophase.

(D) All of the above

(E) B and C, but not A

98. 5-Bromouracil, a thymine analogue, undergoes tautomeric shifts. In its stable keto form, it pairs with adenine; in its excited state and form, it pairs with guanine. If the presence of 5-Bromouracil triggers a transition from a GC to an AT base pair, which of the following is likely to be the mechanism by which the transition occurs?

(A) mistakes in incorporation

(B) mistakes in replication

(C) transversion

(D) wobble

(E) none of the above

99. Which of the following most accurately characterizes the two forms of starch; α-amylose and amylopectin?

   (A) Amylose consists of long, unbranched chains of D-glucose units in α (1 → 4) linkages.

   (B) Amylopectin is highly branched, with its D-glucose units typically branching in α (1 → 6) linkages.

   (C) Amylose forms hydrated micelles in aqueous solution which give a blue color with iodine.

   (D) all of the above

   (E) none of the above

100. The β monomer of maltose may be written as O-α-D-glucopyranosyl (1 → 4)-β-D-glucopyranose. Which of the following Haworth projections illustrates the structure specified?

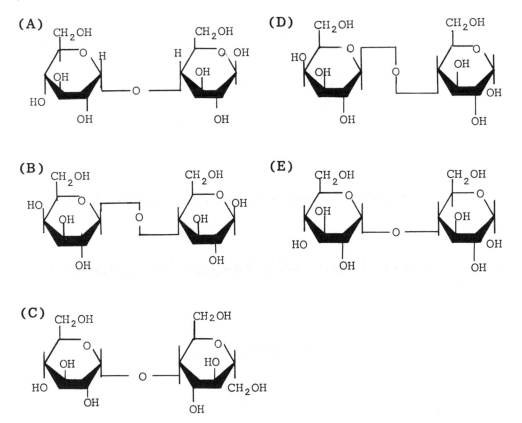

Questions 101 through 105 refer to the diagrams below illustrating early embryonic development in amphioxus. Amphioxus is a small marine chordate; its egg cells are characterized by a very small amount of yolk.

Identify the structures indicated; if there is no specific structure specified then give the appropriate developmental stage.

101.

    (A) blastula   (D) vegetal pole

    (B) morula   (E) gastrula

    (C) animal pole

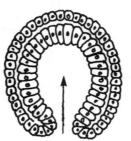

102.

    (A) blastopore   (D) archenteron

    (B) animal pole   (E) blastocoel

    (C) vegetal pole

103.

    (A) early neurula

(B) blastula

(D) gastrula

(C) late neurula

(E) fetula

104.

(A) animal pole

(D) vegetal pole

(B) yolk plug

(E) blastopore

(C) presumptive notochord

105.

(A) blastocoel

(D) blastopore

(B) coelom

(E) archenteron

(C) gut

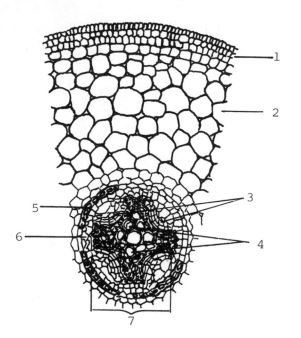

106. Which of the numbered structures corresponds (exclusively) to the phloem?

(A) 3   (D) 7

(B) 6   (E) 5

(C) 4

107. This structure, often composed of a single layer of thin-walled parenchymatous cells, readily adopts meristematic activity to form lateral roots.

(A) 5   (D) 4

(B) 3   (E) 2

(C) 6

108. Water entering the root may traverse large distances via the apoplast network without passing through a plasma membrane. However, water is compelled at some point to enter cells by the presence of the hydrophobic casparian strip, which is associated with the cells indicated by

(A) 3   (D) 5

(B) 4   (E) 7

(C) 2

109. The following numbered list corresponds to characteristics of various plant types. Decide which of the lettered choices best describes the categories of plants to which they apply.

    I. Flower parts arranged in threes.

    II. Principal veins of leaves parallel to each other.

    III. Root system fibrous, no taproot.

    IV. Stem characterized by vascular bundles distributed irregularly through pith tissue.

(A) monocots

(B) II and III characterize monocots, whereas I and IV pertain to dicots

(C) II, III, IV describe manocots but I pertains to dicots

(D) I, III, IV describe dicots, II applies to monocots

(E) I, II and IV describe monocots while III applies to some but not all monocots and some but not all dicots

110. Continuing with the format of question 109.

    I. Cambium adds a new ring each growing season.

    II. 2 seed leaves

    III. Vascular bundles arranged in a single cylinder in stem.

    IV. Leaf vennation pinnate or palmate.

(A) II corresponds to dicots, while the others pertain to monocots

(B) II and IV describe dicots, I and III describe monocots

(C) dicots

(D) I pertains to some but not all monocots, II, III and IV are all features of dicots

(E) monocots

111. Consider what would happen if organisms did not store energy in the form of carbohydrates, but instead maintained an excess of available ATP. What problems would arise? Why is this simpler arrangement not found in nature? If energy was stored as ATP:

    I. Osmotic effects would occur such that the cell would be hypotonic (to plasma or marine environments).

II.   the acid-base equilibrium of the cell would be altered, making the intracellular medium more acidic.

III. Cell walls would have to be much stronger to resist bursting.

IV.   Phosphorus would have to be more widely available than it currently is.

(A) all of the above

(D) all of the above except III

(B) all of the above except I

(E) none of the above

(C) III and IV are correct, II is correct if the word "acidic" is changed to "alkaline"

Questions 112–115 refer to the schematic rendering below of the life cycle of a flowering plant. In each case decide which of the words or phrases best corresponds to the illustrated object or process; unless a line is drawn to indicate a specific element the entire structure is to be indicated.

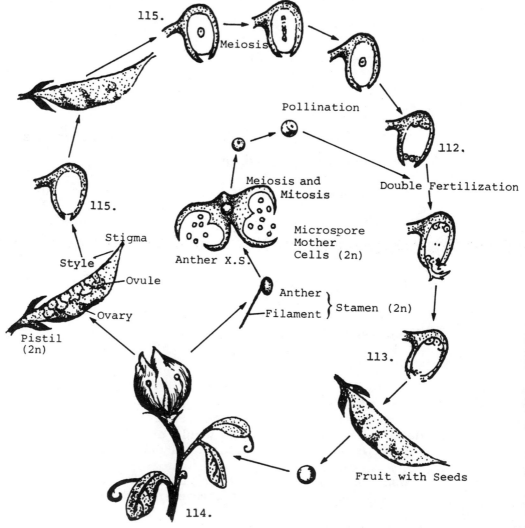

112. (A) egg (n)   (D) megaspore

(B) ovule   (E) polar nuclei (n)

(C) embryo sac

113. (A) sperm nucleus (n)   (D) endosperm nucleus (3n)

(B) polar nucleus (n)   (E) fusion nucleus (2n)

(C) polar nucleus (2n)

114. (A) sporophyte (2n)   (D) sporophyll (2n)

(B) gametophyte (n)   (E) sporophyte (n)

(C) sporangium (2n)

115. (same structure in cross section and outer view)

(A) ovule (2n)   (D) ♀ gametophyte (n)

(B) megaspore mother cell (2n)   (E) none of the above

(C) microsporangium (n)

116. Which of the following lends support to the signal hypothesis of protein translocation? Which tends to refute it, or at least indicate that other mechanisms may exist for inserting proteins into membranes?

I. Of those proteins which span membranes, 50% are found to have their $-CO_2^-$ terminus at the P face and their $-NH_3^+$ terminus at the E face, while 50% have the reverse polarity.

II. The known arrangement of bacteriorhodopsin in bacterial membranes.

III. Proteins like albumen, which does not lose a segment upon entering the endoplasmic reticulum and has no hydrophobic sequence at its N-terminus.

IV. The fact that when free polysomes (or polysomes removed from microsomal membranes) are made to synthesize proteins in a reaction medium to which proteases have been added, the

145

nascent chain may reach a length of 30-40 amino acids, whereas if polysomes bound to microsomes are used the polypeptide can grow much longer.

(A) All support signal hypothesis.

(B) All argue against the hypothesis.

(C) I and II support the hypothesis, III and IV tend to oppose it.

(D) All except II argue for the hypothesis.

(E) All except IV argue against the hypothesis.

117. The schematic diagram below indicates some classifications of marine biomes and oceanic regions. Which letter indicates the type of zone in which the most complex ecological communities would be found?

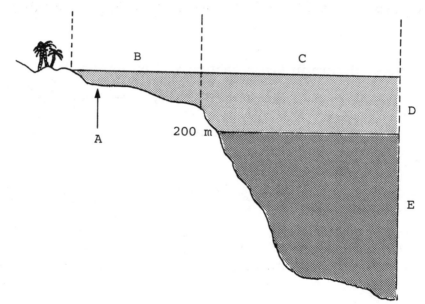

118. Ethylene glycol poisoning may be treated by

(A) quickly giving the victim alcoholic beverages

(B) giving the victim oxalic acid

(C) keeping the victim warm

(D) giving the victim carbonated beverages

(E) none of the above

119. If two populations of a given species are isolated from each other there is an intrinsic tendency for the two populations to begin to diverge from each other. Eventually this can lead to the development of a new species. Which of the following contribute to the underlying tendency of the two populations to deveop divergent characteristics?

 I. Pre-existing variations in gene frequencies between the two groups.

 II. Different mutations in the separated groups.

 III. Different selection pressures upon the two groups after they have been separated.

 IV. Founder effects, genetic drift.

(A) all of the above

(B) I, II and III but not IV

(C) II, III and IV but not I

(D) I, II, III (and IV, but only in the case of small populations)

(E) none of the above

120. Two individuals, A and B, belonging to widely separated populations, are unable to produce viable offspring when brought together. However, it is shown that there exists a series of intermediate populations such that individual A can mate with an individual in the first intermediate population; the progeny of this mating can mate with a member of the next population, and so on down the line to an individual capable of producing viable offspring with individual B. Which statement best describes the relationship between the original individuals A and B?

(A) They are in different species.

(B) They are in different species but the same genus.

(C) They are in the same species.

(D) They are in the same deme.

(E) They are in the same population, but on different ends of a cline.

121. In cell culture, the terms "synchronous growth" and "balanced growth" are used. It can therefore be concluded that

(A) they are synonymous

(B) balanced growth refers to a situation where all routine metabolic and reproductive functions are at a maximum in each cell, so that increases in DNA synthesis are proportional to increases in protein synthesis which are proportional to increases in RNA synthesis

(C) balanced growth defines a situation where all cells are at the same point in the cell cycle at the same time

(D) the description in (B) would be correct if it started with the words "synchroncus growth "

(E) none of the above

122. The side chain amino group of lysine has a pK_a of 10.5. What fraction of the side chain amino groups (ε amino groups) are protonated at a pH of 10?

(A) $10^{\frac{1}{2}}/(10^{\frac{1}{2}} + 1) \approx 3.16/4.16$ (D) 10/11

(B) $10^{\frac{1}{2}}/11 \approx 3.16/11$ (E) none of the above

(C) $(10^{-\frac{1}{2}} + 1)/10^{-\frac{1}{2}}$

123. Which statement most accurately describes the circulatory patterns in humans and fish?

(A) In humans, blood is returned to the heart after being aerated prior to being sent through the tissues; whereas in fish aerated blood is pumped directly to the tissues.

(B) In humans, blood is pumped directly to the tissues after being aerated, whereas in fish the blood is routed through the heart.

(C) In both fish and humans, blood is sent through the heart immediately after aeration and before systemic circulation.

(D) In both fish and humans, blood is routed directly to the tissues after aeration.

(E) In humans, aerated blood is first routed through the heart; in fish either circulatory pattern may occur.

124. Which of the following statements is/are correct?

I. Phosphate bond energy is used during propagation of action potentials.

II. Sodium channels are chemically gated.

III. The density of sodium channels in myelinated and unmyelinated fibers is lowest at the nodes of Ranvier.

IV. The conductance of sodium channels to all permeant cations increases as pH decreases.

(A) all of the above   (D) I and III

(B) III and IV   (E) none of the above

(C) I, III and IV

125. Suppose you made an incision around the entire circumference of a tree trunk (below the branches) such that the phloem was cut all around. The direct result would be that

(A) roots will die because of a lack of nutrients

(B) leaves will not be able to carry out photosynthesis because of a lack of water

(C) buds will not grow above the cut, but roots will continue to grow

(D) the tree will not be affected

(E) none of the above

126. Effective adaptations for gas exchange in plants include

I. thicker cell walls on the side next to the stoma than on the side away from the stoma (for guard cells).

II. a waxy cuticle lining the epidermis of leaves.

III. the location of stomata primarily on the lower surface of the leaf (the side turned away from the sun's rays).

IV. the covering of the lower epidermis with root hairs.

(A) all of the above   (D) I and IV

(B) I, III and IV   (E) none of the above

(C) I only

127. Roots, in contrast to stems and leaves, do not possess structures specialized for gas exchange. Such structures are not needed because

(A) gas exchange does not occur in the roots

(B) most of the cells in the plant stem are dead, and the cells of the leaves obtain oxygen from photosynthesis; therefore, there is very little need for the roots to transport gases to the rest of the plant

(C) unlike the stem and the leaves, which are covered with a waxy cuticle to prevent dehydration, roots branch extensively into a tremendous surface area consisting of a myriad of porous root hairs. Gases can thus diffuse freely across the moist membranes of root hairs and other epidermal structures, making specialized structures unnecessary

(D) the plant obtains all of its needed CO_2 and other gases through the guard cells

(E) none of the above

128 – 131

Consider the following schematic diagram of the tricarboxylic acid cycle. Note that compounds in square brackets are postulated intermediates whose presence is inferred rather than conclusively demonstrated.

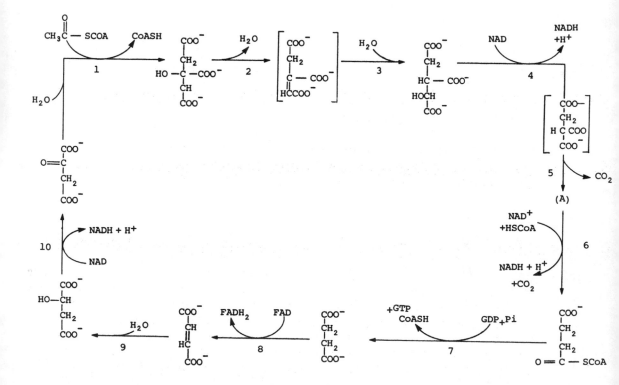

128. The "missing" compound, i.e. the compound that belongs in the position occupied by the letter A is

(A) α-ketoglutarate

(D) citrate

(B) oxalosuccinate

(E) none of the above

(C) fumarate

129. Which of the reactions indicated, if any, are irreversible under typical conditions?

(A) 1 only

(D) 4, 6, and 10

(B) 8 and 10

(E) 1 and 6

(C) 1, 4, 6 and 10

130. Which of the TCA intermediates below is a symmetric molecule which is recognized in asymmetric fashion by an enzyme through a 3 point attachment mechanism?

(A) succinate

(D) L-malate

(B) fumarate

(E) oxaloacetate

(C) citrate

131. Reaction 9 is catalysed by

(A) isocitrate dehydrogenase

(D) hexokinase

(B) malate dehydrogenase

(E) oxalosuccinase

(C) fumarase

132. For many generations, giraffes have stretched their necks in order to reach the leaves on the higher branches of trees. This stretching effort has caused their necks to grow longer; as an individual giraffe acquires a longer neck by stretching it, he is able to pass on this quality of a longer neck to his progeny. After continued small growth in neck length from one generation to the next, giraffes have evolved into exceptionally long-necked animals.

The above reasoning illustrates the theory of

(A) Charles Darwin   (D) Louis Pasteur

(B) Jean Baptiste de Lamarck   (E) none of the above

(C) Gregor Mendel

133. One of the simplest kinds of behavior is the knee jerk, in which a tap below the knee causes the leg to jerk up. This behavior requires as a minimum which of the following combination of structures?

(A) A motor neuron and a muscle

(B) A receptor and at least two segments of the spinal cord

(C) A receptor neuron, a motor neuron, and a muscle

(D) An intact spinal cord and a brain

(E) A receptor neuron connected to a muscle

134. In animal physiology, the latent period is defined as the period

(A) during which synapses are blocked by beta blockers

(B) during which a nerve or muscle recovers from exercise

(C) of maximum fatigue

(D) of repose

(E) between the application of a stimulus and the beginning of response of a nerve or muscle

135. Which of the following is not involved in the process of synaptic transmission?

(A) The release of a neurotransmitter from synaptic vesicles at the pre-synaptic neuron.

(B) The destruction of the post-synaptic membrane after the neurotransmitter has come into contact with it.

(C) Diffusion of the neurotransmitter across the synaptic cleft.

(D) Destruction of the neurotransmitter after transmission of the impulse has taken place.

(E) none of the above

DIRECTIONS: *For each group of questions below, match the numbered word, phrase, or sentence to the most closely related lettered heading and mark the letter of your selection on the corresponding answer sheet. A lettered heading may be chosen as the answer once, more than once, or not at all for the question in each group.*

136 - 140

    (A) 5-Bromouracil   (D) Ultraviolet light

    (B) Nitrous acid   (E) N-nitroso compounds

    (C) Acridines

136. Intercalating agents; give rise to frameshift mutations

137. Base analogue; primarily gives GC=AT transitions

138. Strong alkylating agents; result in GC $\rightleftharpoons$ AT transitions, transversions

139. The damage done to this system is sometimes repaired by a photoreactivation system

140. Deamination, GC $\rightleftharpoons$ AT transitions

141 - 145

    (A) nonsense mutation   (D) competence

    (B) missense mutation   (E) genetic mosaics

    (C) wobble

141. Amber, ochre and opal (UAG, UAA and UGA, respectively)

142. Female mammals

143. Requirement for bacterial transformation

144. Specification of alternate amino acid by a mutant codon

145. AAG
     UUU

146 - 150

    (A) Calciferols   (D) Renin

    (B) Calcitonin   (E) Angiotensin II

    (C) Parathyroid hormone (PTH)

146. Primary regulator of intestinal absorption of Ca^{++}

147. Promotes transfer of calcium and phosphate from bone to blood plasma

148. Promotes secretion of aldosterone, thus stimulating salt and water retention

149. PTH antagonist

150. Stimulates excretion of phosphate by kidney

151 – 155

    (A) T antigen   (D) src gene product

    (B) specialized transduction   (E) Burkitt's lymphoma

    (C) generalized transduction

151. SV-40

152. Occurs through defects in mechanism for cutting and packaging concatamers

153. Occurs through errors in excision mechanism

154. Herpes virus

155. Protein kinase required for oncogenic transformation

156 – 160

    (A) stamen   (D) pollen grain

    (B) pistil   (E) calyx

    (C) embryo sac

156. style, stigma, ovary

157. seven celled, 8 nucleate, haploid

158. haploid, dinucleate, thick walled

159. anther, filament

160. sepals

161 - 165

(A) Causes typhus, Rocky Mountain fever in humans

(B) No cell walls

(C) Exist as dense spore-like cells called elementary bodies

(D) Endospore-forming, anaerobic, gram positive rods

(E) Peritrichously flagellated, gram negative facultatively anaerobic rods

161. Chlamydia

162. Rickettsia

163. Escherichia

164. Mycoplasma

165. Clostridium

166 - 170

(A) colchicine

(B) vinblastine

(C) cytocholasin B

(D) sodium azide

(E) vanadate ions (VO_4^{3-})

166. transition state analog for phosphoryl group hydrolysis

167. inhibits formation of actin filaments

168. uncoupling agent

169. specifically inhibits mitosis at metaphase, prevents distribution of chromosomes to daughter cells without interfering with chromosome replication, and can therefore be used in making polyploid cells

170. inhibits formation of microtubles and spindle fibers by catalyzing intracellular crystallization of tubulin

171 - 175

(A) Taiga   (D) Grasslands

(B) Tundra   (E) Tropical rain forests

(C) Deciduous forests

171. Most pronounced degree of vertical stratification

172. Relatively few trees, but the ground is largely blanketed with mosses, lichens and grasses

173. Dominated by coniferous forests and characterized by small lakes, ponds and bogs

174. Temperate regions with relatively long and warm summers and abundant rainfall

175. Undergoes annual warm-cold or wet-dry cycles in temperate and tropical zones, respectively

176 - 179

(A) cryptic coloration

(D) Müllerian mimicry

(B) aposematic appearance or coloration

(E) geographic isolation

(C) Batesian mimicry

176. Industrial melanism

177. Characterized by gaudy coloration and highly conspicuous appearance

178. Two or more inedible or unpleasant species evolve to resemble each other, each species serving as both model and mimic.

179. A prey species without any characteristics noxious to potential predators takes on the appearance of a species which has such characteristics

180 - 183

(A) cline

(C) species

(B) deme

(D) subspecies (or race)

180. All the bass in a pond

181. Largest unit of population within which gene flow can occur

182. Isolated populations with recognizably different traits but believed to be potentially capable of interbreeding

183. Gradual geographic variation of a given trait within a species

184 - 188

(A) convergent evolution

(D) allopolyploidy

(B) parallel evolution

(E) allopatric

(C) character displacement

184. Exemplified by relationship of Australian marsupials and placental mammals of other continents

185. A mechanism of sympatric speciation

186. Tendency for closely related sympatric species to diverge rapidly

187. Exemplified by relationship between fish and whales

188. Having different ranges

189 - 192

(A) 2,3-diphosphoglycerate

(C) carbonic anhydrase

(B) T quaternary structure ("tense" structure)

(D) R quaternary structure ("relaxed" structure)

189. Binds to deoxyhemoglobin but not oxyhemoglobin

190. Contains intersubunit salt linkages

191. Structure with higher oxygen affinity

192. Stabilized by binding of H^+ and CO_2

193 - 196

Match each example of learning to its appropriate term:

(A) Habituation   (C) Operant conditioning

(B) Classical conditioning   (D) Autonomic learning

E) Instinctive response

193. A rat is placed in a cage equipped with a small lever. As the rat becomes hungry, it begins a random exploration of its cage and accidentally depresses the lever. As the lever is depressed, a food pellet is thrust into the cage. At first the rat shows no signs of associating those two events, but in time his searches become less random and he proceeds more directly to the lever. Eventually, the rat spends most of its time just sitting and pressing the bar.

194. Rats were paralyzed with a curare derivative. Its effect was to allow the animals to remain fully conscious but unable to move any skeletal muscles voluntarily. Positive reinforcement was given to the rats via an electrode implanted in the pleasure center of their brains. Negative reinforcement was given by administering a mild shock.

The minor physiological variations that occur naturally in the body were the starting points for the experiment. If the scientists wanted to teach the animal to slow its heart rate, they stimulated the pleasure center of the brain whenever the heart rate naturally began to slow down. Whenever the heart rate increased, the animal was given a mild shock. Under these circumstances, the researchers found that in some cases the heart rate soon remained slow. Conversely, when a rapid heart rate was rewarded, the heart rate increased.

195. Russian biologist Ivan Pavlov conducted the following experiment.

A dog was harnessed and had a tube attached to its mouth to measure saliva. On successive occasions, a light was flashed five seconds before food was dropped onto a feeding tray under the dog's nose. After a number of such trials, Pavlov discovered that the dog would begin to salivate upon seeing the light.

196. A reef fish, establishing its territory, repeatedly chases after its immediate neighbors, attempting to drive them away. Eventually, it accepts the nearby presence of familiar fish, and only chases after strange fish wandering through the area.

DIRECTIONS: *The following groups of questions are based on laboratory or experimental situations. Choose the best answer for each question and mark the letter of your selection on the corresponding answer sheet.*

197 - 202

Consider what would happen if 100 cells from the following sources were cultured.

    I.   Normal human (N)
   II. Mother of a Lesch-Nyhan child (MLN)
 III. An individual with Lesch-Nyhan syndrome (LN♂)
 IV. An individual whose cells are devoid of thymidine kinase activity (TK⁻)

The following media are used:

DMEM (Dulbecco's Modified Engle's Medium, a nutritionally complete and defined medium)

TG-DMEM with thioguanine added

HAT - Hypoxanthine Aminopterin Thymidine supplemented medium

Budr - medium with 5-bromodeoxyuracil added

The last column heading, AR(HX*), refers to whether the cells can be shown by autoradiography to incorporate labelled hypoxanthine in HAT medium.

Complete the following table by indicating the number of cells surviving out of one hundred plated (or, in the last column, the number giving positive results for incorporation of hypoxanthine in HAT medium).

(A) 0

(A) 0   (D) 75

(B) 25   (E) 100

(C) 50

	DMEM	TG	HAT	Budr	Ar(HX*)
N	100	0	100	0	100
MLN	100	50	Q197	Q198	50
LN♂	100	100	Q199	Q200	///////
TK⁻	100	0	Q201	Q202	///////

203. Consider the following experiment performed using a frog embryo.
In normal development of the zygote, the first cleavage runs
through the grey crescent. If the 2 daughter cells are experimentally separated after the first cleavage each develops into a
normal tadpole. If, on the other hand, the plane of cleavage is
altered (by tying off a portion of the zygote) so that all the grey
crescent material is partitioned into one daughter cell, the cell
with the grey crescent material develops normally while the other
cell gives rise to an undifferentiated ball of cells. This shows that

(A) the first cleavage is determinate

(B) the first cleavage is indeterminate

(C) grey crescent material is required for normal development
within the first few divisions

(D) B and C

(E) A and C

204 - 206

204. An invitro protein synthesizing system was prepared using radioactively labelled amino acids, synthetic components from reticulocytes
and mRNA from human placenta. The predominant transcript was specific for human placental lactogen (HPL), a hormone secreted by the
placenta. Rat liver microsomes derived from either smooth membranes
(SM) or rough ER (RM) were added to the sample; the samples were then
analyzed using SDS-PAGE (sodium dodecyl sulfate polyacrylamide gel

electropheresis), as illustrated in lanes a and b of the figure below. Lanes c and d illustrate the results obtained when the proleolytic enzymes trypsin and chymotrypsin were added to the mixture. The results show that

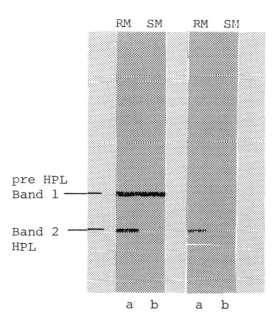

I. the translated product (Pre HPL) is larger than the final, secreted form of the molecule (HPL).

II. The processing of pre HPL to form HPL is dependent on insertion of pre HPL into lumen of endoplasmic reticulum.

III. the processing depends on components exclusively associated with RER and not with SER.

IV. translocation of proteins is mediated by excision of a hydrophobic N-terminal sequence in accordance with the "signal hypothesis."

(A) all of the above   (D) I, III and IV, but not II

(B) I, II and III, but not IV   (E) I and IV

(C) I only

205. Which of the following is the most accurate interpretation of the experimental results?

(A) The experiment supports the signal hypothesis more than it does to the membrane trigger hypothesis.

(B) The experiment supports the trigger hypothesis more than it does to the signal hypothesis.

(C) The experiment supports both hypotheses equally well (or equally poorly).

(D) The experiment shows that the binding of polysomes to endoplasmic reticulum is dependent on a protein receptor in the membrane of RER.

(E) none of the above

206. Suppose another type of mRNA had been used and that this message specified a protein similar in molecular weight and acid-base properties to pre HPL and was native to the cytoplasm of rat liver hepatocytes, you would expect that

(A) a single band would be found in all lines, at approximately the level where HPL was found

(B) a single band would be found in all lines, at approximately the level where pre HPL was found

(C) a single band would be found in lanes a and b at about the level of pre HPL, and no bands would be found in lanes c and d

(D) two bands would be found in lane a as with HPL and pre HPL, one band would be found in lane b at the pre HPL level, one band would be found in lane c at the same level and no bands would appear in lane d

(E) two bands would be found in lane a as above, one in lane b at the HPL level and no bands would be found in lanes c or d

207. Bacterial cells were incubated in a medium containing glucose, lactose and galactose. At regular intervals, cells were removed and plated. Colonies were counted, and the figure below illustrates the fashion in which the population increased over time:

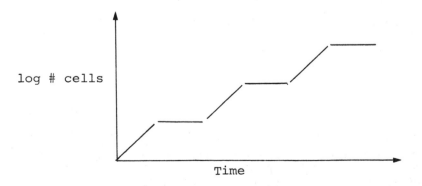

What can you say about the organisms involved?

(A) All enzymes required for fermentation of any of the sugars are adaptive.

(B) All enzymes required for fermentation of each sugar are constituitive.

(C) Lactose hydrolysis is repressible.

(D) Synthesis of β-galactosidase is adaptive.

(E) none of the above

208. Suppose you are trying to determine the purity of an enzyme preparation and using a chromophore as a substrate for the enzyme-catalyzed reaction, you define specific activity as follows:

(μ moles substrate consumed /min /mg protein).

Given that the molar extinction coefficient of the substrate is ϵ, and that A_1 and A_2 are the absorbancies of samples removed from the reaction mixture at times t_1 and t_2, which expression represents enzyme purity in units of specific activity?

Assume that samples removed from the reaction mixture do not need to be diluted before their absorbancies are measured; let L represent the length of the curette used in the spectrophotometer, V the volume of the reaction mixture, and g the amount in grams of enzyme used.

(A) $\dfrac{(1/\epsilon L)(A_2 - A_1)V \times 10^6/(t_2 - t_1)}{10^3 g}$

(B) $\dfrac{(\epsilon L)(A_2 - A_1)V \times 10^6/(t_2 - t_1)}{10^3 g}$

(C) $\dfrac{V \epsilon L/(A_2 - A_1) \times 10^6/(t_2 - t_1)}{10^3 g}$

(D) $\dfrac{(V \epsilon L)}{10^3 g(A_2 - A_1)} + \dfrac{10^3}{g(t_2 - t_1)}$

(E) $\dfrac{(1/\epsilon L)V(t_2 - t_1)10^{-6}}{10^3 g}$

209. Consider the following experiment. Cysteine was incubated with its cognate transfer RNA(tRNA), and cysteine tRNA synthetase. The aminoacyl tRNA produced was then reacted with Raney nickel (which catalyses the removal of the third group from the primary

carbon, ultimately giving a methyl group). The modified amino-acyl tRNA was then used in a cell-free protein synthesizing system with a template comprised of a random copolymer of U and G in 5:1 ratio. Note that such templates normally lead to incorporation of cysteine (UGU) but not alanine (GCU, GCC, GCA or GCG). The peptides obtained from the above

(A) incorporated alanine but not cysteine

(B) incorporated cysteine but not alanine

(C) incorporated both cysteine and alanine

(D) incorporated neither cysteine nor alanine

(E) enough information is not given to properly answer the question

210. Without considering the details of the catalytic processes involved, the elongation of mRNA chains is most appropriately which of the following reaction types?

(A) E_1

(B) E_2

(C) $S_N 1$

(D) $S_N 2$

(E) disproportionation

THE GRADUATE RECORD EXAMINATION BIOLOGY TEST

MODEL TEST II

ANSWERS

1.	D	20.	E	39.	E
2.	E	21.	B	40.	E
3.	B	22.	E	41.	A
4.	D	23.	D	42.	B
5.	C	24.	D	43.	E
6.	E	25.	D	44.	A
7.	E	26.	D	45.	C
8.	D	27.	D	46.	D
9.	E	28.	C	47.	A
10.	C	29.	C	48.	B
11.	D	30.	E	49.	D
12.	E	31.	C	50.	A
13.	D	32.	B	51.	E
14.	C	33.	C	52.	D
15.	D	34.	E	53.	A
16.	C	35.	C	54.	C
17.	B	36.	D	55.	D
18.	D	37.	A	56.	B
19.	A	38.	B	57.	A

58.	B	89.	B	120.	C
59.	C	90.	A	121.	B
60.	A	91.	D	122.	A
61.	E	92.	E	123.	A
62.	E	93.	A	124.	E
63.	E	94.	E	125.	A
64.	C	95.	A	126.	C
65.	E	96.	E	127.	C
66.	E	97.	E	128.	A
67.	A	98.	A	129.	E
68.	C	99.	D	130.	C
69.	B	100.	A	131.	C
70.	D	101.	B	132.	B
71.	C	102.	D	133.	C
72.	D	103.	B	134.	E
73.	E	104.	A	135.	B
74.	A	105.	A	136.	C
75.	E	106.	A	137.	A
76.	E	107.	C	138.	E
77.	E	108.	D	139.	D
78.	B	109.	A	140.	B
79.	C	110.	C	141.	A
80.	B	111.	B	142.	E
81.	C	112.	C	143.	D
82.	C	113.	D	144.	B
83.	B	114.	A	145.	C
84.	E	115.	A	146.	A
85.	C	116.	E	147.	C
86.	D	117.	A	148.	E
87.	E	118.	A	149.	B
88.	A	119.	D	150.	C

151.	A	171.	E	191.	D
152.	C	172.	B	192.	B
153.	E	173.	A	193.	C
154.	E	174.	C	194.	D
155.	D	175.	D	195.	B
156.	B	176.	A	196.	A
157.	C	177.	B	197.	C
158.	D	178.	D	198.	A
159.	A	179.	C	199.	A
160.	E	180.	B	200.	A
161.	C	181.	C	201.	A
162.	A	182.	D	202.	E
163.	E	183.	A	203.	A
164.	B	184.	A	204.	B
165.	D	185.	D	205.	A
166.	E	186.	C	206.	C
167.	C	187.	A	207.	D
168.	D	188.	E	208.	A
169.	A	189.	A	209.	A
170.	B	190.	B	210.	D

THE GRE BIOLOGY TEST

MODEL TEST II

DETAILED EXPLANATIONS
OF ANSWERS

1.   (D)
Phenotypic markers tend to follow the Law of Independent
Assortment if and only if they correspond to loci on separate
chromosomes. It is actually the chromosomes themselves that
assort independently during meiosis.

2.   (E)
An autotroph is by definition capable of manufacturing
organic nutrients from inorganic sources. All heterotrophs
lack this ability, and therefore must ingest carbon in organic
form. Herbivores obtain organic carbon by consuming
autotrophs, while carnivores, one step further up the food
chain, prey upon herbivores or other carnivores.

3.   (B)
The hypothesis of chemiosmotic coupling proposes that the
energy derived from electron transport causes translocation
of protons across the inner mitochondrial membrane, thus
creating a pH and electrochemical gradient across that
membrane. The free energy which results from the flow of

protons back across the membrane is used to drive the formation of ATP from ADP and phosphate. A similar chemiosmotic mechanism is believed to be involved in photosynthetic ATP synthesis.

4.   (D)
The hypothalamus exerts significant control over various aspects of the body by monitoring feedback from the Autonomic Nervous System and by using releasing hormones to regulate the hormonal secretions of the pituitary.

5.   (C)
The NAD^+/NADH system is most important as a carrier of electrons and protons in ATP formation and numerous other reactions. The phosphorylated analogue $NADP^+$/NADPH has many important biosynthetic functions, including production of deoxyribonucleotides and tetrahydrofolate, and is also involved in photosynthetic electronic transport.

6.   (E)
All are clearly involved. The RER is the site to which bound ribosomes are attached; polypeptides produced on bound ribosomes are characteristically destined to be exported from the cell and/or localized in the cell membrane. There is a contrast, in function, between free and bound ribosomes: polypeptides (proteins) produced by free ribosomes are released into the cytoplasm of the cell.

7.   (E)
The lower limits of useful resolution of a light microscope occur at approximately 275 nm. This is more than adequate for resolving individual bacterial cells, but is insufficient for viewing the other structures listed, which are typically far less than 100 nm. Microtubules, for example, are approximately 25 nm in diameter.

8.   (D)
Given that neither allele is dominant, both parents must be homozygous. The progeny which must then be

heterozygous will, as a consequence, be all of uniform length.

This can be demonstrated by assigning values to alleles and charting the cross:

L = long allele
S = short allele
LL = long stemmed
SS = short stemmed
Ls = intermediate

Cross of a long stemmed individual and a short stemmed individual

	L	L
S	Ls	Ls
S	Ls	Ls

9.   (E)
Structure of aminoacyl tRNA complex

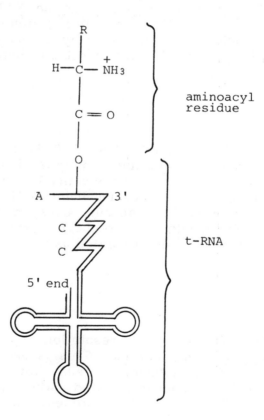

10.   (C)
The four distinct stages of respiration are: glycolysis, the conversion of pyruvate to acetyl-CoA, Kreb's cycle, and oxidative phosphorylation. The tricarboxylic acid and citric

acid cycles are alternative names for the Kreb's cycle. Gluconeogenesis is not a cycle but a process by which glucose is produced from non-carbohydrate molecules.

11.   (D)
There is no requirement that the total number of individuals in the population remain constant; the other four conditions must be met if the Hardy-Weinberg equilibrium is to hold.

The Hardy-Weinberg formulation postulates a binomial distribution of allelic frequencies - two alleles corresponding to a given trait. If we let p designate the frequency of one allele corresponding to a particular trait, and q the frequency of the other allele, then we have

1) $P + q = 1$

2) $(P + q)^2 = P^2 + 2Pq + q^2 = 1$

Given the frequency of one of the alleles, this equation allows us to calculate the frequencies of both types of homozygote as well as the percentage of heterozygotes in the population.

12.   (E)
In general, the overall trend is towards a more complex and more stable community. The greater degree of complexity and specialization permits a more efficient use of energy and support of a greater biomass.

13.   (D)
As is the case with phosphorus and sulphur, nitrogen is absorbed primarily in ionic form.

14.   (C)
Such bacteria are associated with a high frequency of recombination because insertion of the plasmid into the chromosomal DNA allows for transfer of chromosomal DNA as well as sex factor during conjugation.

15.   (D)
    There are numerous examples of organisms not closely linked in a phylogenetic sense taking on similar traits in response to similar environmental pressures. Convergence is exemplified in the way that mammalian, terrestrial-descended whales evolved fins which bear a superficial resemblance to those of fish.

16.   (C)
    Epinephrine, norepinephrine and especially glucagon all oppose insulin in their effect upon blood glucose levels. The net result of all the effects listed is the depleting of blood glucose levels while increasing reserves of glycogen, fat and protein. Glucocorticoids such as cortisone tend to stimulate the formation of polysaccharides from protein and fat, initially raising glycogen levels.

17.   (B)
    Under conditions of rest, a normal breath moves only a small portion of the lungs' total capacity. This small portion of air — 500 ml on the average — is the tidal volume, while the total capacity of the lungs — approximately 3.2 liters for females and 4.5 liters for males — is called the vital capacity.

    The volume of air which can be forcibly inhaled over and above the resting tidal volume is the inspiratory reserve volume; similarly, the volume of air that can be expired beyond the resting tidal volume is called the expiratory reserve volume. Even after forced maximum expiration, a certain volume of air remains in the lungs. This is called the residual volume. The vital capacity is the sum of the tidal volume and the inspiratory reserve volume and the expiratory reserve volume.

18.   (D)
    Kingdom, phylum (division), class, order, genus, species

19.   (A)
    The denaturing conditions disrupt the forces that maintain the double helix. These forces include hydrophobic interactions which keep the largely nonpolar bases in the inside of the molecule, hydrophilic interactions involving the phosphate moieties, hydrogen bonding between paired bases, and the combination of permanent depolar effects and van der Waals forces comprising the so-called "stacking energy" which stabilizes a parallel alignment of the bases.

20. (E)

Bacterial cell wall types form the basis of the most important method of classification of bacteria.

21. (B)

While the subject is complex, there is ample evidence (particularly autoradiography) of the passage of membrane from nuclear envelope to R.E.R. and then S.E.R.; there is also ample evidence to support a flow from golgi apparatus to secretory vesicles to cell membrane.

22. (E)

Prokaryotic cells are found in bacteria and cyanobacteria (members of the Kingdom Monera). They are more primitive than eukaryotic cells, which are found in all Kingdoms except Monera. Prokaryotic cells contain a single circular chromosome made of nucleic acid without associated proteins, have no membrane-bound organelles such as chloroplasts or mitochondria, and their microtubule structure is not the 9.2 structure found in eukaryotic cells.

23. (D)

It is now believed that hydrophobic interactions are driven by an entropy effect. Solvent molecules surrounding exposed hydrophobic groups are thought to be constrained to adopt an ordered, low entropy configuration. It is known that the heat of reaction is slightly positive.

24. (D)

Erythrocytes are produced in red bone marrow.

25. (D)

Tight junctions eliminate the spaces between adjacent cells and thus provide an uninterrupted absorptive surface; desmosomes have been described as analogous to "spot welds" which maintain integrity of the absorptive surface during distention. Gap junctions provide a means for chemical communication between cells.

26.   (D)
The current model of eukaryotic cilia and flagella indicates that their motion is due to the lengthwise sliding of tubules. During this process the nexin arms of a doublet cyclically attach to and detach from the next double, "sliding along its back."

27.   (D)
In a cilium or flagellum of a eukaryotic cell, microtubules are arranged in a specific pattern. Nine peripheral microtubules surround two central microtubules.

28.   (C)
The sarcomere is the area between Z lines of a myofibril. It is the basic, repeating contractile unit containing the A band in which alternating actin and myosin filaments slide together during contractions.

29.   (C)
These techniques take advantage of differences in refractive index and other optical effects (such as phase changes in light waves at the boundaries of different materials). The result is that one obtains a sort of "optical staining" which allows for in vivo observation.

30.   (E)
Auxins can stimulate or inhibit elongation of target cells, depending on cell type and auxin concentration. Auxin transport seems to be highly specific, and auxins are thought to be crucial in plant tropisms.

31.   (C)
Protoplasts are bacterial cells which have been treated to remove the cell wall. Methods of producing protoplasts include treating cells with lysozyme, or growing cells in the presence of penicillin which selectively inhibits cell wall formation. Note that when the peptidoglycan layer is removed from gram negative bacteria but other layers of the cell envelope are left intact, these cells are termed spheroplasts.

32.   (B)
Psychrophiles may grow at 0°C or lower, although optimal temperatures range from 15-30°C.

33.   (C)
All known living organisms are hypertonic with respect to freshwater. Freshwater fish thus tend to take in too much water and have difficulty retaining salt. They are able to compensate for this by seldom drinking, active absorption of salts and the excretion of large volumes of dilute urine.

34.   (E)
By the beginning of anaphase, the centromeres have uncoupled and have begun moving toward opposite poles. In late prophase the chromosomes are not yet lined up along the equator and the nuclear membrane is still visible. The onset of metaphase occurs with the complete disappearance of the nuclear envelope and the alignment of chromosomes along the metaphase plate.

35.   (C)
Most xylem cells are essentially hallowed out cell walls which serve two important functions:

First, they provide a passageway through which water and dissolved substances can be transported up and down the tree; second, they provide structural support for the tree itself (xylem cells comprise what is commonly called wood). Organic materials are transported through the other kind of plant vascular tissue: the phloem.

36.   (D)
In the dicot stem phloem (externally) and xylem (internally) develop in concentric rings around a layer of vascular cambium. A pith is at the core. The dicot root lacks vascular cambium and a pith.

37.   (A)
The clonal selection theory proposes that lymphocytes become specialized early in embryonic development. When these lymphocytes begin to move through the bloodstream, they learn to "recognize" their parent organism, so that they can later distinguish a foreign substance from the body's own molecules.

Later, when the viable individual does encounter a foreign substance - an antigen - the lymphocyte with the antibody specific for that antigen binds to the antigen and begins to proliferate. Thus producing a great number of antigen-specific antibodies to neutralize the invading foreign substance.

38.   (B)
Binding of the inducer to the repressor inactivates it, thus opening up the site on the promoter for binding of polymerase. Review the Lac operon.

39.   (E)
The fact that all of the statements are true indicates the potential usefulness of interferons. Although the mechanism of action is not well understood, interferon is known to bind to the membranes of uninfected cells and prevent their becoming virally infected. Perhaps the most promising use of interferons lies in their use as antitumour agents.

40.   (E)
An outer covering of lipopolysaccharides is associated with gram negative bacteria.

Schematic G[+] cell wall:

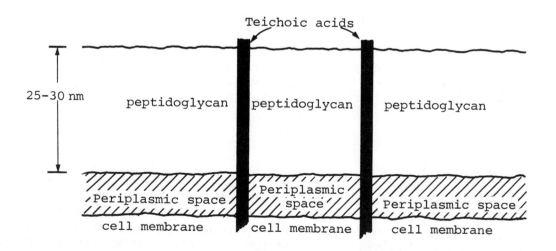

41. (A)

By definition an integral protein cannot be extracted from cell membrane without dissolving the membrane. The fact that the sample is undigested by proteolytic enzymes in an intact cell or sealed membrane but hydrolysed by the same enzymes when in leaky membranes indicates that they are exposed on the cytoplasmic face of the membrane but not on the outer surface.

42. (B)

In a dihybrid cross, one starts with a parental generation of (opposed) homozygous types and obtains a first filial generation composed entirely of heterozygotes which all display the dominant markers. When individuals of the first filial generation are crossed with each other, there are 16 possible combinations of gametes that can result, and these in turn give rise to 4 different phenotypes in the F_2 generation. These phenotypes, arising from the doubly dominant type, the 2 singly dominant recombinant types and the doubly recessive individuals distribute according to the aforementioned 9:3:3:1 ratio.

43. (E)

Trisomies in human beings are generally lethal; the three diseases listed are exceptional insofar as their victims often survive. Down's syndrome, previouly called Mongolianism, is an autosomal trisomy of chromosome 21. Klinefelter's syndrome is the condition in which a male has two X chromosomes and one Y chromosome. An individual affected by Turner's syndrome has only one sex chromosome. The manifested phenoytype of Turner's syndrome is female in humans and male in drosophila.

44. (A)

The experiment relied on the fact that proteins contain sulfur, but no phosphorus; nucleic acids contain phosphorus, but no sulfur. Thus, the proteins and nucleic acids in the bacteriophage labelled themselves by incorporating radioactive sulfur and phosphorus, respectively. Since the infected bacteria was shown to contain only phosphorus, the bacteriophage must have injected only nucleic acids into the bacteria. Given the mechanism of phage reproduction, in which an infected host cell produces hundreds of new phages and then ruptures, the nucleic acid must be the genetic material of the cell.

45.   (C)
As a general rule, the most internal tissues of the body are derived from endoderm; surface tissues and outer body layers are derived from ectoderm while tissues of intermediate depth arise from mesoderm. The exception is neural tissue which is derived from the ectoderm.

46.   (D)
The nucleus and the mitochondria each have both an inner and an outer membrane, so that transport from the nucleus to the internal matrix of a mitochondrion requires passage across four membranes.

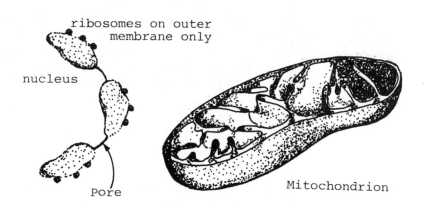

47.   (A)
In this and other questions, "mendelian" refers to genetics without codominance, epistasis, or multiple alleles. "True breeding" indicates "homozygous."

Since the F_1 generation consists entirely of red plants, it is obvious that the red allele is dominant to the white. To solve this problem, it may be helpful to draw a diagram of two punnet squares. As always, indicate dominant traits with uppercase letters and recessive traits with lowercase letters:

Parental Cross RR × rr → F_1 (all red, all Rr)

F_1 cross

	R	r
R	RR	Rr
r	Rr	rr

3:1 red:white

Test Cross Rr × rr   1:1 red:white

	R	r
r	Rr	rr
r	Rr	rr

48.   (B)

On the basis of the results of this experiment Mendel postulated a model of inheritance, the particulate model, even before it was known that chromosomes are involved. Mendel's particulate model proposed that a given trait is determined by two factors, one from each parent, and that one factor - the dominant - is capable of masking the other - the recessive. The masked factors do not disappear but reemerge in subsequent generations, undiminished in intensity. The red and white plants do not give an intermediate pink; the traits remain segregated in their initial types from one generation to the next.

Mendel also proposed that patterns of inheritance can be studied statistically. According to his model of inheritance, an individual's characteristics result from the random inheritance of factors from both parents, in accordance with the laws of probability.

49.   (D)

Proceed as before using a punnet square. Note that each of the two traits follows the same 3:1 ratio as obtained in the monohybrid cross, and the genotypes are 1:2:1 as in the monohybrid cross.

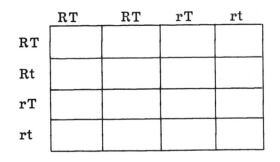

	RT	RT	rT	rt
RT				
Rt				
rT				
rt				

You should try to do problems like this without the punnet square.

50. (A)

Recall that I bands consist of actin filaments alone; the A band consists of the thick myosin filaments between Z lines and includes the actin filaments which overlap those myosin filaments and the H zone is the region between opposing actin filaments in the A band. When muscle contracts, the thin actin filaments slide toward each other along the myosin filaments, causing the I band and H zone to decrease in length, while the A band stays the same.

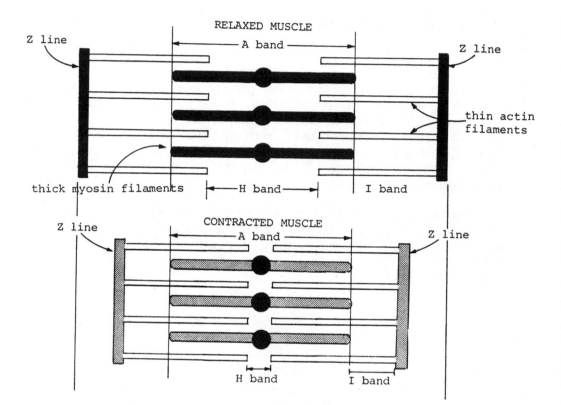

51. (E)

There is a causal relationship between many of these problems. Retention of sodium increases water retention and starts the increase in blood pressure. There is also an increase in excretion of potassium to compensate for the retention of sodium. Potassium is an inhibitor of the heart, so the blood pressure is further increased. Finally the increase in sodium and accompanying decrease of potassium in the blood level significantly alter the equilibrium of the bicarbonate buffer system, resulting in the onset of alkalosis.

52. (D)

Insulin must bind to cell membranes of adipose tissue before glucose can enter such cells, and act on enzymes within the cells to promote synthesis of fats from fatty acids and glycerol. Thus insulin deficiency triggers an accumulation of long chain fatty acids in the blood. These in turn may be oxidized to form ketones which in sufficient concentrations can give rise to dehydration and acidosis.

53. (A)

The convention is to describe genes by proceeding in the 5' → 3' direction along the antisense strand, allowing one to preserve the 5' → 3' orientation of the message while being able to read from left to right. Here we are describing structures along the transcribed strand in the order that a polymerase would encounter them, which is from a 3' to 5' direction.

mRNA → 5'   G   Leader   AUG   Trailer   3'
            A

sense strand
                 C   TAC
3' Promoter   T   Antileader   Gene   Antitrailer   Terminator 5'

54. (C)

Centered at about the -10 region of a DNA template is one of the two known promoter sites in E. coli. The other site is further upstream at about -30 nucleotides. These promoters have an importance extending well beyond E. coli insofar as over 50 prokaryotic promoters have been sequenced and all contain similiar sequences in these two locations.

55. (D)

Promoters function with respect to transcription. The translation-related activities described in (D) are characteristic (in prokaryotes) of the Shine-Dalgarno sequence. The latter is about 5 to 10 base pairs upstream from the AUG codon and is homologous to the 3' terminus of 16 S rRNA. Note that specifying transcriptional direction (E) is essentially the same as choice of sense strand (C).

56.   (B)

In kinetics experiments, those lengths of DNA that reassociate fastest will be those which are most widely repeated throughout the genome and hence have the greatest likelihood of quickly finding complementary sequences to form a duplex with. Although not inevitable, it is generally true that these highly repetitive fragments will have a different percentage of GC from the others in the genome and therefore will generally band in a satellite position in a CsCl gradient. In situ hybridization studies have localized this DNA in those chromosomal regions characterized by constitutive heterochromatin.

57.   (A)

Since I^A and I^B alleles are codominant, the individual produces glycosyl tranferases corresponding to both A and B antigens. The production of both antigens causes the individual to produce neither the two kinds of antibody. Since the primary risk in transfusions arises from agglutination of the transfused erythrocytes by circulating antibodies in the recipient's bloodstream and not by agglutination of the host erythrocytes by injecting antibodies, the AB individual may receive transfusions from any of the bloodtypes (he should not, however, donate to any but other AB recipients). AB types are known as "universal acceptors", O types are known as "universal donars".

58.   (B)

The translated products are isozymes. Allelic forms refer to alternate sequences which occur at the same position along a given chromosome. A gene is epistatic to another gene if gene A can mask or enhance the effects of gene B (i.e. B produces a pigment while A regulates its deposition). Pleiotropy is the translated product of a single gene manifesting itself as more than one phenotypic trait.

59.   (C)

Hemoglobin S results from the substitution of a single amino acid in the β chain. Normal adult hemoglobin (i.e. hemoglobin A) is a tetramer which may be represented as $\alpha_2\beta_2$. Hemoglobin S, the sickling form, is thus α_2S_2. The individual with sickle cell trait is heterozygous for the β producing allele; the hemoglobin is composed of 50% hemoglobin A and 50% hemoglobin S.

60. (A)

The correct choice is A. The other effects listed correspond respectively to Tetracycline, Puromycin, Erythromycin and Streptomycin.

61. (E)

Human embryos haploid for any autosome are nonviable and will not come to term. The following schematic diagram illustrates the different outcomes that are possible if an individual with a translocation of chromosome #21 mates with a normal individual. Note that chromosome #21 is an extremely small chromosome and the translocation fragment is usually almost the entire chromosome.

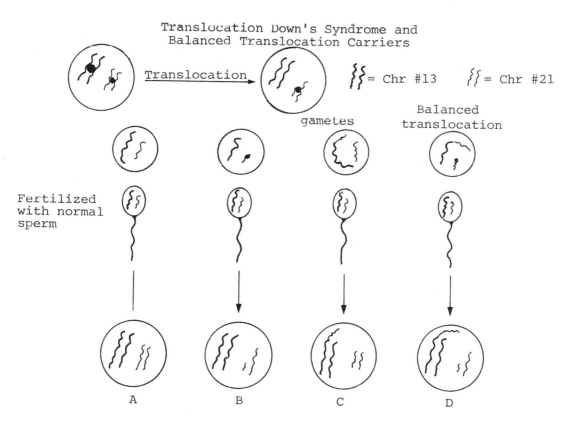

A) Normal individual

B) Virtually haploid for chromosome #21, nonviable

C) Virtually triploid for chromosome #21, Down's syndrome

D) Balanced translocation carrier. Note that this individual produces the same types of gametes so the next generation will be 1/3 translocation carriers, 1/3 Down's syndrome and 1/3 normal.

Note also that Down's syndrome may be caused by other events, such as nondisjunction of chromosome #21. There have been a significant number of cases of institutionalized women with Down's syndrome becoming pregnant; their children have a 50% chance of having Down's syndrome.

62.   (E)
Most if not all of these statements should be familiar. The large amount of basic amino acids in histones causes them to develope a net of positive charge at intracellular pH and become electrostatically bound to the phosphate moieties of DNA. Histones are present in roughly equal weight amounts as DNA, and histone biosynthesis is closely linked to DNA replication: histone synthesis occurs only during the S phase and rapidly ceases if DNA replication is experimentally interrupted. Histones and histone mediated structuring of DNA into nucleosomes represent an evolutionary development unique to eukaryotes.

63.   (E)
Not only are all the above true but there are several further levels of organization present. The solenoidal nucleosome fibers are about 25 nm in diameter, these in turn form "folded loop" structures each running approximately 0.5 μM in length and/or radial loops of 1μM. It is these loops which tend to cluster into the bands, which can be visualized under a light microscope.

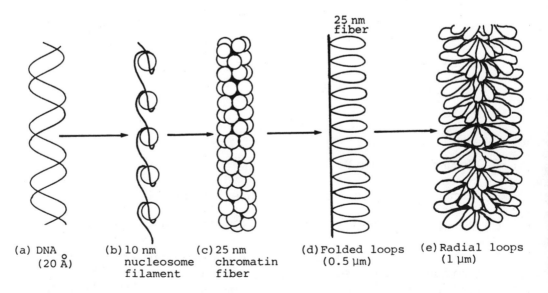

(a) DNA (20 Å)   (b) 10 nm nucleosome filament   (c) 25 nm chromatin fiber   (d) Folded loops (0.5 μm)   (e) Radial loops (1 μm)

64.   (C)

Primase gets its name from its role in forming the RNA primers needed in DNA replication. Unlike RNA polymerization which can proceed de novo, there is a requirement in DNA synthesis for an exposed 3' end to add onto and hence the need for RNA primers. The other functions listed are also necessary and are performed, respectively, by helicase (A), single strand binding protein (B), DNA polymerase III holoenzyme (C) and DNA polymerase I.

65.   (E)

There is no generalized increase; typically only IgM and IgG would be involved in such a sequence. Furthermore, the quantity of IgM would quickly diminish after the rise of IgG levels, and if a further dose of DNP-BSA was given in subsequent months there would be a rapid rise in IgG levels to a further peak without an intervening rise in IgM levels.

66.   (E)

All are correct. Na^+ ions are actively transported out of the ascending limb of the loop of Henle, some of which diffuse back into the descending limb. Water does not diffuse out of the tubule as the filtrate passes through the loop of Henle; therefore the fluid passing through the distal convoluted tubule is less concentrated than the initial filtrate. Whether the urine released by the collecting tubule is dilute or concentrated depends on overall water levels in the body and hormonal signals.

67.   (A)

These compounds are all glycolytic intermediates; clearly the metabolic pathway is thermodynamically favorable under intracellular conditions and so it is only required to write down the order of the compounds as they are formed in glycolysis.

68.   (C)

Recall the definitions of noncompetitive and uncompetitive inhibition. In the latter, the inhibitor does not combine with free enzyme or substrate but combines with the enzyme-substrate complex. A Lineweaver-Burke plot of such a reaction is characterized by retaining a constant slope as inhibitor concentration increases for fixed substrate concentrations. Noncompetitive inhibition occurs when the inhibitor can complex with either free enzyme or enzyme-substrate complex, the inhibitor tends to bind at some site other than the active site and increasing substrate concentration is ineffectual in reversing inhibition. Noncompetitive inhibition may be recognized through Lineweaver-Burke plots of the type shown, i.e. different slopes but a common $\frac{1}{[S]}$ intercept for reactions run in various concentrations of inhibitor.

69.   (B)
Exotoxins are primarily excreted by gram positive organisms, whereas gram negative bacteria are typically the source of endotoxins. The former are proteins, while the latter are lipopolysaccharides. A toxin which cannot be inactivated by either heat treatment or acidification is clearly not a protein.

70.   (D)
The most dramatic result of administering gibberellins, particularly to plants of the type cited, is a radical elongation of the stem.

71.   (C)
As follicles grow they release increasing amounts of estrogen. The high estrogen levels (or possibly the slowdown in estrogen buildup as the peak is reached) elicit a sharp, spiked surge of luteinizing hormone from the pituitary. This, in turn, triggers the opening of the follicle and ovulation.

72.   (D)
In a typical action potential, the first event is a change in the membrane voltage due to Na^+ ions entering the axon. As the Na^+ ions move in, the voltage peaks, triggering the movement of K^+ ions out. The initial change is a change in Na^+ conductivity, the potential is never as positive as the Nernstian Na^+ potential, and there is an increase in the K^+ conductance.

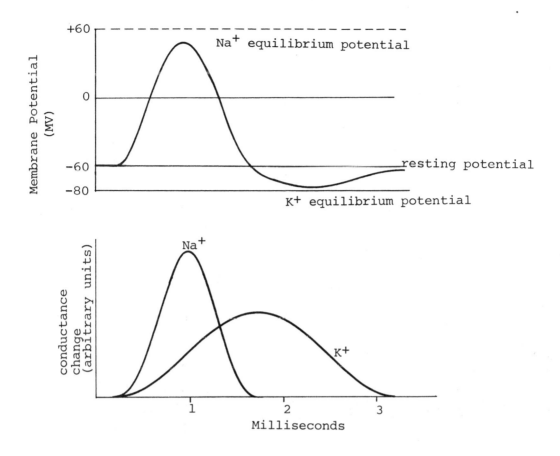

73.   (E)
Chimpanzees and marine animals (the dolphins, the whale) have long been known to possess intelligence advanced enough to allow them to communicate via very simple, basic "language." The case of the bees, though less well known, is equally extraordinary. Through intricate, elaborate dances, bees are able to convey to one another the location of food sources at various and considerable distances from the hive.

74.   (A)
Primer pheromones are distinguished from releaser pheromones insofar as the latter trigger an immediate and reversible change in the behavior of the recipient, while the former produce long term physiological changes and may not give rise to any immediate behavioral response.

75. (E)
At each successive trophic level, there is a loss of energy and productivity from the system. In general, perhaps 10% of the energy at one trophic level can be passed to the next. The result is that it is very rare for a food chain to consist of more than 4 or 5 levels; the productivity of the fifth level is estimated to be no more than 0.0001 that of the first.

76. (E)
All statements are true. A reflex arc consists of a minimum of three cells: a sensory neuron, a motor neuron and an effector cell.

77. (E)
In cyclic photophosphorylation, light energy excites the reactive center of Photosystem I (PI) to the point where it becomes a strong reducing agent and gives up an electron to Ferredoxin, becoming a radical cation in the process. The electron is passed on several times, eventually returning to PI in a lower energy state, with some of the energy having been used to synthesize ATP from ADP and inorganic phosphate. Photosystem II is directly involved in the splitting of water and NADPH is produced only in the noncyclic mode.

78. (B)
Photosynthesis reverses the topology of mitochondrial function - protons are translocated from the outside to the inside - but in each case ATP is synthesized in the lumen of the organelle. Oxygen is evolved only in the noncyclic process, as water is not split in the cyclic process. The stoichiometry is 2 hv:2e⁻:1 Ⓟ ∿

79. (C)
Many processes of active transport fall in the realm of cotransport, in which the diffusion of one solute against a concentration (or electrochemical) gradient is coupled to the diffusion of another solute down its own gradient; phosphate bond energy is only indirectly involved insofar as it is used to maintain that gradient. Both animal and bacterial cells tend to use cotransport processes to take up amino acids and

simple sugars, the process being driven by Na^+ flow in animal cells and proton flow in prokaryotes. Directionality in active transport is provided by the coupling system, which is either located on the cytoplasmic face of the cell membrane or asymetrically oriented within the membrane. Saturation effects distinguish mediated from nonmediated transport, and the persistence of an established direction characterizes active transport.

80. (B)
Territory and home range are concepts from sociobiology, whereas niche and habitat are ecological constructs. A large herbivorous mammal typically has no territory (defined as an area actively defended from conspecifics) but a large home range (total area travelled in the normal course of activities). The large home range is required for sufficient forage.

81. (C)
Habituation is the process of gradually learning not to respond to a stimulus. Intentional movements are component parts of an implied imminent action, such as a man clenching his fist but not striking. Redirected activity is transference of an action from its intended yet unavailable object to an available object. In this case, the dog's anger and desire to bite is redirected from your flesh to the stick.

82. (C)
Some higher plants, especially grasses and trees, pick up quantities of ammonium ion and subsequently incorporate the nitrogen into more complex molecules. However, most flowering plants absorb required nitrogen in the form of nitrates.

83. (B)
Chlamydomonas is characterized by (B). Possession of a macronucleus as well as a micronucleus as noticed in answer (A), is a characteristic of Paramecia and various cellular slime molds such as Dictyostelium. These organisms undergo amoeboid as well as motile multicellular stages, the change from free living amoebae occurring under conditions of diminished local food supply.

84.   (E)
cAMP has been shown to play a role in all these processes. It has a definitie role in the reaction cascade leading to activation of phosphorylase and inactivation of glycogen synthetase, thus accelerating the breakdown of glycogen and retarding its formation. This is accomplished by phosphorylating a protein kinase; phosphorylation of a different kinase stimulates lipolysis so that in some instances cAMP may be regarded as a hunger signal for vertebrate cells. It was therefore a matter of some interest to biologists to find that cAMP is the chemical signal released that leads to the aggregation of amoeboid cells into a plasmodium when food supplies run low.

85.   (C)
Periderm may be immediately eliminated as a possible answer since it is a surface (rather than fundamental) tissue. Parenchyma and collenchyma are both composed of cells which remain alive during most of their functional existence. The former are characterized by thin primary cell walls and no secondary cell walls; the latter are structurally similar except that their cell walls may be irregularly thickened. Endodermal cells typically occur as a single layer surrounding vascular tissues of roots (or sometimes stems). They typically resemble elongate parenchyma cells in their early stages, but in later stages may develop a secondarily thickened wall. Sclerenchyma is a protective or supporting plant tissues which have greatly thickened, lignified secondary walls which confer mechanical support to the plant body.

86.   (D)
The middle lamella is the pectin-based layer formed when the walls of adjoining plant cells come into contact. The primary wall is the first to be laid down, is stretchable, and remains the only wall present as long as the cell continues growing. After ceasing to grow, cells in the more rigid, wooden sections of the plant may lay down a secondary wall, lying internal to the primary wall (and external to the plasma membrane).

87.   (E)
The osmotic pressure is approximately 25 mm Hg at both ends of the capillary bed, while the hydrostatic pressure is 36 mm Hg at the arteriole end and 15 mm Hg at the venule

end. The osmotic pressure, which is always greater in the capillaries than in the tissue fluid, causes the capillaries to take up water and dissolved materials, while the hydrostatic pressure has the opposite effect. The result is a net outward pressure from the capillaries at the arteriole end and a net inward pressure at the venule end. Many lipid soluble materials diffuse through the endothelial cell membranes, while other solutes may either be taken up from the capillary lumen by pinocytotic vesicles and released on the other side of the cell or pass through clefts between adjacent endothelial cells.

88. (A)
The metal-coated structures are impermeable to electrons, while the uncoated areas allow electrons to pass through them. These electrons make light spots on the screen, and dark ones on a photoplate. When positives are printed from the plate, these structures appear as they do on the screen – darker than surroundings.

89. (B)
Osmium tetroxide cross links adjacent lipids at unsaturated moieties (recall first semester organic chemistry of alkenes).

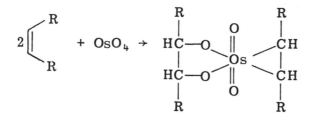

PAS reaction is used to detect polysaccharides; HIO_4 breaks down the polysaccharide to give, among many other products, exposed aldehyde groups which form a colored complex with the Schiff reagent.

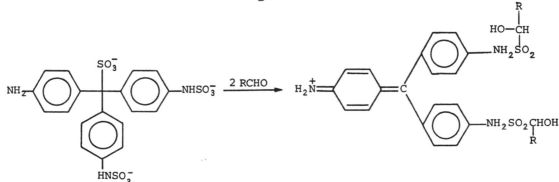

193

Formaldehyde is an appropriate reagent for detection of polypeptides, the latter being cross linked at their N-termini via a methylene bridge (P = Polypeptide)

$$P_1-NH_2 + \overset{\overset{\displaystyle O}{\displaystyle \|}}{H C H} \longrightarrow P_1NHCH_2CH \xrightarrow{P_2NH_2} P_1NHCH_2NHP_2$$

$$+ H_2O$$

Glutaraldehyde works by an analogous method, but forms longer bridges and, having 2 aldehyde moieties, links 4 polypeptides.

90. (A)
In centrifugation, the more massive structures pellet out before lighter or less dense structures. The diagram below shows the order of centrifugates obtained from rat liver cells. Note that rat liver cells do not contain synaptosomes.

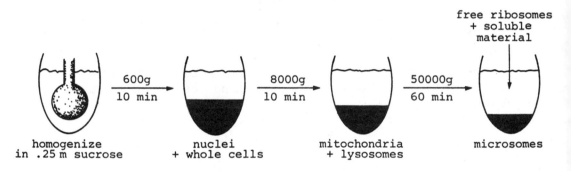

91. (D)
One method, at least, uses all of the above. The technique involves using polyethylene glycol to trigger the fusion of the activated B lymphocytes with mutant tumor B lymphocytes and aminopterin as a selective agent to allow only the hybridomas to survive. The activated B lymphocytes can grow for short periods in the presence of aminopterin, but cannot be maintained in culture for more than a few days, while the mutant cells can grow indefinitely in the right medium but do not survive in the presence of aminopterin. One then allows individual hybridomas to grow into clones, and tests for the presence of desired antibody, after which the appropriate cells may be frozen and grown indefinitely.

92. (E)
Regardless of whether the method is direct or indirect, the cell must be partially fixed and made permeable. The direct method simply involves conjugating anti-P to a florescent dye, incubating the prepared cells with florescent anti-P, washing to remove unbound antibody, and examining the cells under a florescence microscope. The indirect method starts by injecting rabbit immunoglobulins into a goat, using the rabbit antibody proteins as antigens and obtaining goat-anti-rabbit serum (GAR). GAR is then linked to a florescent marker. The prepared cells are incubated with nonflourescent anti-P; unbound anti-P is removed by washing and then the cells are incubated with florescent goat anti-rabbit serum (FGAR). The FGAR binds to the anti-P (unbound FGAR being removed by washing) and so indirectly indicates the location of protein P.

93. (A)
The ATP-ase activity in contraction of skeletal muscle is associated with the myosin thick filaments; more specifically it is localized in the S-1 heavy meromyosin fragment. The ATP-ase activity in microtubule activity is associated with neither of the tubulin subunits but rather with the dynein "arms" that connect adjacent doublets. The MAPs are known to bind to microtubules during polymerization/depolymerization cycles and are presumably functionally involved in initiation, elongation and stabilization of microtubule assembly but possess no known ATP-ase activity.

94. (E)
Most common plants undergo "alternation of generations" with both diploid and haploid multicellular stages. The relative prominence of these stages varies from one plant type to another. The most primitive plants are characterized by the absence of any diploid phase other than the zygote, which meiotically divides to produce spores.

A: Rudimentary Plant

——— haploid
——— diploid

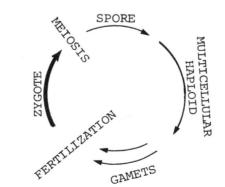

B: More advanced multicellular plant

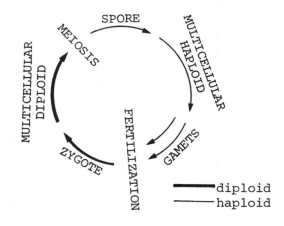

95. (A)
"A" is the only correct answer. A permissive cell permits viral replication - it is one in every 10^4 - 10^5 nonpermissive cells that may be transformed by an oncogenic virus. RNA-DNA hybrids and reverse transcriptases are characteristic of infection by RNA viruses.

96. (E)
In eucaryotes, the group of codons necessary to produce one polypeptide is called a cistron both in mRNA and DNA. In DNA, the cistron is also referred to as the "structural gene," because it determines protein structure.

97. (E)
An equal number of PD and NPD tetrads indicates that two markers are unlinked; the 1:1 ratio reflects the two equally probable alignments of nonhomologous chromosomes at metaphase I. A tetratype results from a single (2-strand) crossover in which one of two unlinked markers is affected. A vastly greater number of parental ditypes than nonparental ditypes is convincing evidence for linkage; in the absence of crossovers between the positions of the two markers on the chromosome, parental ditypes are formed.

98. (A)

Let Bu* represent the excited state and tautomer and consider the following pattern:

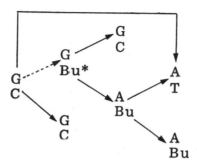

Note that a GC base pair has been replaced by an AT pair; the dotted arrow illustrates the mistaken incorporation of Bu instead of C.

99. (D)

Amylose is long and unbranched with D-glucose in alpha 1-4 linkages. Amylopectin is highly branched with D-glucose in 1-6 linkages. Amylose tests blue in the presence of iodine by forming micelles in water.

100. (A)

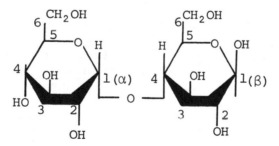

O-α-D-glucopyranosyl - (1 → 4)-β-D-glucopyranose

B is β anomer of lactose
(O-β-D-galactopyranosyl - (1 → 4)-β-D-glucopyranose)

C is sucrose (O-β-D-fructofuranosyl - (2 → 1)-α-D-glucopyranoside)

D and E are, respectively, the α anomers of maltose and lactose.

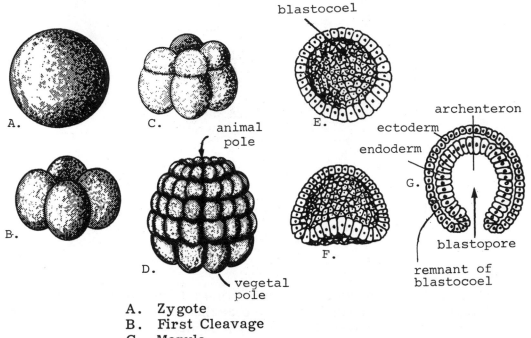

A. Zygote
B. First Cleavage
C. Morula
D. Blastula
E. Longitudinal section through a blastula
F. Longitudinal section through early gastrula
G. Longitudinal section through late gastrula

Note the following characteristics: The early cleavage stages occur without cytoplasmic growth, resulting in cell clusters no larger than original zygote.

Early cleavages are oriented in such fashion that the embryo is composed of an animal and a vegetal hemisphere, with the cells in the former being noticeably smaller. Animals with greater amounts of yolk in the eggs show greater variation in size than amphioxus. In addition the more yolk present, the more the pattern of embryological development diverges from amphioxus. The first cleavages produce a grapelike cell cluster known as the morula. Subsequent cleavages result in a blastula stage. The blastula is a semispherical aggregate of cells (blastomeres) which secrete fluids into a central cavity, the blastocoel.

Gastrulation typically begins when the blastula is composed of approximately 500 cells. The process begins with invagination at the vegetal pole and continues as more cells move to the point of invagination and fold inward, enlarging the invagination. Subsequently the invaginated cell layer comes to lie almost flush against the surface layer, filling in the old blastocoel and forming a 2-tiered cup-like structure. The interior of the cup forms the archenteron and opens to the outside through the blastopore.

106 - 108

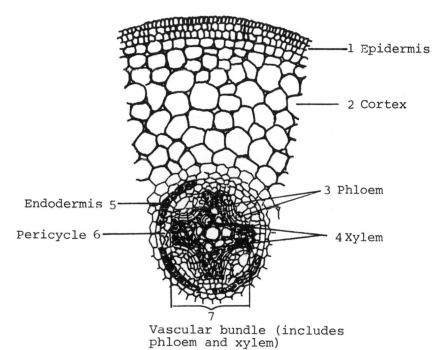

Vascular bundle (includes phloem and xylem)

106. (A)
The use of the term "exclusively" eliminates D as a possibility; the phloem cells intercalate between the arms of x-shaped core of xylem.

107. (C)
The description corresponds to the pericycle.

108. (D)
The Casparian strip is the waterproof band that comprises the radial and end walls of endodermal cells.

	Dicots	Monocots
No. of seed leaves (cotyledons)	2	1
venation of leaves	Pinnate, palmate, i.e. branched network	Parallel
type of root system	Woody, taproot- i.e. large primary root with branch roots growing from it	fibrous, no taproot, all roots approximately equal in size
arrangement of vasculature in stem	Vascular bundles in a single ring	Vascular bundles irregularly distributed through pith tissue
cambium/ growth rings	New growth ring each year or growing season	Stem and root devoid of cambium, no growth rings
arrangement of flower parts	Arranged in twos, fours or most often fives	Arranged in threes

Question 109 listed exclusive traits of monocots, while 110 dealt entirely with dicot features.

111.   (B)

If one considers the complete breakdown of glucose to CO_2 and water through glycolysis, Krebs cycle and respiratory electron transport, one mole of glucose corresponds to 36 ATP equivalents. Furthermore, since glucose is stored in polymer, the energy contained in many thousands of phosphate bonds is osmotically equivalent to a single molecule. If energy was stored as ATP the cells would become so hypotonic that cell walls would have to be phenomenally strong to resist bursting.

Also, the acid-base equilibrium of the cell would be altered, since ATP is an acid (glucose is neutral). Finally, phosphorus would have to be more plentiful, because one glucose molecule corresponds to 34-36 ATP molecules. The body would have to store over one hundred phosphorus atoms for every 1 glucose molecule it stores now.

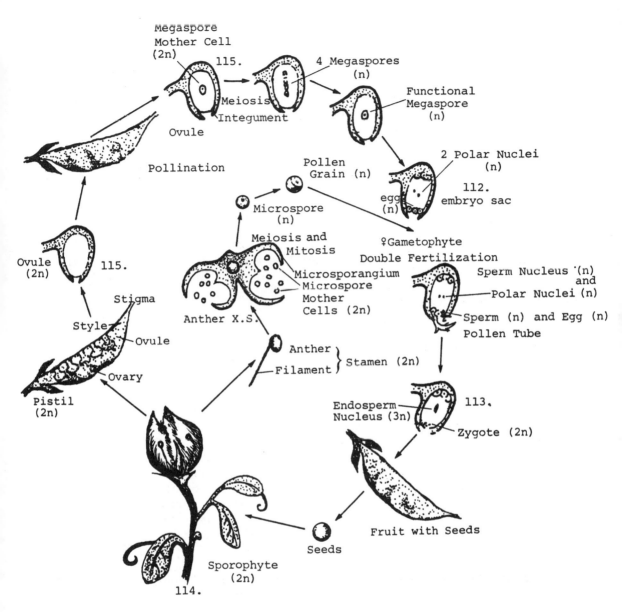

112.   (C)
The femal gametophyte is the embryo sac.

113.   (D)
    The endosperm nucleus indicated is the result of the combination of the diploid fusion nucleus (which arises from the joining of the 2 polar nuclei) with the (second) sperm. The result is thus a triploid nucleus, and after a series of divisions, a triploid endosperm tissue. The other sperm, of course, fertilizes the egg (double fertilization).

114.   (A)
Many flowering plants illustrate the phenomenon of sporophyte dominance - it is the diploid stage that we are customarily aware of as the flowering plant, while the male and female gametophytes are closer to being reproductive intermediates with true independent stages. Sporophyll is derived from the Greek phyllon (leaf) and refers to a leaf modified to bear spores; sporangium is a more general term for any spore producing plant structure.

115.   (A)
The structure indicated is the ovule. If named as a sporangium it would correspond to a megasporangium (and the term meiosis on the figure should indicate a diploid structure).

116.   (E)
Recall that the signal hypothesis proposes that there is a hydrophobic signal sequence at the N-terminus of the nascent chain, that this sequence is cleaved as the product crosses the ER membrane and that proteins must be translocated as they are being translated. Thus if the signal hypothesis were universally valid all transmembrane proteins would have their N-terminus at the E face and their C-terminus at the P face. Bacteriorhodopsin traverses the membrane 7 times, which does not conform to the hypothesis; albumen is another clear example where the hypothesis does not work. Only IV, which indicates that proteins cross the membrane as they are translated, supports the signal hypothesis.

117.   (A)
By far the greatest number of species is to be found in the littoral zone (defined as extending from the beach to a point where the water is sufficiently deep so that it is not completely stirred by waves and/or tides). The zone is characterized by high primary productivity and wide variability in temperature, lighting, salinity and turbulence.

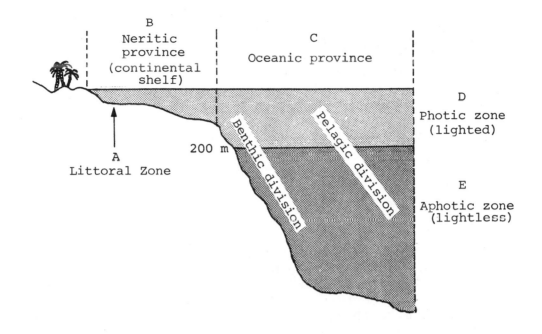

B
Neritic province (continental shelf)

C
Oceanic province

D
Photic zone (lighted)

200 m

Benthic division

Pelagic division

A
Littoral Zone

E
Aphotic zone (lightless)

118. (A)
The danger in ethylene glycol poisoning arises not directly from ethylene glycol but from an oxidation product, oxalic acid.

This oxidation is catalyzed by a dehydrogenase which can also use ethanol as a substrate. Ethanol may therefore competitively inhibit the formation of oxalate.

119. (D)
Since most species exhibit some degree of geographic variation, it is unlikely that a geographic barrier would separate a species into two identical gene pools. Therefore, it is likely that pre-existing variations in gene frequency play a role in the divergence of the two groups.

Mutations, being random, will also cause the two groups to diverge.

Separated populations by definition occupy different ranges; hence different selection pressures are to be expected.

Founder effects, as a form of genetic drift, necessarily pertain to small populations.

120.   (C)

Individuals A and B must be in the same species if A is to be able to produce progeny which after many successive generations may produce viable progeny via individual B. If A and B were genetically dissimilar to the point of being in different species, then future compatible, viable offspring would not be possible. "Species" is sometimes defined as a set of populations among which gene flow can occur.

121.   (B)

As indicated in (B), balanced growth refers to a period where metabolic and reproductive functions are operating at their peak. In fact, such balanced growth is characteristic of a log phase in a standard growth curve.

The term synchronous growth refers to cells being at the same point in their cell cycle; as such it is more commonly associated with eukaryotic cells. Note that such growth is an experimental requirement in a wide variety of cases.

122.   (A)

Recall the Henderson-Hasselbalch equation

$$pH = pk_a + \log \frac{[A^-]}{[HA]} = pk_a + \log \frac{[salt]}{[acid]}$$

In the present case, the conjugate base is the neutral $R-NH_2$ and the conjugate acid is $R-NH_3^+$. Plugging in the values given for pH and pk_a, taking antilogs and rearranging we obtain

$$[R-NH_3^+] = 10^{\frac{1}{2}}[R-NH_2]$$

fraction ionized is given by:

$$\frac{10^{\frac{1}{2}}[R-NH_2]}{[R-NH_2] + 10^{\frac{1}{2}}[R-NH_2]} = \frac{10^{\frac{1}{2}}}{(10^{\frac{1}{2}} + 1)}$$

123.   (A)

Fish retain a primitive type of heart, with one atrium and one ventricle. Thus blood is routed directly from the gills to the tissues in fish.

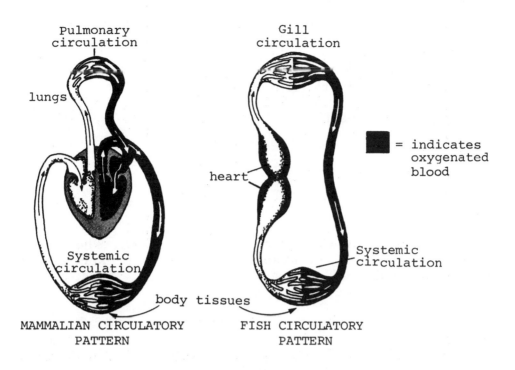

Pulmonary circulation

lungs

Gill circulation

heart

Systemic circulation

body tissues

MAMMALIAN CIRCULATORY PATTERN

Systemic circulation

FISH CIRCULATORY PATTERN

■ = indicates oxygenated blood

124. (E)
Phosphate bond energy is required to establish the underlying electrochemical gradients, but the actual conduction of an impulse is an endergonic process. Sodium channels are voltage gated. The density of sodium channels is higher at the nodes of Ranvier than at any other region of myelinated or unmyelinated fibers. The conductance of sodium channels decreases with decreasing pH; graphs of relative permeability vs. pH actually resemble a titration curve for an acid with a pk_a of 5.2 This suggests the presence of a carboxylate moiety in the active conformation of the channel. It is hypothesized that the selectivity of the channel is due to its small size and the proximity of negative charge.

125. (A)
Recall that phloem serves to transport nutrients from leaves to the rest of the plant, and not to transport water.

126. (C)
In the epidermis each stroma is limited by two epidermal cells called guard cells. These cells have walls of unequal

thickness, the walls next to the stroma are much thicker than the walls on the side away from the stroma. When the guard cells contain much fluid and are turgid, the thin outer walls buckle outward pulling the rest of the cell with it and thus opening the stroma. A waxy cuticle lining the epidermis of leaves, the covering of the lower epidermis with root hairs all make gas exchange more difficult in plants.

127.   (C)
     Roots play a vital role in the absorption and transport of not only gases, but water and nutrients as well. The roots' surfaces are readily permeable; therefore, no specialized structures are needed.

128. (A)   129. (E)   130. (C)   131. (C)

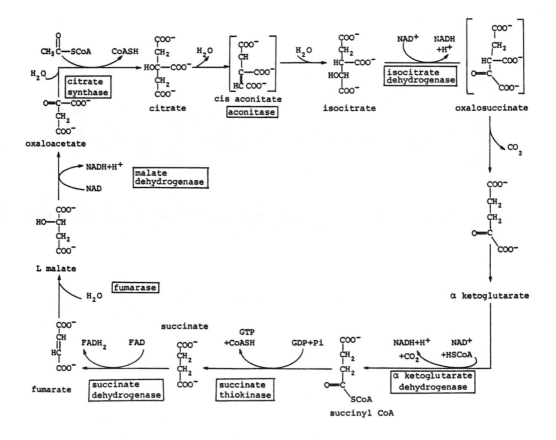

The only irreversible reactions within the cycle are the addition of acetyl-COA and water to oxaloacetate ($\Delta G' = -9$ kcal/mol) and the reaction producing succinyl COA ($\Delta G' - -8$ kcal/mol).

The figure below illustrates, schematically, a type of 3 point binding of citrate which could lead to, among other effects, asymmetric distributions of radioactive labels in TCA intermediates obtained from incubating actively respiring tissue with $CH_3^{14}COOH$.

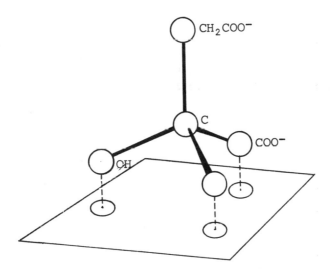

132. (B)

Lamarck believed that evolution occurred through physical changes during the lives of individuals who struggled to adapt to their environments. This theory is now known to be wrong. We now know that DNA directs the formation of the phenotype, and that DNA does not accept instruction back from the phenotype. Lamarck's theory has been replaced by Darwin's theory of natural selection, which proposes that changes in the phenotype of a certain type of organism occur as individuals with characteristics which favor survival pass those characteristics on to successive generations, while individuals with characteristics less suited to survival are not as successful at passing on their genes; hence, characteristics which are less suited to survival fade out and eventually disappear.

133. (C)

Highly developed animals can react extremely quickly to certain types of stimuli. This can be important when the animal comes into contact with harmful stimuli, such as fire or sharp objects piercing the skin. The faster the animal responds, the less damage is done.

These kinds of responses do not involve the brain, for conscious input is unnecessary and would slow the response. Instead, the impulse is carried from the receptor neuron directly to the spinal cord and then back out through a motor neuron to the appropriate muscle. Thus, the receptor neuron, the motor neuron, and the muscle are the only structures necessary for the reflex action to occur.

134.   (E)
There is a time lag or delay between the point of nerve stimulation and when it begins to fire an impulse. There is also a latent period between the point at which a skeletal muscle is stimulated and when it begins to contract.

135.   (B)
When a nerve impulse reaches the synaptic cleft, neurotransmitters diffuse across the cleft and contact receptor sites on the next neuron. If enough combining sites are filled, an action potential will be created. None of the membranes (presynaptic or postsynaptic) are destroyed in this process.

136. (C) 137. (A) 138. (E) 139. (D) 140. (B)

5-Bromouracil is a thymine analogue; it mispairs with guanine and can cause transitions in both directions between GC and AT base pairs. Nitrous acid is a deaminating agent, transforming cytosine to uracil (and adenine to hypoxanthine). The predominant mutagenic effects are again GC ⇌ AT transitions. Acridines, of the type illustrated below, function as intercalating agents. These complex with a portion of a DNA chain and mask it, thus giving rise to frameshift mutations.

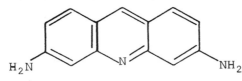

Proflavin, a simple acridine

Ultraviolet light primarily causes dimerization of adjacent pyrimidines; since the dimers formed cannot fit into the double helix, replication and/or transcription are blocked until the defect is repaired.

This may occur through excision repair, which involves the following: sequential nicking by a UV-specific endonuclease, DNA synthesis by a polymerase, and 5'→3' excision by that same polymerase followed by DNA ligase-catalysed sealing of the nick. Alternatively, most cells contain a photoreactivating enzyme which, upon absorption of blue light, can recognize a pyrimidine dimer and catalyse its cleavage back to monomers.

N-nitroso compounds are among the most potent alkylating agents known. Guanine is particularly susceptible to these reagents, whose predominant mutagenic effects are GC → AT transitions.

141. (A)   142. (E)   143. (D)   144. (B)   145. (C)

A nonsense mutation refers to an alteration of the base sequence such that a terminator codon (UAG, UAA or UGA) appears at an internal point in a structural gene. This is a much more serious lesion than a missense mutation because the chances of obtaining a functional polypeptide in spite of premature termination are minimal. Several strains of E. coli have been isolated with amber suppressor mutations (i.e. a t-RNA is specified which misreads the termination codon and inserts an amino acid, resulting in a translated product with at least partial activity).

As the above suggests, the converse type of point mutation is a missense mutation which causes an alternative amino acid to be incorporated into the peptide. If the residue is not at a critical site in the product, and if the substituted amino acid is simlar to the orginal (such as Asp incorporation for Glu) then protein function should show little or no impairment.

Wobble refers to the flexibility that exists in codon/anticodon recognition at the third codon base and results in the capacity of some tRNA molecules to bind several distinct codons. The pattern of degeneracies in the genetic code is such that XYU and XYC always specify the same amino acid, whereas XYA and XYG are not as consistent; this led Crick to hypothesize that the steric constraints on pairing of the third (3' codon, 5' anticodon) base would be less rigorous than for the first two. This is illustrated in the diagram:

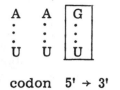

anticodon 3' ← 5'

$$\begin{array}{ccc} A & A & \boxed{G} \\ \vdots & \vdots & \vdots \\ U & U & \boxed{U} \end{array}$$

codon 5' → 3'

Competence refers to the ability of cells to take up fragments of naked DNA from the medium. If, for example, linkage or mapping in E. coli is to be studied via transformation experiments the cells are treated with concentrated $CaCl_2$ to make their membranes permeable to DNA (the cells must also be devoid of exonuclease I and V to function).

An organism is a genetic mosaic if it contains cells of more than one genotype. Lyarization is the facultative heterochromatization undergone by one X chromosome in female mammalian cells. The process is universal insofar as it occurs in all cells aside from those making up the germ line and, more importantly for the purposes of this question, the choice of whether the maternal or paternal X chromosome is to be inactivated is made at random.

146. (A)   147. (C)   148. (E)   149. (B)   150. (C)

The absorption of dietary calcium is primarily regulated by calciferol. The latter is derived cholecalciferol, a fat-soluble vitamin (vitamin D).

Calcitonin, a peptide secreted by extrafollicular cells in the thyroid gland, is released in greatest amounts around the time of food intake. It promotes transfer of calcium and phosphate from blood to bone. Calcitonin seems to exert its function by decreasing the reabsorption of calcium and phosphate.

Parathyroid hormone is released when the plasma concentration of Ca^{++} falls below a threshold level, travels to bone and other sites and stimulates increased transfer of calcium and phosphate from bone to blood. In this sense, PTH and calcitonin form an agonist/antagonist pair. PTH also acts on the kidneys to increase the rate of excretion of phosphate; thus the net result of PTH is to increase plasma levels of calcium while decreasing phosphate concentrations.

Angiotensin II is an octapeptide which exerts its effects by stimulating cells of the adrenal cortex to increase secretion of aldosterone. This promotes increased salt and water retention. Angiotensin II also stimulates contraction of smooth muscle cells lining the blood vessels; when coupled with the aldosterone mediated rise in blood volume the net result is an increase in blood pressure.

Renin is a proteolytic enzyme released by the juxtaglomerular cells of the kidney in response to decreases in blood pressure, blood volume, or sodium levels. Its substrate is produced in the liver and the cleavage product is the decapeptide Angiotensin I. Angiotensin II is formed by the action of a lung enzyme in cleaving two further residues.

151. (A) 152. (C) 153. (E) 154. (E) 155. (D)

In the early region of the SV-40 genome (so named for its being the first portion to be transcribed) there is a length of DNA which codes for 2 products: t antigen and T antigen. The latter is a protein kinase which activates the replication of viral DNA by host cell machinery. The former becomes associated with the cell membrane and may be involved in the loss of contact inhibition. Both are thought to be required for the establishment and maintenance of the transformed state.

Specialized transduction is that in which fragments of chromosomal DNA break off with viral DNA when the prophage is induced to come out of its lysogenic state. The included host DNA will then be transferred along with the viral genome; the transfer of genetic information is specialized insofar as only sequences close to phage attachment site can be transduced.

The converse to specialized transduction, generalized transduction, occurs by an entirely different mechanism. Recall that viral DNA is often replicated in the form of long concatamers; the latter is then spliced into fragments of the appropriate length and packaged into a capsid. Transducing particles are formed when pieces of host DNA, generally of the same length as the phage chromosome, are mistakenly packaged into a phage coat and subsequently released with the phage progeny when the cell lyses. Note that such a transducing particle contains only host specific DNA instead of bacterial and viral DNA as in specialized transduction; note also the lack of constraints on gene loci transported - hence the apellation "generalized".

The src gene is a component of the RNA oncovirus ASV (Avian Sarcoma Virus). A protein kinase, it is significant as the sole proteinaceous product found in cells transformed by ASV but not in cells which are ASV infected and nontransformed. Hence, expression of this gene seems to be the critical step in the oncogenic transformation mediated by ASV.

Burkitt's lymphoma is a cancer found in African children. It is thought to be caused by the Epstein-Barr virus, which is a type of herpes virus. E.B. virus is also implicated in mononucleosis and nasopharyngeal carcinomas.

156. (B) 157. (C) 158. (D) 159. (A) 160. (E)

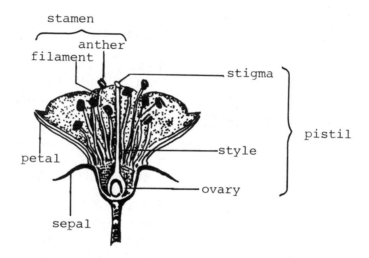

The stamen comprises the male reproductive organs (sporophylls), of which the 2 primary structures are the stalk-like filament and the terminal pollen producing structure known as the anther.

The female reproductive organs correspond to the pistil, which, as illustrated, is composed of the ovary, the style and the stigma. The embryo sac is the female gametophyte; the pollen grain, of course, corresponds to the male gametophyte.

The calyx is the structure made up of all the sepals (corolla is the term for the corresponding petal structure).

161. (C) 162. (A) 163. (E) 164. (B) 165. (D)

Leaky membranes are associated with obligate intracellular parasites such as rickettsias and chlamydias. Rickettsias may occur as short rods, cocci or pleomorphic coccobacillary forms; they are gram negative, non-motile and do not produce spores. Because ATP, NADH and other compounds seep through their cell membranes, they tend to be difficult to culture out of tissue. Arthropods provide the natural habitat for rickettsias. They are generally nonpathogenic to humans but can cause Rocky Mountain Spotted Fever and epidemic typhus.

Mycoplasmas are unique among bacteria. They are the smallest free-living organisms (approximately 3 μM in diameter) and are filterable. They are able to squeeze through the pores of the filter because they (uniquely) lack cell walls.

Chlamydias are obligate intracellular parasites which cannot grow outside a host cell because they lack the necessary enzymes for generating compounds needed for energy transfers. In this aspect they are similar to rickettsias; they however differ from rickettsias by the fact that they exist as dense spore-like cells called elementary bodies outside of the host cell.

Bacteria of the genus Clostridium are endospore forming rods, generally gram positive, obligate anaerobes. Species which are pathogenic to humans and conform to the above description include Cl. tetani, Cl. botulinum and Cl. septicum (causes gas gangrene in humans).

Finally, Escherichia are gram negative facultative anaerobic rods that are peritrichously flagellated (i.e. have numerous flagellae along the sides).

166. (E) 167. (C) 168. (D) 169. (A) 170. (B)

Colchicine and vinblastine are both specific inhibitors of mitosis and both exert their effects by interacting with tubulin. Colchicine binds tubulin dimers and forms colchicine-tubulin dimers that associate with the (+) end of growing microtubule structures (i.e. at the fast growing, fast disassembling end). This association prevents the continued formation of microtubules; thus spindle fibers are not formed and mitosis is arrested at the metaphase stage. More stable microtubule structures, such as those in flagellae, do not seem to be affected.

Vinblastine arrests mitosis by the crystallization of tubulin, as mentioned (i.e. problem 170) and so prevents formation of spindle microtubules.

Cytocholasin B is a specific inhibitor of cytokinesis and has no effect upon mitosis. Eukaryotic cells incubated with the alkaloid show altered shapes and inhibited motility. The effects are produced by a specific interaction with one end of actin microfilaments which prevents further assembly of actin filaments.

Sodium azide (NaN_3) is a metabolic poison which inhibits respiratory ATP synthesis. It prevents energy produced by oxidation of Krebs cycle intermediates from being used to drive the phosphorylation of ADP.

The presence of vanadaterons disrupts numerous processes that require ATPase activity (such as sodium-potassium pumping). The pentavalent vanadium ion binds a bipyramidal array of ligands in a conformation similar to that around a phosphorus atom during the hydrolysis of a phosphate bond.

171. (E) 172. (B) 173. (A) 174. (C) 175. (D)

Taiga (also known as boreal forest) comprises a wide zone in North America and Eurasia, south of the tundra. It is characterised by large coniferous forests interspersed with small bodies of fresh water.

Although the subsoil is frozen for much of the year and the winters are quite cold, the subsoil thaws during the summers during which time there is an abundance of vegegation.

In contrast to the above, the primary characteristic of tundra is the permanently frozen subsoil. The word "tundra" apparently is derived from a Siberian expression for "North of the timberline," and trees are in fact relatively rare. The dominant vegetation consists of mosses and grasses.

Deciduous forests cover large areas of the Eastern U.S. and are characterized by summers that are warmer and longer than those of the taiga as well as abundant rainfall.

Grasslands occur both in temperate and tropical zones, and are typically regions where relatively low annual rainfall or seasonally uneven rainfall prevent the establishment of forests but allow for abundant grasses. Those grasslands in tropical zones (savannas) are characterized by wet-dry cycles as opposed to the annual temperature cycles of those in the temperate zones.

Tropical rain forests contain tall, closely-packed trees whose foliage forms continuous canopies which absorb large quantities of sunlight and leave the forest floor relatively dark throughout the day. These trees catch rain during storms and allow the water to continue to percolate down to the ground long after the rain has ceased. Furthermore, lower levels are shielded from wind, which results in a decreased rate of evaporation and leaves the forest floor more humid than the upper levels. All these factors lead to an abundant variety of life forms found at different elevations.

176. (A) 177. (B) 178. (D) 179. (C)

"Cryptic coloration" refers to the similarity in appearance between certain organisms and their natural environment. This similarity enhances the organisms' survival by making it difficult for predators to see them.

A classic example of cryptic coloration occurred in the 19th century in Manchester, England, where a species of moth composed almost entirely of white individuals evolved into a species composed almost entirely of black individuals. This change in color, which was called industrial melanism, was brought about by the industrial revolution, in which Manchester was centrally involved. Before the industrial buildup, the trees around Manchester were covered with spotty white lichens; hence, spotty white moths were difficult to see on the trees, and therefore had a better chance of escaping death from the moths' main predators – the birds. Black moths, on the other hand, stood out conspicuously against the lichens and therefore had a lesser chance of survival. A black moth was considered a rare and valuable catch to the avid British moth collector.

With the industrial revolution came profuse pollution, and the trees, like everything else, became covered with dark soot. This gave the black moth the survival advantage, and eventually it achieved a predominant frequency of 98% in Manchester and its immediate surrounding areas. In rural areas, the white moth remained predominant. Note that this evolutionary process of cryptic coloration follows precisely the principles of Darwin's natural selection.

Aposematic appearance is the highly noticeable, flashy appearance of certain organisms which possess qualities noxious to predators. While most organisms wish to hide from predators, these noxious organisms rely upon their unpleasant characteristics being remembered and subsequently avoided by predators. For this reason, it is advantageous for the species as a whole to possess physical characteristics which are easily distinguished from the appearance of animals without the same noxious qualities.

Batesian mimicry is the adaptation of an innocuous species to the appearance of a similar but noxious species. This is highly advantageous to the innoxious organism, for it escapes the predation of those animals which have learned to avoid the noxious species.

Müllerian mimicry is the mutually beneficial evolution of a similar appearance by two or more species possessing qualities which are noxious to their predators. Each of the species involved benefits because the predator has only one avoidance response to learn, instead of two or more. Each species is afforded greater protection from predators because the repellent qualities are more frequently advertised.

180. (B) 181. (C) 182. (D) 183. (A)

The synonymous terms "subspecies" and "race" raise a certain amount of controversy. They are sometimes applied to geographically isolated populations which, despite apparent genetic divergencies, are nevertheless able to interbreed. They are also used to designate populations which occur on different sides of a sharp break in a cline. For example, if stem length in some species of flower varies gradually over distance, but the flowers on one side of a mountain range have an average stem length significantly greater than those on the other side, then the flowers on different sides of the mountain may be considered two subspecies.

A basic definition which incorporates both usages: subspecies (or races) are populations within a species which possess genetic differences and are to some extent isolated from each other reproductively because they occupy different ranges; nevertheless, they are capable of interbreeding.

184. (A) 185. (D) 186. (C) 187. (A) 188. (E)

Convergent evolution denotes a situation whereby two groups which are not closely related develop similar characteristics in response to similar selection pressures. As an example, note the similarities between Australian marsupials and unrelated placental mammals. They are similar because their environments are similar.

Convergent evolution is equally well exemplified by the development of flippers by marine mammals such as the whale. Fish also employ flippers to maneuver through the water but are not closely related to any mammal.

Contrasting terms to convergent evolution are divergent evolution and parallel evolution. In the latter, two related species evolve in similar ways over an extended period of time. Divergent evolution denotes a splitting of one group into two groups which continually become more unlike over time.

Character displacement describes the tendency for closely related sympatric species to develop in such fashion as to decrease both the competition between them and the possibilities of interbreeding. This occurs because of the adaptive advantages of reducing competition for a single niche and because of selective hybrid elimination: if two closely related sympatric species produce hybrids as well as or better adapted than themselves, then the original populations will soon become indistinct and will not properly represent different species. If, on the other hand, the hybrids are less well adapted than the parental populations they (the hybrids) will die out, and any genetic factors in the parental population that promote correct mate selection will increase in frequency at the expense of those promoting incorrect selection.

Allopolyploidy denotes a multiplication in chromosome number in an intraspecific hybrid. If, for example, two diploid individuals in distinct species produce a hybrid which (through a process of nondisjunction, for example) has its complement of chromosomes doubled, then the latter will have received a full diploid set from each parent. Thus there is no requirement for pairing of chromosomes from different species in meiosis, and the hybrid is thus able to produce normal gametes. In all the above, there is no requirement for allopatry. Note that this is an important and frequent type of speciation among plants.

Remember that allopatric refers to groups with different ranges; sympatric denotes populations with the same range.

189. (A) 190. (B) 191. (D) 192. (B)

Briefly, the allosteric effects shown by hemoglobin are thought to depend on an equilibrium between two quaternary structures: tense and relaxed. The former is constrained by the presence of 8 salt links between the 4 subunits and a low oxygen affinity; in the latter the salt links are absent and the oxygen affinity is higher.

The two forms are thought to be in equilibrium; allosteric modulators exert their effect by shifting this equilibrium to one side or the other. The presence of 2,3-diphosphoglycerate tends to decrease the oxygen affinity of hemoglobin by binding only to deoxyhemoglobin (in fact oxygenation and 2,3-DPG binding are to a first approximation mutually exclusive, the latter being released from hemoglobin upon oxygenation). Protons and CO_2 are also involved in the formation of the salt links characterizing the T form; increased acidity and CO_2 concentrations favor decreased oxygen binding.

193.   (C)
In operant conditioning, reinforcers follow the behavioral response. It is based on the fact that if a pattern is rewarded, the probability of that pattern reappearing is increased. In this example, the rat continues to press the lever because of the reward in the form of food.

194.   (D)
Autonomic learning is the ability of the body to learn to control bodily functions which are generally regulated by the autonomic nervous system. This is also called "visceral learning."

195.   (B)
Pavlov's experiments are very well known. The kind of learning illustrated in this example is called "classical conditioning" because the dogs' salivation became conditional on the light. The light became associated with the original stimulus – the meat – and eventually became the stimulus itself.

196.   (A)
Habituation is, in a sense, learning not to respond to a stimulus. While it is generally beneficial for an animal to be wary and to respond to any stimulus which presents a possible threat, the animal would waste a great deal of time and energy if it continued to respond to stimuli which consistently did not attack, threaten or compete. Thus the animal learns to ignore stimuli which has proved to be harmless by experience and instead reserves his energy to respond to extraordinary stimuli which may pose real danger.

	DMEM	TG	HAT	Budr	Ar(Hx*)
N	100	0	100	0	100
MLN	100	50	50	0	50
LN ♂	100	100	0	0	///////
TK⁻	100	0	0	100	///////

197. (C) 198. (A) 199. (A) 200. (A) 201. (A) 202. (E)

Two genes are involved: the gene for hypoxanthine guanine phosphoribosyl transferase (HGPRT) and for thymidine kinase (TK).

HGPRT is necessary for salvage synthesis of purines, catalysing the coupling of free purines with phosphoribosyl pyrophosphate (PRPP). The lack of HGPRT activity in human (males) is an inborn error of metabolism referred to as Lesch-Nyhan syndrome. As may be inferred from the sex linkage, the locus for the HGPRT gene is on the X chromosome. A Lesch-Nyhan patient is hemizygous for HGPRT and lacks the functional product; his mother doesn't show symptoms but her cells are a mosaic showing only half HGPRT activity.

The second gene, thymidine kinase, catalyses the phosphorylation of deoxythymidine to form TMP. Note that 5-bromodeoxyuracil is also a substrate for TK, and thioguanine is also a substrate for HGPRT. In both cases the incorporation of sufficient quantities of either Budr or TG is fatal to cells.

We come now to the constituents of the media used. Aminopterin blocks the de novo synthesis of purine nucleotides, IMP is not formed and salvage synthesis becomes the sole source of purines. Cells may survive if and only if hypoxanthine is added to the medium and they have functional HGPRT activity. Aminopterin also inhibits dihydrofolate reductase (DHFR), the result is that tetrahydrofolate cannot be regenerated and the de novo synthesis of deoxythymidine is also blocked. Cells exposed to aminopterin can survive only if they have functional thymidine kinase activity and if thymidine is added to the medium.

The results require no further background information. The normal cells (HGPRT$^+$, TK$^+$) grow in the control and are poisoned by the uptake of TG or Budr since both HGPRT and TK activity are present. The HGPRT activity causes them to survive in HAT and give positive results for uptake of HX*.

The MLN are TK$^+$ and therefore are killed in Budr. These are only half HPRT$^+$, therefore half survive in HAT medium and take up HX*, the same half is killed by the incorporation of thioguanine.

LN♂ cells die in HAT and survive in TG because of their lack of HGPRT (but are TK$^+$ and die in Budr medium). TK$^-$ cells do the exact opposite.

203. (A)
In the present context an indeterminate cleavage is one where the daughter cells produced retain their full spectrum of developmental potentialities. A determinate cleavage is one where asymmetric partitioning of cytoplasmic components limits the development of one or more daughter cells.

204. (B) 205. (A) 206. (C)

The results clearly indicate that preHPL is of greater molecular weight than HPL as the latter travels further in the gel than the former.

Clearly, some factor in the membrane of RER which is not present in SER is required for processing, as shown by the fact that only the unprocessed band appears in lane b. It is also obvious that the processing is dependent upon entry into the ER, as shown by lanes c and d.

However, the experiment shows nothing that indicates that the excised portion is a "hydrophobic, N-terminal sequence". The experiment supports the signal hypothesis more than the trigger hypothesis because it does show that some sequence is being excised and that translocation occurs during translation. The experiment does not attempt to show what factor(s) cause binding to be limited to RER.

If the experiment was repeated with mRNA specifying a protein that had an electrophoretic mobility similar to preHPL but was found in the cytoplasm, the protein would not be processed or inserted into the microsomes. Hence, single bands would be found in lanes a and b at the preHPL level but no protein would escape digestion in lanes c and d.

207. (D)
The (triauxic) growth pattern indicates that one fermentation pathway is constituitive, while the other two are adaptive. Since lactose is the disaccharide composed of glucose and galactose, fermentation of glucose or galactose must be constituitive, but both cannot be constituitive. If both were constituitive, there would be a maximum of one growth plateau (which would occur if lactose hydrolysis was adaptive). The presence of two lag phases indicates that lactose hydrolysis and at least one of the fermentation pathways from the monosaccharides are adaptive. Note that this sort of pattern would occur in an organism with constituitive glucose catabolism and inducible lactose and galactose utilization.

208. (A)
The Beer-Lambert Law relates the absorbance to the concentration of an absorbing species as

$$A = \varepsilon c L$$

so that if ε is given in liters mol^{-1} cm^{-1} and L in cm, c will be in moles per liter (A is dimensionless). Since ε and L are constants, the equation

$$\left(\frac{1}{\varepsilon L}\right)(A_2 - A_1)V$$

gives the change in the number of moles of absorbing species. Thus the expression below,

$$\frac{(1/\varepsilon L)(A_2 - A_1) \times 10^6/(t_2 - t_1)}{10^3 g}$$

in which $(t_2 - t_1)$ is given in minutes, gives the desired

$$\frac{\mu moles\ consumed/minute}{mg\ protein}$$

209. (A)
The polypeptide incorporated alanine in place of the usual

cysteine. The experiment is important because it demonstrates that codon recognition is mediated exclusively by the anticodon rather than by some interaction involving the activated amino acid.

210. (D)
The reaction is obviously bimolecular, with the 3' terminus of the growing chain reacting with a free nucleoside triphosphate. Furthermore, the reaction proceeds through nucleophilic attack by a 3' oxygen on the α phosphorus atom of the triphosphate.

GRE

BIOLOGY TEST

MODEL TEST III

THE GRADUATE RECORD EXAMINATION

BIOLOGY TEST

ANSWER SHEET

1. Ⓐ Ⓑ Ⓒ Ⓓ Ⓔ
2. Ⓐ Ⓑ Ⓒ Ⓓ Ⓔ
3. Ⓐ Ⓑ Ⓒ Ⓓ Ⓔ
4. Ⓐ Ⓑ Ⓒ Ⓓ Ⓔ
5. Ⓐ Ⓑ Ⓒ Ⓓ Ⓔ
6. Ⓐ Ⓑ Ⓒ Ⓓ Ⓔ
7. Ⓐ Ⓑ Ⓒ Ⓓ Ⓔ
8. Ⓐ Ⓑ Ⓒ Ⓓ Ⓔ
9. Ⓐ Ⓑ Ⓒ Ⓓ Ⓔ
10. Ⓐ Ⓑ Ⓒ Ⓓ Ⓔ
11. Ⓐ Ⓑ Ⓒ Ⓓ Ⓔ
12. Ⓐ Ⓑ Ⓒ Ⓓ Ⓔ
13. Ⓐ Ⓑ Ⓒ Ⓓ Ⓔ
14. Ⓐ Ⓑ Ⓒ Ⓓ Ⓔ
15. Ⓐ Ⓑ Ⓒ Ⓓ Ⓔ
16. Ⓐ Ⓑ Ⓒ Ⓓ Ⓔ
17. Ⓐ Ⓑ Ⓒ Ⓓ Ⓔ
18. Ⓐ Ⓑ Ⓒ Ⓓ Ⓔ
19. Ⓐ Ⓑ Ⓒ Ⓓ Ⓔ
20. Ⓐ Ⓑ Ⓒ Ⓓ Ⓔ
21. Ⓐ Ⓑ Ⓒ Ⓓ Ⓔ
22. Ⓐ Ⓑ Ⓒ Ⓓ Ⓔ
23. Ⓐ Ⓑ Ⓒ Ⓓ Ⓔ
24. Ⓐ Ⓑ Ⓒ Ⓓ Ⓔ

25. Ⓐ Ⓑ Ⓒ Ⓓ Ⓔ
26. Ⓐ Ⓑ Ⓒ Ⓓ Ⓔ
27. Ⓐ Ⓑ Ⓒ Ⓓ Ⓔ
28. Ⓐ Ⓑ Ⓒ Ⓓ Ⓔ
29. Ⓐ Ⓑ Ⓒ Ⓓ Ⓔ
30. Ⓐ Ⓑ Ⓒ Ⓓ Ⓔ
31. Ⓐ Ⓑ Ⓒ Ⓓ Ⓔ
32. Ⓐ Ⓑ Ⓒ Ⓓ Ⓔ
33. Ⓐ Ⓑ Ⓒ Ⓓ Ⓔ
34. Ⓐ Ⓑ Ⓒ Ⓓ Ⓔ
35. Ⓐ Ⓑ Ⓒ Ⓓ Ⓔ
36. Ⓐ Ⓑ Ⓒ Ⓓ Ⓔ
37. Ⓐ Ⓑ Ⓒ Ⓓ Ⓔ
38. Ⓐ Ⓑ Ⓒ Ⓓ Ⓔ
39. Ⓐ Ⓑ Ⓒ Ⓓ Ⓔ
40. Ⓐ Ⓑ Ⓒ Ⓓ Ⓔ
41. Ⓐ Ⓑ Ⓒ Ⓓ Ⓔ
42. Ⓐ Ⓑ Ⓒ Ⓓ Ⓔ
43. Ⓐ Ⓑ Ⓒ Ⓓ Ⓔ
44. Ⓐ Ⓑ Ⓒ Ⓓ Ⓔ
45. Ⓐ Ⓑ Ⓒ Ⓓ Ⓔ
46. Ⓐ Ⓑ Ⓒ Ⓓ Ⓔ
47. Ⓐ Ⓑ Ⓒ Ⓓ Ⓔ
48. Ⓐ Ⓑ Ⓒ Ⓓ Ⓔ

49. Ⓐ Ⓑ Ⓒ Ⓓ Ⓔ
50. Ⓐ Ⓑ Ⓒ Ⓓ Ⓔ
51. Ⓐ Ⓑ Ⓒ Ⓓ Ⓔ
52. Ⓐ Ⓑ Ⓒ Ⓓ Ⓔ
53. Ⓐ Ⓑ Ⓒ Ⓓ Ⓔ
54. Ⓐ Ⓑ Ⓒ Ⓓ Ⓔ
55. Ⓐ Ⓑ Ⓒ Ⓓ Ⓔ
56. Ⓐ Ⓑ Ⓒ Ⓓ Ⓔ
57. Ⓐ Ⓑ Ⓒ Ⓓ Ⓔ
58. Ⓐ Ⓑ Ⓒ Ⓓ Ⓔ
59. Ⓐ Ⓑ Ⓒ Ⓓ Ⓔ
60. Ⓐ Ⓑ Ⓒ Ⓓ Ⓔ
61. Ⓐ Ⓑ Ⓒ Ⓓ Ⓔ
62. Ⓐ Ⓑ Ⓒ Ⓓ Ⓔ
63. Ⓐ Ⓑ Ⓒ Ⓓ Ⓔ
64. Ⓐ Ⓑ Ⓒ Ⓓ Ⓔ
65. Ⓐ Ⓑ Ⓒ Ⓓ Ⓔ
66. Ⓐ Ⓑ Ⓒ Ⓓ Ⓔ
67. Ⓐ Ⓑ Ⓒ Ⓓ Ⓔ
68. Ⓐ Ⓑ Ⓒ Ⓓ Ⓔ
69. Ⓐ Ⓑ Ⓒ Ⓓ Ⓔ
70. Ⓐ Ⓑ Ⓒ Ⓓ Ⓔ
71. Ⓐ Ⓑ Ⓒ Ⓓ Ⓔ
72. Ⓐ Ⓑ Ⓒ Ⓓ Ⓔ

73. Ⓐ Ⓑ Ⓒ Ⓓ Ⓔ
74. Ⓐ Ⓑ Ⓒ Ⓓ Ⓔ
75. Ⓐ Ⓑ Ⓒ Ⓓ Ⓔ
76. Ⓐ Ⓑ Ⓒ Ⓓ Ⓔ
77. Ⓐ Ⓑ Ⓒ Ⓓ Ⓔ
78. Ⓐ Ⓑ Ⓒ Ⓓ Ⓔ
79. Ⓐ Ⓑ Ⓒ Ⓓ Ⓔ
80. Ⓐ Ⓑ Ⓒ Ⓓ Ⓔ
81. Ⓐ Ⓑ Ⓒ Ⓓ Ⓔ
82. Ⓐ Ⓑ Ⓒ Ⓓ Ⓔ
83. Ⓐ Ⓑ Ⓒ Ⓓ Ⓔ
84. Ⓐ Ⓑ Ⓒ Ⓓ Ⓔ
85. Ⓐ Ⓑ Ⓒ Ⓓ Ⓔ
86. Ⓐ Ⓑ Ⓒ Ⓓ Ⓔ
87. Ⓐ Ⓑ Ⓒ Ⓓ Ⓔ
88. Ⓐ Ⓑ Ⓒ Ⓓ Ⓔ
89. Ⓐ Ⓑ Ⓒ Ⓓ Ⓔ
90. Ⓐ Ⓑ Ⓒ Ⓓ Ⓔ
91. Ⓐ Ⓑ Ⓒ Ⓓ Ⓔ
92. Ⓐ Ⓑ Ⓒ Ⓓ Ⓔ
93. Ⓐ Ⓑ Ⓒ Ⓓ Ⓔ
94. Ⓐ Ⓑ Ⓒ Ⓓ Ⓔ
95. Ⓐ Ⓑ Ⓒ Ⓓ Ⓔ
96. Ⓐ Ⓑ Ⓒ Ⓓ Ⓔ

97. Ⓐ Ⓑ Ⓒ Ⓓ Ⓔ
98. Ⓐ Ⓑ Ⓒ Ⓓ Ⓔ
99. Ⓐ Ⓑ Ⓒ Ⓓ Ⓔ
100. Ⓐ Ⓑ Ⓒ Ⓓ Ⓔ
101. Ⓐ Ⓑ Ⓒ Ⓓ Ⓔ
102. Ⓐ Ⓑ Ⓒ Ⓓ Ⓔ
103. Ⓐ Ⓑ Ⓒ Ⓓ Ⓔ
104. Ⓐ Ⓑ Ⓒ Ⓓ Ⓔ
105. Ⓐ Ⓑ Ⓒ Ⓓ Ⓔ
106. Ⓐ Ⓑ Ⓒ Ⓓ Ⓔ
107. Ⓐ Ⓑ Ⓒ Ⓓ Ⓔ
108. Ⓐ Ⓑ Ⓒ Ⓓ Ⓔ
109. Ⓐ Ⓑ Ⓒ Ⓓ Ⓔ
110. Ⓐ Ⓑ Ⓒ Ⓓ Ⓔ
111. Ⓐ Ⓑ Ⓒ Ⓓ Ⓔ
112. Ⓐ Ⓑ Ⓒ Ⓓ Ⓔ
113. Ⓐ Ⓑ Ⓒ Ⓓ Ⓔ
114. Ⓐ Ⓑ Ⓒ Ⓓ Ⓔ
115. Ⓐ Ⓑ Ⓒ Ⓓ Ⓔ
116. Ⓐ Ⓑ Ⓒ Ⓓ Ⓔ
117. Ⓐ Ⓑ Ⓒ Ⓓ Ⓔ
118. Ⓐ Ⓑ Ⓒ Ⓓ Ⓔ
119. Ⓐ Ⓑ Ⓒ Ⓓ Ⓔ
120. Ⓐ Ⓑ Ⓒ Ⓓ Ⓔ

121. Ⓐ Ⓑ Ⓒ Ⓓ Ⓔ
122. Ⓐ Ⓑ Ⓒ Ⓓ Ⓔ
123. Ⓐ Ⓑ Ⓒ Ⓓ Ⓔ
124. Ⓐ Ⓑ Ⓒ Ⓓ Ⓔ
125. Ⓐ Ⓑ Ⓒ Ⓓ Ⓔ
126. Ⓐ Ⓑ Ⓒ Ⓓ Ⓔ
127. Ⓐ Ⓑ Ⓒ Ⓓ Ⓔ
128. Ⓐ Ⓑ Ⓒ Ⓓ Ⓔ
129. Ⓐ Ⓑ Ⓒ Ⓓ Ⓔ
130. Ⓐ Ⓑ Ⓒ Ⓓ Ⓔ
131. Ⓐ Ⓑ Ⓒ Ⓓ Ⓔ
132. Ⓐ Ⓑ Ⓒ Ⓓ Ⓔ
133. Ⓐ Ⓑ Ⓒ Ⓓ Ⓔ
134. Ⓐ Ⓑ Ⓒ Ⓓ Ⓔ
135. Ⓐ Ⓑ Ⓒ Ⓓ Ⓔ
136. Ⓐ Ⓑ Ⓒ Ⓓ Ⓔ
137. Ⓐ Ⓑ Ⓒ Ⓓ Ⓔ
138. Ⓐ Ⓑ Ⓒ Ⓓ Ⓔ
139. Ⓐ Ⓑ Ⓒ Ⓓ Ⓔ
140. Ⓐ Ⓑ Ⓒ Ⓓ Ⓔ
141. Ⓐ Ⓑ Ⓒ Ⓓ Ⓔ
142. Ⓐ Ⓑ Ⓒ Ⓓ Ⓔ
143. Ⓐ Ⓑ Ⓒ Ⓓ Ⓔ
144. Ⓐ Ⓑ Ⓒ Ⓓ Ⓔ

145. Ⓐ Ⓑ Ⓒ Ⓓ Ⓔ	167. Ⓐ Ⓑ Ⓒ Ⓓ Ⓔ	189. Ⓐ Ⓑ Ⓒ Ⓓ Ⓔ
146. Ⓐ Ⓑ Ⓒ Ⓓ Ⓔ	168. Ⓐ Ⓑ Ⓒ Ⓓ Ⓔ	190. Ⓐ Ⓑ Ⓒ Ⓓ Ⓔ
147. Ⓐ Ⓑ Ⓒ Ⓓ Ⓔ	169. Ⓐ Ⓑ Ⓒ Ⓓ Ⓔ	191. Ⓐ Ⓑ Ⓒ Ⓓ Ⓔ
148. Ⓐ Ⓑ Ⓒ Ⓓ Ⓔ	170. Ⓐ Ⓑ Ⓒ Ⓓ Ⓔ	192. Ⓐ Ⓑ Ⓒ Ⓓ Ⓔ
149. Ⓐ Ⓑ Ⓒ Ⓓ Ⓔ	171. Ⓐ Ⓑ Ⓒ Ⓓ Ⓔ	193. Ⓐ Ⓑ Ⓒ Ⓓ Ⓔ
150. Ⓐ Ⓑ Ⓒ Ⓓ Ⓔ	172. Ⓐ Ⓑ Ⓒ Ⓓ Ⓔ	194. Ⓐ Ⓑ Ⓒ Ⓓ Ⓔ
151. Ⓐ Ⓑ Ⓒ Ⓓ Ⓔ	173. Ⓐ Ⓑ Ⓒ Ⓓ Ⓔ	195. Ⓐ Ⓑ Ⓒ Ⓓ Ⓔ
152. Ⓐ Ⓑ Ⓒ Ⓓ Ⓔ	174. Ⓐ Ⓑ Ⓒ Ⓓ Ⓔ	196. Ⓐ Ⓑ Ⓒ Ⓓ Ⓔ
153. Ⓐ Ⓑ Ⓒ Ⓓ Ⓔ	175. Ⓐ Ⓑ Ⓒ Ⓓ Ⓔ	197. Ⓐ Ⓑ Ⓒ Ⓓ Ⓔ
154. Ⓐ Ⓑ Ⓒ Ⓓ Ⓔ	176. Ⓐ Ⓑ Ⓒ Ⓓ Ⓔ	198. Ⓐ Ⓑ Ⓒ Ⓓ Ⓔ
155. Ⓐ Ⓑ Ⓒ Ⓓ Ⓔ	177. Ⓐ Ⓑ Ⓒ Ⓓ Ⓔ	199. Ⓐ Ⓑ Ⓒ Ⓓ Ⓔ
156. Ⓐ Ⓑ Ⓒ Ⓓ Ⓔ	178. Ⓐ Ⓑ Ⓒ Ⓓ Ⓔ	200. Ⓐ Ⓑ Ⓒ Ⓓ Ⓔ
157. Ⓐ Ⓑ Ⓒ Ⓓ Ⓔ	179. Ⓐ Ⓑ Ⓒ Ⓓ Ⓔ	201. Ⓐ Ⓑ Ⓒ Ⓓ Ⓔ
158. Ⓐ Ⓑ Ⓒ Ⓓ Ⓔ	180. Ⓐ Ⓑ Ⓒ Ⓓ Ⓔ	202. Ⓐ Ⓑ Ⓒ Ⓓ Ⓔ
159. Ⓐ Ⓑ Ⓒ Ⓓ Ⓔ	181. Ⓐ Ⓑ Ⓒ Ⓓ Ⓔ	203. Ⓐ Ⓑ Ⓒ Ⓓ Ⓔ
160. Ⓐ Ⓑ Ⓒ Ⓓ Ⓔ	182. Ⓐ Ⓑ Ⓒ Ⓓ Ⓔ	204. Ⓐ Ⓑ Ⓒ Ⓓ Ⓔ
161. Ⓐ Ⓑ Ⓒ Ⓓ Ⓔ	183. Ⓐ Ⓑ Ⓒ Ⓓ Ⓔ	205. Ⓐ Ⓑ Ⓒ Ⓓ Ⓔ
162. Ⓐ Ⓑ Ⓒ Ⓓ Ⓔ	184. Ⓐ Ⓑ Ⓒ Ⓓ Ⓔ	206. Ⓐ Ⓑ Ⓒ Ⓓ Ⓔ
163. Ⓐ Ⓑ Ⓒ Ⓓ Ⓔ	185. Ⓐ Ⓑ Ⓒ Ⓓ Ⓔ	207. Ⓐ Ⓑ Ⓒ Ⓓ Ⓔ
164. Ⓐ Ⓑ Ⓒ Ⓓ Ⓔ	186. Ⓐ Ⓑ Ⓒ Ⓓ Ⓔ	208. Ⓐ Ⓑ Ⓒ Ⓓ Ⓔ
165. Ⓐ Ⓑ Ⓒ Ⓓ Ⓔ	187. Ⓐ Ⓑ Ⓒ Ⓓ Ⓔ	209. Ⓐ Ⓑ Ⓒ Ⓓ Ⓔ
166. Ⓐ Ⓑ Ⓒ Ⓓ Ⓔ	188. Ⓐ Ⓑ Ⓒ Ⓓ Ⓔ	210. Ⓐ Ⓑ Ⓒ Ⓓ Ⓔ

THE GRE BIOLOGY TEST

MODEL TEST III

Time: 170 Minutes
      210 Questions

DIRECTIONS: *Choose the best answer for each question and mark the letter of your selection on the corresponding answer sheet.*

1. Which of the following best describes the society of a social insect?

   (A) The members of the society are very industrious.

   (B) Each society cooperates with other societies.

   (C) There is a division of labor among the various members.

   (D) There exists a social hierarchy.

   (E) Dominance hierarchy is practiced.

2. When two or more traits produced by genes located on two or more different chromosome pairs are expressed independently, there has been

   (A) mutation.

   (B) independent segregation.

   (C) crossing over.

   (D) independent assortment.

   (E) cross linkage.

3. Which of the following is not characteristic of heterotrophic organisms?

(A) They obtain their energy from the oxidation of organic molecules.

(B) There is an interdependent relationship between them and photosynthetic organisms.

(C) Most heterotrophic organisms use aerobic respiration.

(D) They can manufacture their own food.

(E) They have well developed enzyme systems.

4. Increase in population density of most vertebrates results in

(A) more aggressive behavior.   (D) hoarding of food.

(B) increased cooperation be-   (E) more efficient distribu-
    tween individuals.   tion of food.

(C) increased care of the
    young.

5. During the proliferative phase of the female reproductive cycle there is increased secretion from pituitary of

(A) luteinizing hormone.   (D) estrogen.

(B) follicle-stimulating hormone. (E) adrenocorticotropic hor-
                                      mone.
(C) progesterone.

6. Phosphorylation refers to

(A) transfer of electrons to oxygen by different electron carriers.

(B) formation of ATP from ADP.

(C) excretion of phosphates by the kidney.

(D) breakdown of proteins.

(E) oxidation of carbohydrates.

7. Which of the following explains why evergreen trees do not have to shed their leaves in winter?

   (A) Low transpiration rates

   B) Conical shape

   (C) High transpiration rates

   (D) Only one species of trees in each forest

   (E) Great height

8. The medulla oblongata contains centers which regulate

   (A) respiration.

   (B) posture.

   (C) sleep.

   (D) appetite.

   (E) body temperature.

9. All of the following are involved in hydrogen transfer except

   (A) NAD (DPN).

   (B) FAD.

   (C) RNA.

   (D) oxygen.

   (E) NADP.

10. The use of the light microscope in the study of fine cellular structure is limited due to its

   (A) small size.

   (B) type of lenses.

   (C) difficulty of preparing materials.

   (D) lack of contrast.

   (E) relatively low power of resolution.

11. When certain bacteria are removed from the human gut, which of the following cannot be synthesized?

   (A) Riboflavin

   (B) Vitamin B_{12}

   (C) Thiamine

   (D) Glucosamine

   (E) Myosin

12. Which of the following is concerned mainly with cellular respiration?

(A) Cell membrane

(B) Golgi apparatus

(C) Ribosomes

(D) Rough endoplasmic reticulum

(E) Mitochondria

13. The force of attraction between non-polar (uncharged) molecules is referred to as

(A) covalent bond.

(B) hydrogen bond.

(C) van der Waals forces.

(D) ionic bond.

(E) molecular bond.

14. Which of the following is the most important monosaccharide?

(A) Glucose

(B) Fructose

(C) Sucrose

(D) Galactose

(E) Lactose

15. Lactose consists of

(A) two glucose molecules.

(B) one glucose molecule and one galactose molecule.

(C) one fructose molecule and one glucose molecule.

(D) one glucose molecule and one maltose molecule.

(E) one galactose molecule and one fructose molecule.

16. Which of the following cannot be digested by humans?

(A) Glycogen

(B) Amylose

(C) Starch

(D) Fructose

(E) Cellulose

17. Centrioles are found in the cells of all of the following except

(A) animals.

(D) higher plants.

(B) protists.

(E) lower plants.

(C) fungi.

18. All microtubules are made of a common protein. This protein is

(A) tubulin.

(B) myosin.

(C) pectin.

(D) actin.

(E) actinomyosin.

19. During protein synthesis, information is taken from the DNA to the ribosomes. Which of the following takes the information from the DNA?

(A) Ribosomal RNA (rRNA)

(D) Nucleotides

(B) Messenger RNA (mRNA)

(E) Endoplasmic reticulum

(C) Transfer RNA (tRNA)

20. Sometimes it is found that viruses can transfer genetic material from one bacterial strain to another. This process is called

(A) transduction.

(D) transmission.

(B) recombination.

(E) mutation.

(C) conjugation.

21. Certain strains of E. Coli are unable to metabolize lactose due to a deficiency of β-galactosidase, but when grown continuously on a medium containing lactose they are able to synthesize it. This is because

 (A) lactose is able to diffuse easily across the cell wall.

 (B) lactose suppresses the metabolism of other substrates.

 (C) at first E. Coli did not require lactose but later developed the need for it.

 (D) β-galactosidase is not responsible for lactose metabolism.

 (E) lactose acted as an enzyme inducer.

22. The kangaroo rat is able to live successfully in desert areas because it

 (A) has a small surface to volume ratio.

 (B) has very few predators.

 (C) is only active at night.

 (D) excretes a very concentrated urine.

 (E) lives close to waterholes.

23. The functional unit of the human excretory system is the

 (A) kidney.   (D) urinary bladder.

 (B) nephron.   (E) urethra.

 (C) glomerulus.

24. When the blood of an individual becomes hypertonic the kidney will then

 (A) secrete more antidiuretic hormone (ADH).

 (B) excrete more water.

 (C) decrease its rate of filtration.

 (D) excrete a smaller volume of, but more concentrated urine.

 (E) increase its rate of filtration.

Questions 25-30 refer to the following diagram of the human heart.

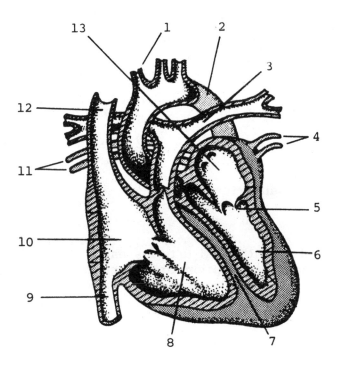

25. The pacemaker of the heart is found at

   (A) 7.   (D) 10.

   (B) 13.   (E) 8.

   (C) 5.

26. Number 5 represents the

   (A) tricuspid valve.   (D) aortic valve.

   (B) pulmonary valve.   (E) semilunar valve.

   (C) bicuspid valve (mitral
       valve).

27. Deoxygenated blood returns to the heart by

   (A) 12 and 4.

   (B) 11 and 4.

(C) 9 and 13.

(D) 3 and 9.

(E) 9 and 12.

28. Number 1 is the

(A) right subclavian artery.

(B) right common carotid artery.

(C) left subclavian artery.

(D) brachiocephalic artery.

(E) left common carotid artery.

29. It is a general rule that all arteries carry oxygenated blood. Which artery is the exception to this rule?

(A) 2

(B) 12

(C) 4

(D) 3

(E) 1

30. The blood pressure is maximum in which vessel?

(A) 3

(B) 2

(C) 9

(D) 1

(E) 12

31. Marine ameba lack a contractile vacuole. This is because

(A) they have alternative means of osmoregulation.

(B) their internal environment is isotonic to their surroundings.

(C) their internal environment is hypotonic to their surroundings.

(D) during the course of evolution they have lost the need for a contractile vacuole.

(E) their internal environment is hypertonic to their surroundings.

32. The taste buds of the tongue are what kind of receptors?

(A) Proprioceptors

(D) Mechanoreceptors

(B) Tastereceptors

(E) Chemoreceptors

(C) Osmoreceptors

33. In certain parts of the Middle-East where agriculture is done on a very limited scale, many people suffer from poor bone growth and poor healing of wounds. This has been traced to a severe deficiency of fresh fruits. The underlying biochemical cause is a lack of

(A) vitamin K.

(D) vitamin C.

(B) calcium.

(E) folic acid.

(C) vitamin E.

34. It is found that when pregnant mothers consume insufficent amounts of dairy products, their babies develop rickets. This is due to lack of

(A) iron.

(D) thiamine.

(B) calcium.

(E) zinc.

(C) vitamin E.

35. The digestive action of bile salts is to

(A) emulsify fats.

(B) act as an enzyme in the breakdown of lactose.

(C) hydrolyse fatty acids.

(D) esterify cholesterol.

(E) hydrolyse fats to fatty acids and glycerol.

36. Humans have developed different mechanisms to conserve body heat. Which of the following is not a means of heat conservation?

(A) Vasodilation of the blood vessels of the skin

(B) Vasoconstriction of the blood vessels of the skin

(C) Shivering

(D) Piloerection

(E) Increased metabolism

37. Which of the following statements is correct?

(A) Root tips show positive geotropism and positive phototropism.

(B) Shoots show positive phototropism and positive geotropism.

(C) Root tips show positive phototropism and negative geotropism.

(D) Shoots show positive geotropism and negative phototropism.

(E) Root tips show positive geotropism and negative phototropism.

38. Excision of the parthyroid glands would result in

(A) muscle convulsions.   (D) decalcification of bones.

(B) strengthening of muscles.   (E) hypermobility.

(C) weakening of bones.

39. In males, which of the following hormones is responsible for secondary sexual characteristics?

(A) Estrogen   (D) Prolactin

(B) Progesterone   (E) Testosterone

(C) Growth hormone

40. Which of the following hormones would be expected to promote an increase in blood pressure?

(A) Insulin

(B) Oxytocin   (D) Histamine

(C) Epinephrine   (E) Prolactin

41. The ability of plants such as the giant redwood to raise water several hundred feet above the earth is due to

(A) root pressure.   (D) all of the above.

(B) transpiration pull.   (E) none of the above.

(C) capillary action.

42. Which of the following is true about alternation of generations in plants?

(A) The haploid, sexual generation follows the diploid, asexual generation.

(B) The sporophyte is haploid.

(C) The gametophyte is diploid.

(D) All of the above.

(E) None of the above.

43. The passage of water through the porous cell walls into the xylem is prevented by

(A) the cortex.   (D) all of the above.

(B) the casparian strip. (E) none of the above.

(C) the endodermis.

44. In an experiment designed to demonstrate the effect of auxin on root growth, a pea seedling was laid on its side after germination. After some time the stem became erect while the root turned downward and penetrated the soil. On examination of the root, the concentration of auxin was found to be highest in the lower side of the root. It can be concluded that

(A) while certain concentrations of auxin in a stem cause elongation, the root may react differently to the same concentrations so that elongation is inhibited.

(B) root cells are more sensitive to inhibition by higher concentrations of auxin than stem cells.

(C) root cells and stem cells react differently to auxin.

(D) all of the above.

(E) none of the above.

45. Gibberelins were injected into cabbage plants. After a few weeks the observed effects would be

(A) cabbage plants with short stems and bushy leaves.

(B) plants which failed to produce normal flowers.

(C) plants which were more resistant to the fungus <u>Gibberella fujikuroi</u>.

(D) all of the above.

(E) none of the above.

46. The neurotransmitter acetylcholine is found at

(A) all autonomic ganglia.   (D) all of the above.

(B) neuromuscular junctions.   (E) none of the above.

(C) presynaptic sympathetic nerve endings.

47. Repolarization of the membrane of a neuron is accomplished by

(A) a sudden influx of sodium ions (Na^+).

(B) a rapid increase in intracellular calcium (Ca^{2+}).

(C) a rapid efflux of potassium ions (K^+).

(D) all of the above.

(E) none of the above.

48. A man was involved in an accident and suffered extensive damage to the cerebellum. Which of the following functions would he be unable to perform?

(A) Recalling facts prior to his accident

(B) Driving his car

(C) Reading for long periods

(D) All of the above

(E) None of the above

49. The lateral line organs of bony fishes are what type of receptors?

(A) Distance receptors

(B) Contact receptors

(C) Auditory receptors

(D) All of the above

(E) None of the above

50. The rattlesnake is able to detect its prey by means of

(A) high-frequency vibration.

(B) high-frequency sound.

(C) polarized light.

(D) all of the above.

(E) none of the above.

51. What is the fundamental difference between active and passive transport across cell membranes?

(A) Active transport occurs more rapidly than passive transport.

(B) Passive transport is never selective.

(C) Passive transport requires a concentration gradient across the cell membrane as the driving force, while active transport needs energy expenditure to transport substances regardless of concentration gradient.

(D) Passive transport occurs only among gases.

(E) None of the above

52. Many single cells have evolved the following method for excreting cellular wastes:

A vacuole containing material to be expelled travels to the cell membrane and fuses with it. After this fusion has been completed, the site of contact opens up and the contents of the vacuole are jettisoned out of the cell. This process is known as

(A) endocytosis.

(D) exocytosis.

(B) phagocytosis.

(E) none of the above.

(C) pinocytosis.

53. The main function of leucoplasts is to

(A) carry out photosynthesis.

(B) prevent the loss of water through the stem.

(C) store starch after it is formed from glucose.

(D) store enzymes which are active in glycolysis.

(E) none of the above.

54. A given cell has a membrane which is permeable to water but impermeable to solute. What will happen to that cell if it contains 1% solute and is placed in a solution containing 3% solute?

(A) The cell will swell.

(B) The cell will shrink.

(C) The cell will remain the same size but will change shape to minimize surface area.

(D) The cell will remain the same size but will change shape to maximize surface area.

(E) None of the above.

55. A universally accepted scientific principle states that in a closed system, energy can neither be created nor destroyed. This principle is called

(A) the second law of thermodynamics.

(B) the first law of thermodynamics.

(C) the Law of Conservation of Energy.

(D) B and C, but not A.

(E) none of the above.

56. In a certain breed of rabbit, neither the allele for red color nor the one for white color shows dominance: the resulting heterozygote from a mating of these two alleles is pink. Which of the following progenies would be expected from the cross of a red and a white individual?

(A) 100 pink

(B) 100 white

(C) 75 red and 25 white

(D) 50 pink and 50 white

(E) 25 red, 50 pink, and 25 white

57. Wolves in the Mohave desert were observed by two scientists to display peculiar behavior. When hunting a hare which can outrun them, one wolf would chase the hare in a large circular pattern for approximately ten minutes. It would then stop to rest and another wolf would continue the chase. These two wolves would alternate chasing the hare in this fashion until it was exhausted. They would then catch and kill the hare and share the meat. This type of behavior displayed by the wolves is typical of

(A) communication.

(B) recognition.

(C) altruism.

(D) cooperation.

(E) none of the above.

58. Which of the following statements is incorrect with regard to the physiological basis of phototropism?

(A) Phototropism takes place as a result of unequal rates of growth.

(B) Phototropism results from a difference in cell division and cell elongation.

(C) Auxins are known to hinder cell growth in actively growing regions.

(D) All of the above.

(E) None of the above.

59. When platelets rupture on the rough edges of torn blood vessels, they initiate blood clotting by releasing an enzymatic substance called

(A) thrombokinase.

(D) fibrinogen.

(B) prothrombin.

(E) none of the above.

(C) thrombin.

60. A resting, non-stimulated neuron is electrically _____ along the outside of its surface membrane and electrically _____ along the inside.

(A) negative, positive

(D) neutral, negative

(B) negative, neutral

(E) positive, negative

(C) neutral, positive

61. Factors that play a role in determining an individual's rank in a dominance hierarchy include which of the following?

I. physical condition
II. experience
III. gender

(A) I only

(D) II and III only

(B) II only

(E) I, II, and III

(C) I and II only

62. Diabetes mellitus, which results from the body's inability to convert glucose to glycogen, and from the inability of cells to absorb glucose is due to a deficiency of which of the following hormones?

(A) Glucogon

(E) Parathormone

(B) Adrenalin

(E) Insulin

(C) secretin

63. Among married couples, complications can arise in pregnancy due to the Rh factor. These complications, however, can be avoided by taking a simple precautionary measure. This involves

   (A) treating Rh negative mothers of Rh positive babies with an injection of antibodies against the Rh positive antigen after the first pregnancy.

   (B) treating Rh positive mothers of Rh negative babies with an injection of antibodies against the Rh negative antigen before the first pregnancy.

   (C) treating Rh negative mothers of Rh positive babies with an injection of antibodies against the Rh negative antigen after the first pregnancy.

   (D) treating Rh positive mothers of Rh negative babies with an injection of antibodies against the Rh negative antigen after the first pregnancy.

   (E) none of the above.

64. Proteins differ from polypeptide chains in that they have well-defined three-dimensional structures which confer biological activities. Which one of the following enables polypeptide chains to reverse their direction?

   (A) α helix   (D) β -turns

   (B) γ helix   (E) None of the above

   (C) β pleated sheet

65. Both hemoglobin and myoglobin are oxygen carriers in vertebrates. Hemoglobin is found in red blood cells while myoglobin is located in muscles. The functional differences between these two carrier proteins arise from the fact that

   (A) myoglobin has a heme prosthetic group which binds with oxygen.

   (B) hemoglobin has a diminished affinity for carbon monoxide.

   (C) hemoglobin is an allosteric protein.

   (D) the oxygen affinity of myoglobin depends on pH.

   (E) the oxygen affinity of myoglobin is regulated by diphosphoglycerate.

66. Which of the following are important biological properties of water?

   I. Water is a polar molecule.

   II. Water can act as interacting molecules.

   III. Water has the ability to form a solvent shell around an ion.

   IV. Water tends to drive the hydrophobic interactions of non-polar molecules.

   (A) I, II

   (B) II, III

   (C) III, IV

   (D) I, II, III

   (E) I, II, III, IV

67. An enzyme facilitates a reaction by

   (A) increasing the free energy difference between reactants and products.

   (B) decreasing the free energy difference between reactants and products.

   (C) lowering the activation energy of the reaction.

   (D) raising the activation energy of the reaction.

   (E) none of the above.

68. The most significant adaptation of succulent plants to arid environments is the

   (A) low surface-to-volume ratio of above-ground organs.

   (B) accumulation of H_2O in the cells.

   (C) formation of highly dissected leaves.

   (D) absence of stomata.

   (E) production of non-pigmented cuticles.

69. Neutrophils in the human body are:

   (A) small phagocytes involved in the immune response.

   (B) lymphocytes generated from the bone marrow.

(C) lymphocytes generated from the thymus.

(D) large macrophages involved in the immune response.

(E) none of the above.

70. In 1928, bacteriologist Fred Griffith performed what is now considered a classic experiment. He had been studying the virulence of two strains of Pneumonococcus, the bacteria that cause pneumonia. One strain was dangerous and one was harmless. When grown on agar in petri dishes, the virulent strain produced "smooth" colonies; the harmless strain produced rough colonies.

When the virulent strain was injected into mice, the mice died. When the rough strain was injected into mice, the mice lived. When the smooth strain was boiled first and then injected into mice, the mice lived. However, when boiled smooth strain and live rough strain, both harmless, were mixed together and then injected into mice, the mice died.

The rough strain had somehow been converted to smooth. In subsequent experiments, various materials from the dead smooth bacteria were isolated and purified and then injected to see if they were the transforming substance. Which of the following has been determined to be the substance responsible for the transformation?

(A) Ribonucleic acid (RNA)

(D) Glucose polymers

(B) Deoxyribonucleic acid (DNA)

(E) Adenosine triphosphate (ATP)

(C) Twenty different amino acids

71. Which of the following is true of both photosynthetic phosphorylation and oxidative phosphorylation?

(A) The primary electron donor is a sugar molecule.

(B) The ultimate electron acceptor is oxygen.

(C) Both take place in the mitochondria.

(D) Both produce chemical energy in the form of ATP.

(E) None of the above.

72. The light reactions of photosynthesis include cyclic photophosphorylation and non-cyclic photophosphorylation. Of these two
kinds of reactions, it can be said that:

(A) cyclic and non-cyclic photophosphorylation each produce
both ATP and NADPH necessary for the dark reactions of
photosynthesis.

(B) both cyclic and non-cyclic photophosphorylation involve
photosystem I and photosystem II.

(C) cyclic photophosphorylation involves only photosystem I
and produces only ATP; non-cyclic photophosphorylation
involves photosystem I and photosystem II and produces
both ATP and NADPH.

(D) cyclic photophosphorylation involves the reduction of NADP
and the liberation of oxygen; non-cyclic photophosphorylation involves the reduction of NADP but not the liberation of oxygen.

(E) cyclic photophosphorylation involves the splitting of water,
whereas non-cyclic photophosphorylation does not involve
the splitting of water.

73. Which of the following growth movements is responsible for
deepening of the plant roots?

(A) Geotropism   (D) Phototropism

(B) Trigmotropism   (E) All of the above

(C) Chemotropism

74. The most vital function of gibberellins in plants is to

(A) lengthen stems and stimulate pollen germination.

(B) accelerate the rate of division of cells.

(C) stimulate the elongation of cells.

(D) act as a growth inhibitor of stems and leaves.

(E) promote the shedding of branches and leaves.

75. The motile and secretory activity of the gastrointestine are
regulated by a large number of chemical messengers which include

(A) neurotransmitters and endocrine hormones.

(B) neurotransmitters and paracrine hormones.

(C) endocrine hormones and paracrine hormones.

(D) neurotransmitters, endocrine hormones and paracrine hormones.

(E) none of the above.

76. Plant cells are able to withstand a much wider fluctuation in the osmotic pressure of the surrounding medium than animal cells. This is due to the plant cells'

(A) stomata.

(B) lipid membrane.

(C) cell wall.

(D) chloroplasts.

(E) all of the above.

77. In humans, the large bone extending from the hip to the knee is called the

(A) tibia.

(B) fibula.

(C) patella.

(D) humerus.

(E) femur.

78. Procaryotic cells differ from eucaryotic cells in that the former lack

(A) ribosomes.

(B) a plasma membrane.

(C) endoplasmic reticulum.

(D) a cell wall.

(E) all of the above.

79. Which of the following contributes to the "powerhouse" properties of the mitochondria?

(A) The presence on the mitochondrion's inner membrane of the enzymes necessary for the functioning of the electron transport system.

(B) The presence of cristae, which increase the surface area of the mitochondrion's inner membrane.

(C) The ease with which pyruvate can travel through the outer and inner membrane into the mitochondrion's inner matrix.

(D) None of the above.

(E) All of the above.

80. An amphoteric molecule is one which

(A) always acts as an electron donor.

(B) forms covalent bonds only.

(C) contains at least one double bond between two carbon atoms.

(D) is capable of acting either as an acid or a base in oxidation-reduction reactions.

(E) always contains both a hydrophobic end and a hydrophilic end.

81. The citric acid cycle is the major degradative pathway that generates ATP. It is also important in

(A) removing excessive glucose in our bodies.

(B) providing intermediates for biosynthesis.

(C) maintaining the proper pH of the body fluid.

(D) keeping a desirable turnover rate of enzymes participating in the citric acid cycle.

(E) none of the above.

82. Activated carriers mediate the interchange of activated groups in many biochemical reactions. Coenzyme A is a universal carrier of

(A) acyl groups.   (D) methyl groups.

(B) phosphoryl groups.   (E) phenyl groups.

(C) glucose.

83. After incubation of two bacteria cultures for 132 minutes, the number of bacteria present is 20,000. What is the generation time?

(A) 2 minutes

(D) 30 minutes

(B) 5 minutes

(E) Greater than 30 minutes

(C) 10 minutes

84. One of the three major classes of hormones are the steroids, which are produced in the

(A) adrenal cortex.

(D) thyroid.

(B) hypothalamus.

(E) anterior pituitary.

(C) liver.

85. First, second and third order refer to what aspect of neuro-endocrine hormonal arrangement?

(A) The length of time that a certain hormone remains in the bloodstream after being released, first order being the shortest length of time, third order being the longest.

(B) The quantity of a particular hormone released into the bloodstream, first order being the smaller quantity, third order being the larger.

(C) The number of organs in the body directly affected by the release of a single hormone, first order meaning that one organ is affected, second order meaning that two organs are affected, third order meaning that three or more organs are affected.

(D) The number of hormones involved in a regulation process, first order meaning that one hormone is released, second order being when two hormones are used, third order being when three hormones are used.

(E) The areas of behavior affected: first order being those hormones which affect sexual behavior, second order being those which affect aggressive/submissive behavior, third order being those which affect feeding behavior.

86. A kinase is an enzyme that

(A) transfers a phosphate from a donor to an acceptor.

(B) is often activated by increased levels of cyclic AMP (protein kinases).

(C) often is involved in the transformation of ATP and ADP.

(D) all of the above.

(E) none of the above.

87. The valve(s) located between the left atrium and the left ventricle of the human heart is (are) called the

    I. left atrioventricular valve

   II. tricuspid valve

  III. mitral valve

(A) I only   (D) I and II

(B) II only   (E) I and III

(C) III only

88. Which of the following is the correct sequence for blood flow through the human circulatory system?

(A) Aorta, right atrium, right ventricle, pulmonary artery, pulmonary vein, left atrium, left ventricle, superior or inferior vena cava

(B) Superior or inferior vena cava, right atrium, right ventricle, pulmonary vein, pulmonary artery, left atrium, left ventricle, aorta

(C) Superior or inferior vena cava, left atrium, left ventricle, right atrium, right ventricle, pulmonary artery, pulmonary vein, aorta

(D) Superior or inferior vena cava, left atrium, left ventricle, pulmonary artery, pulmonary vein, right atrium, right ventricle, aorta

(E) Superior or inferior vena cava, right atrium, right ventricle, pulmonary artery, pulmonary vein, left atrium, left ventricle, aorta

89. If a sperm cell in which nondisjunction of the sex chromosomes

has occurred fertilizes a normal human female egg, how many chromosomes will the resulting zygote contain?

(A) 47

(D) 40

(B) 24

(E) 22

(C) 23

90. A hemophiliac man is married to a woman who has no genes for hemophilia. What is the probability that their first child will be a boy who is hemophilic?

(A) 0%

(D) 75%

(B) 25%

(E) 100%

(C) 50%

91. For which combination is there the greatest need for the overproduction of egg cells so that the species may survive?

(A) Internal fertilization and internal development

(B) Internal fertilization and external development

(C) External fertilization and internal development

(D) External fertilization and external development

(E) No combination has a greater need than another

92. When large numbers of roan cattle are interbred, percentages occur as follows: 25% red, 50% roan, and 25% white. These results illustrate

(A) independent assortment

(D) natural selection

(B) blending inheritance

(E) genetic mutation

(C) dominance

93. Which organism(s) does/do not carry out extracellular digestion?

(A) Paramecium

(B) Earthworm

(C) Hydra

(D) Grasshopper

(E) All of the above

94. Calcium plays a role in all of the following EXCEPT:

(A) blood clotting

(B) nerve transmission

(C) bone development

(D) muscle contraction

(E) ovulation

95. In a reaction controlled by enzymes, the initial increase in rate with the increase in temperature is due to

(A) greater numbers of molecules in the system acquiring enough energy to overcome the activation energy barrier and react

(B) an increasingly favorable equilibrium position for a key reaction

(C) the conversion of an enzyme from an inactive state to an active one

(D) greater numbers of cells taking part in the catalyzed reaction

(E) greater numbers of enzymes taking part in the catalyzed reaction

96. The rate of an enzyme-controlled reaction reaches a peak and then falls off after a certain temperature because of

(A) competition between the activation of enzymes and the inactivation of enzymes

(B) the antagonism of two opposing reactions, one whose rate increases with temperature, the other whose rate decreases with temperature

(C) the opposition caused by the increased temperature, which makes the reaction go faster, and on the other hand, inactivates enzymes

(D) a destruction of key vitamins with increasing temperature

(E) enzymatic degeneration into their original building block proteins

97. Nerves controlling the contraction of smooth muscle are derived from the

(A) autonomic nervous system.

(B) somatic nervous system.

(C) sympathetic nervous system.

(D) parasympathetic nervous system.

(E) A, C and D only.

98. In the nervous system, certain branches of the afferent neurons activate interneurons which inhibit the motor neurons of antagonistic muscles whose activity would oppose a reflex flexion. This inhibition of antagonistic motor neurons is called

(A) dual innervation.   (D) A and C only

(B) reciprocal innervation.   (E) A, B, and C

(C) mutul innervation.

99. Which of the following is a correct explanation for the feeling of dizziness after being rotated in a swivel chair?

(A) The movement of hair cells with respect to the movement of perilymph in the cochlea of the inner ear causes the hair cells to send impulses to the cerebellum.

(B) The movement of hair cells with respect to the movement of endolymph in the ampulla of the inner ear causes the hair cells to send impulses to the cerebellum.

(C) The movement of perilymph with respect to the movement of hair cells in the ampulla of the inner ear causes the hair cells to send impulses to the cerebrum.

(D) The movement of endolymph with respect to the movement of perilymph in the ampulla of the inner ear causes the former to send impulses to the cerebellum.

(E) The movement of hair cells with respect to the movement of perilymph in the cochlea of the inner ear causes the hair cells to send impulses to the cerebrum.

100. If both the dorsal and the ventral root of a peripheral nerve to the foot of a dog were severed, what would be observed upon touching the foot of the dog with a hot iron?

(A) The dog would show signs of experiencing pain, but would be unable to move his foot from the iron.

(B) The dog would show signs of experiencing pain, and accordingly would withdraw his foot from the iron.

(C) The dog would show no signs of experiencing pain nor would he be able to move his foot away from the iron.

(D) The dog would show no sign of experiencing pain, but would be able to move his foot away from the iron.

(E) The dog would experience pain on the opposite foot, and would withdraw this foot as if it were in contact with the iron.

101. Which of the following statements gives the basic physiological reason for the occurrence of the "bends" in some deep sea divers?

(A) The solubility of a gas in our blood increases with a decrease in barometric pressure.

(B) Regardless of depth, nitrogen in our blood remains at the same solubility.

(C) The solubility of a gas in our blood decreases with a decrease in barometric pressure.

(D) Our lungs cannot be expanded at depths 200 feet below sea level.

(E) Quick changes in barometric pressure can cause the body's bones to bend.

102. If the genes for the different traits are located on the same autosome, the two traits would be expected to be

(A) inherited together

(B) visible in the offspring

(C) recessive in the off-spring

(D) sex-linked

(E) dominant in the offspring

103. Striated muscle includes:

(A) skeletal muscle

(B) cardiac muscle

(C) smooth muscle

(D) A and B only

(E) B and C only

104. In a muscle cell, the space between adjacent thick and thin filaments is seen to be bridged by projections at intervals a-long the filaments. These projections are portions of myosin molecules which extend from the surface of the thick filaments to the thin filaments. These projections which extend from myosin to actin filaments are called

(A) cross bridges

(B) myofibrils

(C) muscle fibers

(D) cilia

(E) pseudopods

105. Consider a biochemical reaction X ⇄ Y. Which of the graphs below correctly describes the changes in free energy during both the catalyzed and the uncatalyzed conversion of X to Y?

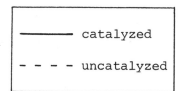

(A)

(B)

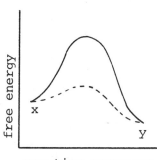

(C)

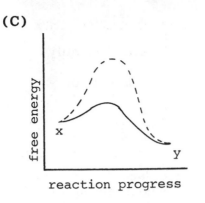

free energy

reaction progress

(D)

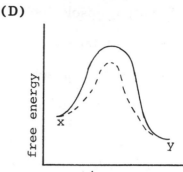

free energy

reaction progress

(E)

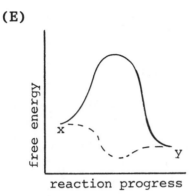

free energy

reaction progress

106. A person afflicted with a complete obstruction of the common bile duct would have a decreased ability to digest

(A) proteins

(D) fats

(B) carbohydrates

(E) none of the above

(C) nucleic acids

107. Animal cells and plant cells differ in which of the following way(s)?

  I. Only animal cells contain centrosomes.

  II. Spindle formation occurs in animal cell division only, not in plant cell division.

  III. Animal cells divide by a cleavage furrow, i.e. the outside cell membrane grows inward; plant cells, by contrast, divide by the formation of a cell plate.

(A) I only   (D) II and III only

(B) II only   (E) I and III only

(C) III only

108. Vitamin D is essential in the control of calcium and phosphorus metabolism. The precursor of Vitamin D is

(A) glycocholate.   (D) sphingosine.

(B) cholesterol.   (E) none of the above.

(C) phosphatidate.

109. Small animals and plants which live in an aquatic environment lack transport systems because

(A) they evolved from ancestors that originated in dry habitats.

(B) they have little need for the nutrient contained within water.

(C) they can meet their nutritional and waste disposal needs by diffusion.

(D) all of the above.

(E) none of the above.

110. John is injured and badly needs a blood transfusion. Carla volunteers, and the physician uses her blood. A few months later, Carla needs a transfusion. John volunteers but the physician turns him down. Given that the physician would accept any blood which was compatible, which of the following could be the blood types of John and Carla?

(A) John is type B and Carla is type O.

(B) John is type B and Carla is type A.

(C) John is type AB and Carla is type O.

(D) A and C only

(E) A, B, and C

111. Capillaries play a vital role in the exchange of substances between blood and intercellular tissue fluid because

   (A) certain types of lymphocytes leave the bloodstream by way of the capillaries.

   (B) the blood is purified by the homeostatic action of capillaries.

   (C) materials leave and enter the bloodstream through the capillaries.

   (D) capillaries control blood pressure.

   (E) capillaries join arteries and veins.

112. Luteinizing hormone (LH) and follicle stimulating hormone (FSH) are secreted by which gland?

   (A) Anterior pituitary   (D) Adrenal medulla

   (B) Posterior pituitary   (E) Adrenal cortex

   (C) Thyroid

113. Which of the following is not one of the four nitrogenous bases of DNA?

   (A) Uracil

   (B) Guanine

   (C) Adrenine

   (D) Cytosine

   (E) Thymine

114. When a virus infects a bacterium, the material that enters the bacterium from the virus is

   (A) sulfur   (D) protein

   (B) nucleic acid   (E) none of the above

   (C) a mutagen

115. The sequence of bases found in a strand of DNA which served as a template for the synthesis of mRNA is adenine - guanine - cytosine - thymine. Thus the sequence of bases found in the newly-synthesized mRNA is

(A) guanine - cytosine - thymine - adenine

(B) uracil - cytosine - guanine - adenine

(C) thymine - cytosine - guanine - adenine

(D) uracil - adenine - cytosine - guanine

(E) none of the above

116. Which gland releases STH (growth hormone)?

(A) Anterior pituitary   (D) Adrenal cortex

(B) Posterior pituitary   (E) Thyroid

(C) Adrenal medulla

117. A man has a rare dominant X-chromosome determined disease called "vitamin D resistant rickets." He is married to a normal woman. If affected daughters have the disease, to which of their progeny will they transmit it?

(A) To the sons only.

(B) To the daughters only.

(C) To either the daughters or the sons, or both.

(D) To neither the daughters nor the sons (woman cannot pass on the disease).

(E) Women cannot be affected by the disease.

118. Listed in order of dominance, four alleles in rabbits are: c^+, colored; c^{ch}, chincilla; c^h, himalayan; and c, albino. What phenotypes and ratios would be expected from the following cross: $c^+c^h \times c^{ch}c$?

(A) 1 colored:1 chincilla:1 himalayan:1 albino

(B) 2 colored:1 chincilla:1 himalayan

261

(C) 2 colored:2 chincilla:1 himalayan:1 albino

(D) 1 colored:2 chincilla:1 himalayan:2 albino

(E) 2 colored:2 chincilla:1 himalayan

119. DNA and RNA are different from each other in several important respects. Among the differences are their nitrogenous
bases. In DNA, the bases are adenine, guanine, cytosine,
and thymine. The nitrogenous base not found in RNA is

(A) adenine   (D) cytosine

(B) guanine   (E) uracil

(C) thymine

120. A tall pea plant with axial flowers when crossed with a dwarf
pea plant having terminal flowers produces pea plants that are
tall and have terminal flowers. If the F_1 are crossed with a
plant homozygous for both tall and axial flowers, what phenotypic ratio can be expected in their offspring?

(A) Half tall; half dwarf; half terminal; half axial

(B) All tall; all terminal

(C) Half tall; half dwarf; all axial

(D) All tall; all axial

(E) All tall; half terminal; half axial

121. In humans, blond hair is dominant to red hair, and tasting of
PTC is dominant to non-tasting. What would be the expected
F_2 phenotypic ratio when homozygous blond tasters marry homozygous blond non-tasters?

(A) All blond tasters.

(B) 75% blond tasters; 25% blond non-tasters.

(C) One-quarter blond tasters; one-quarter blond non-tasters;
one-quarter red tasters; one-quarter red non-tasters.

(D) One-half blond tasters; one-half red tasters.

(E) One-half blond tasters; one-quarter red tasters; one-
quarter red non-tasters.

122. Ectrodactyly or lobster claw is an inherited physical defect in humans, occurring in homozygous recessive individuals. If two normal parents have a daughter affected with this condition, and a normal son, what is the probability that the son will be a carrier of the recessive allele?

(A) 25%

(D) 67%

(B) 33%

(E) 75%

(C) 50%

123. Some animal behavior recurs on a cyclical basis. Of the following, which three are the most common cycles of animal behavior?

   I. Circadian (approximately 24 hours)
   II. Annual
   III. Monthly
   IV. Hourly
   V. Weekly

(A) I, II, and III

(D) II, III, and IV

(B) II, IV, and V

(E) I, II, and V

(C) I, III, and V

124. Which of the following general statements concerning biological succession are true?

   I. The species composition changes continuously during the succession, but the change is usually more rapid in the earlier stages than in the later ones.

   II. The total number of species represented increases initially and then becomes fairly stabilized in the later stages.

   III. Both the total biomass in the ecosystem and the amount of non-living organic matter increase during the succession until a more stable stage is reached.

   IV. The food web becomes more complex, and the relations between species in them better defined.

   V. Although the amount of new organic matter synthesized by the producers remains approximately the same, except at the beginning of succession, the percentage utilized at the various trophic levels rises.

(A) I only

(D) II, III, and IV only

(B) I and II only

(E) I, II, III, IV, and V

(C) III, IV, and V only

125. The "10 percent rule" of ecological efficiency refers to

(A) the percentage of similar species that can coexist in one ecosystem

(B) the average death total of all mammals before maturity

(C) the percent of animals resistant to insecticides

(D) the level of energy production present in a given trophic level and used for production by the next higher level

(E) none of the above

DIRECTIONS: *For each group of questions below, match the numbered word, phrase, or sentence to the most closely related lettered heading and mark the letter of your selection on the corresponding answer sheet. A lettered heading may be chosen as the answer once, more than once, or not at all for the question in each group.*

Questions 126–130.

(A) Stoneworts

(D) Dicotyledon

(B) Ferns

(E) Red algae

(C) Monocotyledon

126. Have certain structures called sori

127. They usually have no vascular cambium

128. Have netlike veins in their leaves

129. Live only in aquatic environments

130. Their cell walls contain calcium salts

Questions 131–135

    (A) Blood group A   (D) Blood group O

    (B) Blood group B   (E) Rhesus factor

    (C) Blood group AB

131. Universal donor

132. No serum antibodies (agglutinins)

133. Universal recipient

134. No antigens (agglutinogens) on the surface of its erythrocytes

135. Plays integral role in erythroblastosis fetalis

Questions 136 through 143 refer to the diagram of the human eye below. For each question, select the structure which corresponds to the question number in the diagram.

Questions 136–143

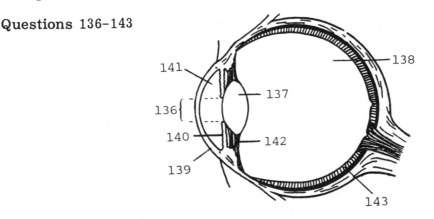

265

136. (A) Pupil   (D) Cornea

    (B) Retina   (E) Muscle

    (C) Iris

137. (A) Vitreous humor   (D) Retina

    (B) Iris   (E) Lens

    (C) Cornea

138. (A) Retina   (D) Lens

    (B) Muscle   (E) Pupil

    (C) Vitreous humor

139. (A) Vitreous humor   (D) Cornea

    (B) Retina   (E) Iris

    (C) Aqueous humor

140. (A) Iris   (D) Vitreous humor

    (B) Lens   (E) Aqueous humor

    (C) Pupil

141. (A) Pupil   (D) Retina

    (B) Cornea   (E) Muscle

    (C) Aqueous humor

142. (A) Cornea   (D) Pupil

    (B) Muscle   (E) Vitreous humor

    (C) Lens

143. (A) Lens   (D) Cornea

    (B) Retina   (E) Aqueous humor

    (C) Iris

Questions 144-146.

   Consider the following diagram of a nephron:

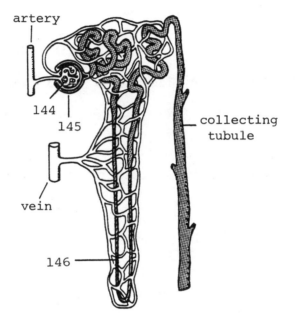

144. This structure is called the

    (A) loop of Henle.   (D) inner renal medulla.

    (B) glomerulus.   (E) outer renal cortex.

    (C) nephric capsule.

145. The name of this structure is the

    (A) loop of Henle.   (D) inner renal medulla.

    (B) glomerulus.   (E) outer renal cortex.

    (C) Bowman's capsule.

146. This structure is the

(A) loop of Henle.

(D) inner renal medulla.

(B) glomerulus.

(E) outer renal cortex.

(C) nephric capsule.

Questions 147-150. For each of the following, choose from the list below the behavior pattern described.

(A) Instinct

(D) Tropism

(B) Insight learning

(E) Taxis

(C) Habit

147. Flight pattern of bees gathering pollen.

148. The "fight or flight" response.

149. A hungry child stands on a chair to reach an apple resting on the table.

150. Direction of plant growth toward the sun.

Questions 151-155 refer to the following graph that depicts the logarithmic growth rate of E. coli in hospitable medium.

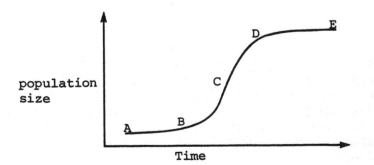

151. Point at which the population of E. coli has reached its carrying capacity.

152. Point at which the population of E. coli has reached its optimal yield.

153. Area of lag phase of growth.

154. Area of stationary phase of growth.

155. Area of logarithmic phase of growth.

Questions 156-160. Match each of the situations described in questions 156 through 160 with one of the five terms below.

(A) Competition

(B) Mimicry

(C) Mutualism

(D) Parasitism

(E) Allelopathy

156. Two species of aquatic life both feed off of salmon, although each species occupies a distinct ecological niche.

157. Two species of plant appear equally uninviting to predators; one species is toxic and the other is not.

158. A worm called Schistosoma mansoni is found in the Nile Valley of Egypt. The extensive irrigation canals fed by the Aswan Dam have allowed aquatic snails to flourish; these snails serve as intermediate hosts to the worm. Schistosoma mansoni enters the human body right through the skin when people wade in the water. The worm then feeds on the nutrients in the human blood.

159. A certain relationship exists between rhinoceri and a species of small birds. The birds perch on the rhino's back and feed on the small insects which infect the rhino's skin. This relationship is beneficial both to the rhinoceri and to the birds.

160. Yeast stop growing when their own waste product, ethanol, reaches 12%.

Questions 161-165. For questions 161 through 165, find a lettered word that corresponds to the description in each question.

(A) Mitochondria

(D) Nucleus

(B) Endoplasmic reticulum

(E) Plasma membrane

(C) Golgi bodies

161. Contains the chromosomes, and controls protein synthesis and transmission of hereditary information.

162. The site for nearly all ATP synthesis in the cell.

163. A complex of membranes that traverses the cytoplasm, providing a means of transport for lipids and proteins and containing enzymes that play an important role in the metabolic process.

164. Involved in the condensation and concentration of protein, and also contains enzymes involved in the synthesis of complex carbohydrates and in the coupling of these carbohydrates to protein.

165. Important for its discriminating permeability and its role in cell adherence.

166-170.

(A) Selective precipitation

(B) Ion-exchange chromatography

(C) Affinity chromatography

(D) Gel filtration

(E) Electrophoresis

166. This technique is used to purify proteins. It relies upon the fact that the solubility of a protein depends on the relative balance between protein-solvent interactions, which tend to keep it in solution; and protein-protein interactions, which tend to cause it to aggregate and separate from the solution. To extract a protein from solution, a salt is added. The salt increases the ionic strength of the solution, which decreases the solubility of the protein, thus causing it to separate from solution.

167. This technique depends on the ionic association of proteins with charged groups bound to an inert supporting material. The proteins leave the solvent and become associated with an alternative, immobile structure, such as filter paper or charged resin.

168. This technique takes advantage of the unique structural properties of a protein whereby that protein can be specifically withdrawn from solution while all other molecules remain behind in the dissolved state. An example of this technique is as follows:

    A person wants to separate the hormone (a protein) receptor that is specific for insulin from solution containing numerous membrane bound receptor proteins. This technique would utilize a column containing an inert material (a matrix) to which insulin molecules are covalently attached. When the impure membrane bound receptor solution is passed through the column, the receptor for insulin combines with the hormone; while the rest of the solution, including all other proteins, pass out through the bottom of the column. The receptor molecules are then displaced from the matrix, and a one-step purification of receptor has been accomplished.

169. This is a technique that separates proteins by molecular weight but does not involve the electric charge of the molecules.

170. This is a technique that separates proteins according to the ability of charged molecules to migrate toward an anode or cathode when placed in an electric field.

171-175. The fundamental contractile unit of striated muscle myo-fibrils is the sarcomere, which consists of actin and myosin filaments. Questions 171-175 concern the structure of a sarcomere.

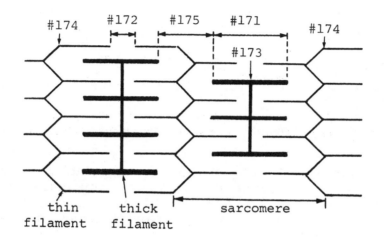

(A) M line

(B) H zone

(C) A band

(D) Z line

(E) I band

171. The region consisting of myosin filaments and overlapping actin filaments

172. The region between the ends of opposing actin filaments, where only thick (myosin) filaments are found

173. A thin dark band produced by linkages between the myosin filaments. By cross-linking the thick filaments, this structure keeps all these in a single sarcomere in parallel alignment

174. This structure defines the limit of a sarcomere on either end. The thin filaments (actin) are attached to this structure at either end of the sarcomere; these structures interconnect the thin filaments from two adjoining sarcomeres and this provides an anchoring point for the thin filaments

175. This region lies between the ends of the dark bands of myosin filaments of two adjacent sarcomeres

176-180.

(A) Troponin   (D) ATP

(B) Tropomyosin   (E) Myoglobin

(C) Calcium

176. A molecule which partially covers the myosin binding site on actin molecules, thereby preventing them from binding to the myosin cross bridges. This molecule functions to prevent muscles from being in a continuous state of contraction

177. A molecule which holds the blocking molecule of question 176 in place on the actin filament. This molecule is bound both to actin and to the blocking molecule of question 176, and contributes to the relaxed state of muscles

178. This molecule acts upon the inhibitory agent of question 177, changing it so that the blocking molecule of question 176 is pulled aside, exposing the cross bridge binding sites on actin This facilitates contraction.

179. A protein which binds oxygen and increases the rate of oxygen diffusion into the muscle cell, as well as provides a small store of oxygen within the fiber. Fibers containing large amounts of this protein have a dark red color which distinguishes them from the paler fibers which lack appreciable amounts of the protein

273

180. This provides the energy for cross bridge movement and detachment, thereby providing the energy for muscle contraction and relaxation

Questions 181-186. Select one of the five choices that best corresponds to each statement below.

(A) Cardiac output

(B) Stroke volume

(C) End-diastolic volume

(D) End-systolic volume

(E) None of the above

181. The total amount of blood pumped by both ventricles in one minute

182. The volume of blood ejected by each ventricle during each heartbeat

183. The volume of blood pumped by each ventricle in one minute

184. The volume of blood in the atrium just prior to systole.

185. The volume of blood in the ventricle immediately following systole

186. The volume of blood in the ventricle just prior to systole

Questions 187-191.

(A) Tetanus

(B) Summation

(C) Recruitment

(D) Pacemaker potential

(E) Hypertrophy

187. The increase in size of muscle fibers due to increased neural activity, such as accompanying repeated exercise.

188. The increase in mechanical response of a muscle to action potentials occurring in rapid succession.

189. The greatest tension that a muscle can develop.

190. An increase in the number of motor neurons that are actively discharging action potentials to a given muscle.

191. In some muscles, action potentials are generated spontaneously in the absence of any neural or hormonal input. This occurs because of spontaneous depolarization to threshold.

DIRECTIONS: *The following groups of questions are based on laboratory or experimental situations. Choose the best answer for each question and mark the letter of your selection on the corresponding answer sheet.*

Questions 192 and 195 refer to the following graph that depicts the level of three hormones secreted during human pregnancy.

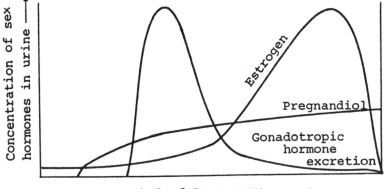

192. The rise in estrogen concentration near the end of gestation is due to

   (A) the decrease in the concentration of gonadotrophic hormone.

   (B) increased secretion of estrogen by the ovaries.

   (C) increased secretion of pregnanediol by the corpus luteum.

   (D) increased secretion of estrogen by the placenta.

   (E) decreased steroid secretion by the placenta.

193. The removal of the subject's ovaries after the sixth month of pregnancy would affect the graphs of which of the following hormones?

   (A) Estrogen

   (B) Pregnanediol

   (C) Gonadotrophic hormone

   (D) Estrogen and Pregnanediol

   (E) None of the above

194. The placenta and fetus are maintained even though the concentration of human chorionic gonadotropin hormone falls after the first $5\frac{1}{2}$ weeks of pregnancy. This is because:

   (A) human chorionic hormone is no longer needed

   (B) estrogen and progesterone levels are rising

   (C) the placenta is secreting steroids

   (D) the placenta is secreting human chorionic hormone

   (E) leuteinizing hormone is being produced

195. Human chorionic gonadotropin functions to:

   (A) maintain the placenta

   (B) maintain the fetus

   (C) maintain the endometrium

   (D) maintain the corpus luteum

   (E) maintain implantation

In the tricarboxylic acid cycle (Krebs cycle) succinic dehydrogenase catalyzes the conversion of succinic acid to fumaric acid. The following experiment was conducted to illustrate this process:

Tube	0.1% Methylene Blue (ml)	Distilled Water (ml)	Phosphate Buffer (ml)	Sodium Succinate (ml)	Heart Homogenate (ml)
1	1.0	3.0	2.0	2.0	2.0
2	1.0	3.5	2.0	2.0	1.5
3	1.0	4.0	2.0	2.0	1.0
4	1.0	4.5	2.0	2.0	.5

Following the addition of the homogenate the reactions were timed, in seconds, and % absorbance for each tube was recorded at 530 μm.

196. A control for this experiment might be:

(A) 1.0 ml methylene blue, 5.0 ml distilled water, 2.0 ml phosphate buffer, 2.0 ml sodium succinate, 0 ml homogenate

(B) 1.0 ml methylene blue, 5.0 ml distilled water, 2.0 ml phosphate buffer, 0 ml sodium succinate, 2.0 ml homogenate

(C) 5.0 ml distilled water

(D) (A) or (B)

(E) (A) or (C)

197. The heart homogenate contributes:

(A) oxidative enzymes   (D) succinic acid

(B) reductive enzymes   (E) ATP

(C) hydrolytic enzymes

198. The reagent which acted most like FAD was:

(A) heart homogenate   (D) distilled water

(B) sodium succinate   (E) methylene blue

(C) phosphate buffer

199. If 2.0 ml malonic acid were added to each tube, the rate of reaction would

(A) increase.

(D) decrease then increase.

(B) decrease.

(E) remain the same.

(C) increase then decrease.

Questions 200-203 refer to the metabolic process given below:

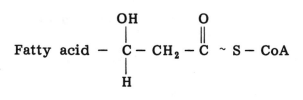

Fatty acid $-$ CH$_2$ $-$ CH$_2$ $-$ C $\overset{\displaystyle O}{\underset{\displaystyle OH}{\diagup}}$

ATP
CoA
AMP + PPi

Fatty acid $-$ CH$_2$ $-$ CH$_2$ $-$ $\overset{\displaystyle O}{\overset{\|}{C}}$ ~ S $-$ CoA

FAD$_{red}$
FAD$_{ox}$

Fatty acid $-$ $\overset{H}{\overset{|}{C}}$ = $\overset{H}{\overset{|}{C}}$ $-$ $\overset{O}{\overset{\|}{C}}$ ~ S $-$ CoA

H$_2$O

Fatty acid $-$ $\overset{OH}{\underset{H}{\overset{|}{\underset{|}{C}}}}$ $-$ CH$_2$ $-$ $\overset{O}{\overset{\|}{C}}$ ~ S $-$ CoA

NAD$_{red}$
NAD$_{ox}$

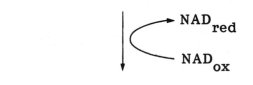

Fatty acid $-$ $\overset{O}{\overset{\|}{C}}$ $-$ CH$_2$ $-$ $\overset{O}{\overset{\|}{C}}$ ~ S $-$ CoA

Acetyl CoA

278

200. If the initial fatty acid chain is 16 carbons long, how much ATP is produced by the reduced products of this process?

(A) 32

(D) 80

(B) 35

(E) 96

(C) 64

201. What would the total number of ATP produced be, if each Acetyl-CoA molecule produced then entered the Krebs cycle?

(A) 96

(D) 160

(B) 128

(E) 192

(C) 119

202. The condition in which the body would resort to this metabolic pathway is:

(A) lactic acid buildup

(D) acidosis

(B) muscle development

(E) hypoglycemia

(C) ketosis

203. This pathway is commonly known as:

(A) the citric acid cycle

(D) gluconeogenesis

(B) the fatty acid cycle

(E) glucogenesis

(C) β-oxidation

Carla, curious as to the cause of disease in her mother's tobacco plants, designed the following experiment. She crushed the infected leaves with sterile water. Then, the solution was filtered once through a funnel filter and once through a 0.4-micron filter. Both the filtrate and residue of each filter were kept and tested by rubbing each, individually, onto an uninfected plant. The table below shows the results.

	Infection Ensues	No Infection
initial filtrate	x	
initial residue	x	
second filtrate	x	
second residue		x

204. Which of the following is the most likely pathogen?

(A) virus

(D) fungus

(B) bacteriophage

(E) none of the above

(C) bacterium

205. The second filtrate was used to infect a healthy plant. This plant then underwent the same procedure as the plant that yielded the first set of filtrates and residues. The process was repeated four times; the degree of infection remained constant each time. It is assumed that the cause of disease is a pathogen and not a toxin because:

(A) the initial filtrate remains infectious

(B) the initial residue remains infectious

(C) the second filtrate remains infectious

(D) the second residue remains uninfectious

(E) none of the above

206. Which of the following tests corroborates the diagnosis given in question 204?

(A) isolation and growth of the pathogen on essential media similar in chemical constitution to that of the plant

(B) isolation and growth of the pathogen on agar (nutrient-deficient) media

(C) lack of growth of the pathogen on essential media similar in chemical constitution to that of the plant

(D) lack of growth of the pathogen on agar (nutrient-deficient) media

(E) There is no test available for corroboration.

The pancreases of four rats were operated on as follows:

    #1 removal of the entire pancreas

    #2 tying of the pancreatic ducts

    #3 destroying only the β-cells

    #4 cut open but pancreas left untouched

207. Following surgery, two of the rats begin to ingest large quantities of food and water but remain dehydrated and underweight. They are most likely:

(A) 1 and 2   (D) 2 and 4

(B) 1 and 3   (E) 3 and 4

(C) 2 and 3

208. The rats are fed a postoperative diet which consists primarily of celery and lettuce. The rat who is most affected by this diet as a result of its surgery is:

(A) 1   (D) 4

(B) 2   (E) They would be equally affected.

(C) 3

209. Tying off the pancreatic ducts ties off:

(A) the endocrine function

(B) the neural function

(C) the digestive function

(D) (A) and (B)

(E) (B) and (C)

210. Rats numbered 1 and 3 will most likely develop:

(A) lactose intolerance

(B) goiter

(C) hypertension

(D) diabetes insipidus

(E) diabetes mellitus

THE GRADUATE RECORD EXAMINATION BIOLOGY TEST

MODEL TEST III

ANSWERS

1.	C	20.	A	39.	E
2.	D	21.	E	40.	C
3.	D	22.	D	41.	D
4.	A	23.	B	42.	A
5.	B	24.	D	43.	B
6.	B	25.	D	44.	D
7.	A	26.	C	45.	B
8.	A	27.	E	46.	D
9.	C	28.	D	47.	C
10.	E	29.	D	48.	B
11.	B	30.	B	49.	A
12.	E	31.	B	50.	E
13.	C	32.	E	51.	C
14.	A	33.	D	52.	D
15.	B	34.	B	53.	C
16.	E	35.	A	54.	B
17.	D	36.	A	55.	D
18.	A	37.	E	56.	A
19.	B	38.	A	57.	D

58.	C	89.	A	120.	E
59.	A	90.	A	121.	B
60.	E	91.	D	122.	D
61.	E	92.	B	123.	A
62.	E	93.	A	124.	E
63.	A	94.	E	125.	D
64.	D	95.	A	126.	B
65.	C	96.	C	127.	C
66.	E	97.	E	128.	D
67.	C	98.	B	129.	E
68.	A	99.	B	130.	A
69.	A	100.	C	131.	D
70.	B	101.	C	132.	C
71.	D	102.	A	133.	C
72.	C	103.	D	134.	D
73.	A	104.	A	135.	E
74.	A	105.	C	136.	A
75.	D	106.	D	137.	E
76.	C	107.	E	138.	C
77.	E	108.	B	139.	D
78.	C	109.	C	140.	A
79.	E	110.	D	141.	C
80.	D	111.	C	142.	B
81.	B	112.	A	143.	B
82.	A	113.	A	144.	B
83.	C	114.	B	145.	C
84.	A	115.	B	146.	A
85.	D	116.	A	147.	E
86.	D	117.	C	148.	A
87.	E	118.	B	149.	B
88.	E	119.	C	150.	D

151.	E	171.	C	191.	D
152.	C	172.	B	192.	D
153.	A	173.	A	193.	E
154.	D	174.	D	194.	C
155.	B	175.	E	195.	D
156.	A	176.	B	196.	A
157.	B	177.	A	197.	A
158.	D	178.	C	198.	E
159.	C	179.	E	199.	B
160.	E	180.	D	200.	B
161.	D	181.	E	201.	C
162.	A	182.	B	202.	E
163.	B	183.	A	203.	C
164.	C	184.	E	204.	A
165.	E	185.	D	205.	D
166.	A	186.	C	206.	C
167.	B	187.	E	207.	B
168.	C	188.	B	208.	E
169.	D	189.	A	209.	C
170.	E	190.	C	210.	E

THE GRE BIOLOGY TEST

MODEL TEST III

DETAILED EXPLANATIONS
OF ANSWERS

1.   (C)
There is a division of labor among the various members in which each member has a particular task to perform. For example, in the honey bee society which is known as a hive, there is one reproductive female, the queen; several hundred males, the drones, whose main functions are reproductive; and many thousands of immature females, the workers.   The workers build the hive, collect food, feed the queen and the drones, nurse the young and protect the colony from strange bees and enemies.

When a hive becomes overcrowded, the queen, some drones, and several thousand workers form a "swarm" and migrate to a new location to establish a new hive. In the old hive a new queen takes over.

2.   (D)
The law of independent assortment is Mendel's second law.   Phrased in modern terms, this law states: the inheritance of a gene pair located on a given chromosome pair is unaffected by simultaneous inheritance of other gene pairs located on other chromosome pairs.

In other words, two or more traits produced by genes located on two or more different chromosomes pairs assort independently, each trait being expressed independently as if no other traits were present.

3. (D)
Heterotrophic organisms are unable to manufacture their own food because they are unable to synthesize the complex organic molecules they need from the simple ones found in nature.   Hence they must obtain these compounds in an already manufactured form.

4. (A)
An increase in population density of most vertebrates would result in a more aggressive behavior due to competition among the members.   There will be competition for food, territory, leadership, and mates.   This means that those which are stronger will get the better of these things while the others will need to defend them; all of which leads to aggressive behavior.

5. (B)
In the proliferative phase, the pituitary, under the influence of the hypothalamus, releases more F.S.H., which acts on the immature ova of the germinal epithelium of the ovary.   Under the influence of increasing levels of F.S.H., the cell clusters around several immature ova will grow, forming follicles.

6. (B)
Phosphorylation occurs whenever inorganic phosphate is added to another molecule to yield a high energy bond. In the presence of inorganic phosphate and energy, ADP is converted to the high energy ATP molecule, a process of phosphorylation.

7. (A)
Evergreen trees have adapted to their environments by reducing their leaves to needles which are covered by a

waxy secretion. This results in very low transpiration rates, thereby conserving water.

8. (A)
The medulla oblongata is concerned primarily with the vegetative functions of the body. It contains the nerve centers controlling heartbeat, vasomotion and respiration.

9. (C)
Ribonucleic acid (RNA) is involved in protein synthesis. It can work as mRNA which transfers information from DNA to the ribosomes.

10. (E)
Theoretically an object can be magnified indefinitely but in practice this is not so. At very high magnification clarity of the image is lost. This is due to the limits of resolution - that is, objects lying close to one another cannot be distinguished as separate objects if the distance between them is less than one-half the wavelength of the light being used. Hence light microscopes can only magnify objects to 1000-1500 times the actual size.

11. (B)
There are certain microorganisms in the small intestine of humans which synthesize vitamin B_{12}. Destruction of these microorganisms results in a deficiency of this vitamin.

12. (E)
The mitochondria contain mainly respiratory enzymes and coenzymes which are responsible for carrying out respiration. They are the chief "factories" of cellular respiration.

13. (C)
The weak attraction between non-polar molecules is called

van der Waals force. Uncharged molecules tend to attract each other, but they do not overlap electron shell orbitals, since their orbitals are already filled and stable. This lack of orbital overlap is what makes van der Waals attractions weaker than other types of molecular bonding.

14. (A)
The most important monosaccharide is glucose. This six-carbon sugar is fundamentally involved in energy metabolism and photosynthesis and is also the building block of starch, glycogen and cellulose.

15. (B)
Lactose consists of one glucose molecule covalently linked to one galactose molecule.

16. (E)
Cellulose is a linear polymer of glucose subunits joined by 1-4 linkages. Humans lack the enzyme cellulase which can break the 1-4 linkages. So they are unable to digest cellulose.

17. (D)
The centrioles are important in cell division where they assist in spindle formation. Higher plants appear to have lost their centrioles during the course of evolution. When the cells of higher plants divide there is no spindle formation due to the absence of the centrioles.

18. (A)
Microtubules are small hollow tubes, which in cross-section, appear as tiny circles. Each microtubule is made of a spiral arrangement of spherical bodies of tubulin protein.

19. (B)
The DNA sub-unit contains all the genetic material for

the synthesis of proteins. During protein synthesis, the sense strand of the DNA duplex acts as a template for the messenger RNA (mRNA). The mRNA then acts as a template for tRNA and arranges itself on the correct reading frame of ribosomes to synthesize the corresponding protein.

20. (A)
When a virus infects a bacterium, one of its fate is to have its DNA integrated into the bacterial genome which corresponds to the lysogenic phase of its life cycle. When the viral DNA emerges from the bacterial genome, it usually takes up some of the bacterial DNA through an imprecise excision process. The virus then becomes a transducing phage which acts as a vector to promote the exchange of bacterial genes.

21. (E)
In lower organisms such as bacteria, when a particular substrate is not metabolized, the bacteria will be deficient in the enzyme responsible for metabolizing that substrate. When this substrate is included in the medium on which the bacteria grow, after some time they will acquire the necessary enzyme which is said to have been induced by the substrate.

22. (D)
The kangaroo rat, which is an inhabitant of the Southwestern deserts of the USA, has kidneys which are remarkably efficient at concentrating urine so that little water is lost. This rodent is also remarkable in that it obtains water from its own body metabolism. All of this makes the kangaroo rat adaptable to its environment.

23. (B)
The functional unit of the mammalian excretory system is the nephron which consists of the Bowman's capsule, proximal convoluted tubule, loop of Henle, distal convoluted tubule and the collecting duct.

24. (D)
Hypertonic blood is caused by a lack of sufficient water in the body. Under these conditions, water must be conserved. To this end, the hypothalamus secretes antidiuretic hormone (ADH) which makes the cells of the collecting duct more permeable to water. There is also inhibition of the release of aldosterone. These two effects enable the kidney to excrete more sodium and less water – urine of greater concentration but lesser volume.

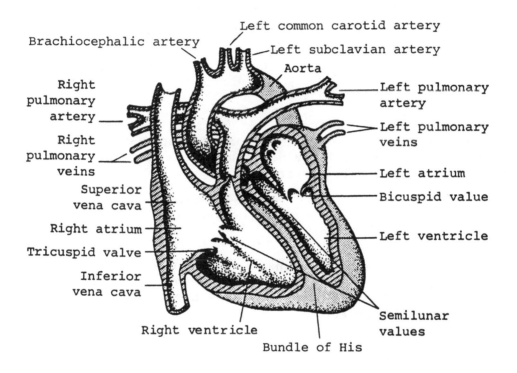

25. (D)
The pacemaker tissue of the heart consists of special fibers having their own automaticity. These fibers are located in the walls of the right atrium and it is from them that the heartbeat originates.

26. (C)
The bicuspid (mitral) valve consists of two flaps which separate the left atrium from the left ventricle. It prevents the backflow of blood from the left ventricle to the left atrium during ventricular systole.

27. (E)
Deoxygenated blood from the head, neck and upper limbs returns to the heart through the superior vena cava. Blood from the rest of the body returns through the inferior vena cava. Both venae cavae open into the right atrium.

28. (D)
As the aorta arches around the pulmonary artery it gives off its first set of branches. On the right, it gives off the brachiocephalic artery which then branches into the right common carotid and subclavian arteries. The left common carotid and subclavian arteries arise directly from the aorta.

29. (D)
The arteries generally carry oxygenated blood except for the pulmonary arteries which pump deoxygenated blood to the lungs where it is oxygenated.

30. (B)
In a normal individual, when the ventricles contract, there is a tremendous build up of pressure which reaches a maximum of about 120 mmHg. This pressure is transmitted to the walls of the aorta.

31. (B)
Marine amoebae lack a contractile vacuole because they are isotonic to the surrounding seawater. There is a balanced exchange of water and other substances with the environment, making a specialized transport structure unnecessary. Exchange occurs freely across the cell membrane.

32. (E)
The taste buds contain very specialized chemoreceptors which can distinguish certain chemicals. In this way, one is able to distinguish sweet, salty, sour and bitter tastes.

33. (D)
Vitamin C is abundant in fresh fruits, expecially citrus, tomatoes and potatoes. A deficiency of this vitamin leads to scurvy which is characterized by poor bone growth, hemorrhages and slow healing of wounds.

34. (B)
Dairy products are rich in calcium which is an essential component of bones and teeth. Calcium also plays an integral part in muscle contraction. When pregnant mothers do not take enough calcium, an insufficient amount of calcium crosses the placenta to the fetus. The newborn infant therefore develops rickets.

35. (A)
Bile salts have no enzymatic action as they do not change any substrate into another component. Their function is to emulsify fats so that lipases can hydrolyse them.

36. (A)
Vasodilation of the blood vessels of the skin is a means of losing heat and not a means of conserving heat.

37. (E)
Root tips show positive geotropism because they grow downward, i.e. with gravity. On the other hand, they show negative phototropism as they tend to grow away from light.

38. (A)
Excision of the parathyroid glands would result in a complete lack of parathyroid hormone (PTH), which would lead to low levels of calcium in the blood. Calcium plays an integral role in muscle contraction and its deficiency results in muscle convulsions and eventually death.

39. (E)
Testosterone belongs to the group of male sex hormones known as androgens. It is produced by the interstitial cells that lie outside the seminiferous tubules of the testes. It is vital to the development of the male's secondary sex characteristics, including body hair and baritone voice.

40. (C)
Epinephrine is one of the hormones released by the adrenal medulla. Its action is to increase the heart rate, cause vasodilation in skeletal muscle, and vasoconstriction in the splanchnic and skin circulations.

41. (D)
There is no single explanation for the ability of very tall plants to raise water to their leaves. Root pressure is thought to be an indirect effect of the active transport of inorganic nutrients into the root, which then sets up an osmotic gradient, allowing water to enter. Root pressure alone cannot produce enough pressure to force water up through tall trees.

Capillary action is due to the adhesive and cohesive properties of water.

Transpiration pull occurs through the evaporation of water from the cell surface, thereby setting up an osmotic gradient. Each depleted cell can generate a pulling pressure equal to 12 atmospheres, which is enough to take water to the top of the tallest tree.

42. (A)
In the alteration of generations in plants, the sporophyte begins with the fertilized egg cell. The egg cells mature and grow into the diploid sporophyte. The sporophyte generation ends with meiosis and the production of haploid spores, marking the beginning of the gametophyte generation.

43. (B)
The Casparian strip consists of partly suberized (waxy) cell walls in the endoderm. It blocks the passage of water

through the porous cell walls into the xylem, routing the water through the cytoplasm of the endodermal cells.

44. (D)
Auxin is believed to promote cell elongation in the growing tips of stems, while in root cells it may inhibit elongation. This leads to one hypothesis that roots and stems react differently to auxin. It can also be postulated that root cells are much more sensitive to inhibition by higher concentrations of auxin. As there is no actual proof for these hypotheses, some botanists prefer a simpler hypothesis: namely, that the root tip is producing some as yet unidentified auxin-like growth factor that moves upward.

45. (B)
Gibberellins received their name from the fungus Gibberella fujikuroi. Their action is to produce stem elongation and prevent the plant from producing normal flowers. They also act as chemical messengers to stimulate the synthesis of α-amylase in some seeds such as barley and corn.

46. (D)
The neurotransmitter acetylcholine is the chemical transmitter found at all the autonomic ganglia, presynaptic sympathetic nerve endings, all postsynaptic parasympathetic nerve endings, neuromuscular junctions and in some parts of the brain.

47. (C)

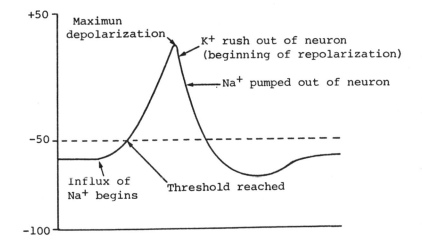

When a stimulus above the threshold value of a neuron is received, it sets in motion an action potential. The action potential begins with depolarization of the membrane which is brought about by a sudden influx of sodium ions (Na^+). As soon as Na^+ influx begins there is a leakage of potassium ions (K^+) to the outside of the membrane. The potassium efflux stays longer than the sodium influx and therefore slows the rise of the potential which finally returns to its initial level, i.e. repolarization. Active pumping of Na^+ and K^+ into the cell follows repolarization to restore the original intracellular ionic concentrations.

48. (B)

The cerebellum is found towards the back of the brain above the medulla. It controls balance, equilibrium and coordination. Damage to it would lead to loss of these functions. Since driving a motor car requires a great deal of coordination, one would expect that this ability would be impaired when the cerebellum is extensively damaged.

49. (A)

The lateral line organs of bony fishes are used to determine distant movements. They are sensitive to water motion. As animals move through the water, they create disturbances that are picked up in the lateral line canal where sensory endings detect them, sending impulses to the cranial nerves and brain.

50. (E)

The rattlesnake is able to detect its prey by infrared radiation. In the depressions near their eyes are heat-sensing devices which tell the snake when it is facing a living thing that is generating metabolic heat.

51. (C)

Essentially, the only difference between these two kinds of diffusion is their source of energy. Active transport requires that energy be expended by the cell often, because diffusion is taking place across an unfavorable concentration gradient. Passive diffusion occurs across a favorable concentration gradient, and does not cost the cell any energy.

52. (D)

The process described is exocytosis. Endocytosis is just the opposite – moving material from outside the cell to the inside of the cell. This is accomplished by buckling the cell membrane inward, forming a depression which is eventually surrounded by the membrane and then becomes pinched off to form a new vacuole inside the cell, holding contents which were previously outside the cell. These contents can then be digested and used for energy.

Endocytosis is the name given to the general process; phagocytosis and pinocytosis are specific forms of endocytosis. Phagocytosis occurs when visible, solid material is engulfed; pinocytosis occurs when only dissolved materials, such as proteins, are consumed.

53. (C)

Leucoplast is a form of plastids that specializes in the storage of starch. Chloroplast and chromoplast are also plastids. All plastids share the characteristics of having their own small genome, a double membraned envelope and an origin from proplastids which are small organelles found in meristematic cells.

54. (B)

Water, like all other substances, diffuses from regions of greater concentration to regions of lower concentration. The diffusion of water through a semipermeable membrane is called osmosis. In this case, the concentration of water is higher inside the cell than out, so water will diffuse out of the cell, causing the cell to shrink.

55. (D)

This law, known by both names, is fundamental to our understanding of energy and the myriad chemical reactions involved in sustaining life. In a closed system, energy is neither created nor destroyed, but it can and does change form. The second law of thermodynamics applies to this change in form: entropy always increases. This means that a closed system changes only to a lower energy state: from a state of lesser to greater randomness.

56. (A)

Since neither allele shows dominance, all heterozygotes are pink. Furthermore, all red and all white rabbits are homozygous.

    Let R = the red allele, and
        W = the white allele.

Then your genotypes and corresponding phenotypes are:

    RR = Red
    RW = Pink
    WW = White.

The cross is:

	R	R
W	RW	RW
W	RW	RW

all pink.

57. (D)

Cooperative behavior occurs when two or more animals act toward their mutual benefit. When the animals belong to the same species, it is called intraspecific cooperation; when they belong to different species, it is known as interspecific cooperation. In the case of the two wolves (intraspecific), each wolf benefits by helping the other catch the rabbit, and each sacrifices one-half the rabbit to the other in return for the help.

Altruism differs from cooperation insofar as it involves an activity which benefits another organism but at the individual's own expense. Altruism often is performed by parents protecting their offspring: many species of birds, for example, will allow themselves to be attacked while sitting on a nest rather than fleeing to safety and leaving the vulnerable young exposed.

Communication occurs in many forms and in all kinds of behavior - friendly, hostile, aggressive.

Recognition is the ability of individuals within a population to recognize and distinguish other individuals in the population. This is especially important in mate recognition, for each member of a pair wants to be certain that its efforts toward raising the young are directed to its own offspring, and not the offspring of other individuals.

58.   (C)
The bending of a plant shoot toward light in phototropism takes place as a result of unequal rates of growth between the side facing the light and the side shielded from the light.   Since growth of an organism involves cell division and enlargement, the difference in the rate of growth in the two sides of the plant reflects a difference in the rate of cell division or enlargement or both. Auxins are known to accelerate cell growth in actively growing regions, such as stem tips and vascular cambium.   In response to unidirectional light, it is found that there is a differential distribution of auxin in the stem; the side of the stem facing light receives a lower concentration of auxin than the side away from the light.   The shaded side elongates faster than the illuminated side, and the effect is that the tip and the top part of the stem curve toward the source of the light.

59.   (A)
Broken platelets release thrombokinase, which interacts in the plasma with calcium ions and the blood protein prothrombin.   Prothrombin is an inactive precursor of the catalyst thrombin.   In the presence of calcium ions and thrombokinase , prothrombin becomes thrombin, which reacts with fibrinogen, another of the blood proteins.   This reaction yields fibrin, an insoluble coagulated protein which forms the blood clot.

60.   (E)
The separated charges of a resting neuron are said to compose an electric potential.   These electric charges are carried by ions that are part of, or are attached to, the two sides of the cell membrane of the neuron.   In the rest state the positive and negative charges are prevented from coming together, and the membrane is said to be polarized electrically.   When an impulse sweeps along a nerve fiber, the permeability of the membrane charges at successive points along the fiber.   As this happens at any one point, an avenue is created through which the positive and negative ions of an adjacent point can pass, thus depolarizing that region.   In this manner, the impulse is propagated wave-like along the fiber.   Shortly after an impulse has passed a given point, the membrane at that point regains both its original state of permeability and polarization.

61. (E)

All three characteristics play an important role in dominance hierarchy. Physical condition is important because much of dominance position is attained through physical confrontation. Sex is a factor because females are usually weaker than males, and also are somewhat vulnerable and incapacitated during pregnancy and birth. Therefore females tend to be subordinate to males. Experience is a factor because it can contribute favorably to an individual's survival: an experienced individual may be more adept at finding food and avoiding danger than a less experienced individual.

62. (E)

Insulin has two methods of regulating blood glucose. First, it promotes absorption of glucose by the cells, possibly by making the plasma membranes more permeable to glucose. Second, insulin promotes the absorption of glucose by the liver and the conversion of glucose to glycogen within the liver. Both of these actions tend to reduce the glucose level of the blood to normal.

63. (A)

When an Rh-negative woman is made pregnant by an Rh-positive man, the fetus she carries may be Rh-positive. This can result in serious and even fatal complications, not in the first pregnancy, but in subsequent pregnancies.

The incompatibility does not occur in the first pregnancy, because the Rh-negative woman does not build up antibodies against the Rh-positive antigen until Rh-positive red blood cells enter her bloodstream. If this happens, it takes place during birth, therefore, the first infant is unaffected.

If, at the time of delivery, small quantities of the baby's blood cells enter the mother's bloodstream, the mother may build up antibodies against the Rh-positive antigen.

A subsequent pregnancy with an Rh-positive baby could then be disastrous, because the mother's anti-Rh-positive antibodies could enter the baby's bloodstream and destroy its Rh-positive red blood cells.

To prevent this, the mother is given an injection of antibodies against the Rh-positive antigen shortly after her first pregnancy. This destroys the Rh-positive cells before they can trigger the immune response. Subsequent pregnancies are as normal as the first, because the mother has no anti-Rh-positive antibodies.

64. (D)
Proteins accomplish their globular shapes by frequent reversals of the direction of polypeptide chains. This is done by β-turns which have the CO group of residue n of a polypeptide hydrogen bonded to the NH group of residue (n + 3).

α helix and β-pleated sheet also contribute to the conformation of proteins. α helix is a coiled rodlike structure with a polypeptide main chain forming the backbone; its structure is stabilized by hydrogen bonds between NH and CO groups in the same polypeptide chain. β-pleated sheet is an extended sheet-like structure stabilized by hydrogen bonds between NH and CO of different polypeptide strands.

65. (C)
Allosteric interaction refers to the interaction between spatially distinct sites. Hemoglobin is the best studied of the allosteric proteins. Its allosteric behavior is manifested by its cooperative binding propensity for oxygen, its pH dependent affinity for oxygen and its organic phosphates regulated affinity for oxygen.

The presence of a heme prosthetic group and a diminished affinity for carbon monoxide are properties of both hemoglobin and myoglobin.

66. (E)
The polarity and hydrogen bonding capabilities of water serves as a major driving force in hydrophobic and hydrophilic interactions. These two interactions participate in the folding of macromolecules, the formation of a close membrane system and many interactions between biological molecules.

67. (C)
The following is a graph showing the function of an enzyme:

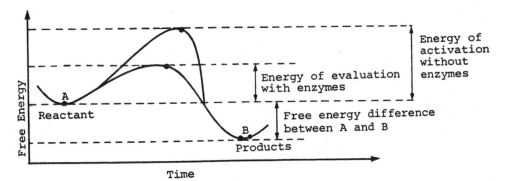

301

The enzyme decreases the energy barrier which must be overcome in order for the reaction to proceed, thus allowing the reaction to take place more readily. The difference in free energy between reactants and products does not change.

68. (A)
The first priority of succulent plants adapting to arid environments is conservation of water. Since most water loss will occur in the hot, dry air above ground, the plant must adapt by reducing the area exposed to this air. The less surface area exposed to the air, the less water the plant will lose.

69. (A)
Neutrophils are phagocytes that engulf foreign material, including bacteria and other phagocytes that have been destroyed at infection sites. Neutrophils work in cooperation with T-lymphocytes and B-lymphocytes, both of which produce antibodies in response to invading antigens. Neutrophils, T-lymphocytes, and B-lymphocytes are all involved in the immune response, whereas the larger macrophages which arise from monocytes, are primarily involved in response to physical injury (wounds which have cut through the skin).

70. (B)
After years of research, we know that DNA is the genetic material. In Griffith's experiment, the harmless rough cells took in smooth-cell DNA and were converted.

The demonstration that pure DNA extracted from smooth strain pneumonococcus could transform rough strain bacteria was made in 1944. However, for many years thereafter a good number of scientists remained unconvinced. They had thought that proteins were the genetic material, and old theories often die hard.

71. (D)
Both photosynthetic phosphorylation and oxidative phosphorylation produce ATP as the final energy-rich

product. They differ, however, in their source of energy, site of reaction, and in the nature of their electron donor and acceptor.

Photosynthetic phosphorylation uses light energy to synthesize ATP from ADP and inorganic phosphate. When light strikes a green plant it is absorbed by the chlorophyll. The chlorophyll molecule then ejects a high energy electron which passes down an electron transport chain. During this passage, the electron returns to its original energy level, producing ATP molecules in the process. The chlorophyll molecule serves as both the electron donor and electron acceptor, and the entire process takes place in the chloroplasts.

Unlike photosynthetic phosphorylation, which uses neither oxygen nor organic substrates, oxidative phosphorylation relies on both, because animals are incapable of utilizing sunlight as their energy source. Oxidative phosphorylation, which takes place in the mitochondria, uses the energy released when certain reduced substrates are oxidized. The electrons released are passed in a series of redox reactions down the electron transport chain in a system of electron acceptors of decreasing reduction potential. In oxidative phosphorylation, the ultimate electron acceptor is oxygen and the primary electron donor is sugar or some other organic substrates. As in photosynthetic phosphorylation, the final product is energy-rich ATP.

72. (C)

In cyclic photophosphorylation, light striking a chlorophyll a molecule excites one of the electrons to an energy level high enough to allow it to leave the molecule. The chlorophyll$^+$ molecule, having lost an electron, is now ready to serve as an electron acceptor because of its net positive charge. However, the ejected electron does not return to its ground state and the chlorophyll$^+$ molecule directly; instead it is taken up by ferredoxin and passed along an electric transport chain. As the electron passes from ferredoxin to the cytochromes and finally back to chlorophyll$^+$, two ATP molecules are produced. No oxygen is liberated, since water is not split; no NADP is reduced since it does not receive electrons. ATP is formed during the electron flow, and light energy is thus converted into chemical energy in the ATP molecules. However, since NADPH$_2$ is not formed, cyclic photophosphorylation is not adequate to bring about CO_2 reduction and sugar formation, processes which require the energy of NADPH$_2$.

Non-cyclic photophosphorylation produces both ATP and NADPH$_2$ molecules necessary for the dark reactions of photosynthesis. In this process, electrons from excited chlorophyll a molecules are trapped by NADP in the formation of NADPH$_2$ and do not cycle back to chlorophyll a. Electrons are ejected from chlorophyll b to be donated to chlorophyll a through a series of electron carriers. To restore chlorophyll b to its ground level, water is split into protons, electrons, and oxygen. The electrons are picked up by chlorophyll b, the hydrogens are used to form NADPH$_2$, and oxygen escapes to the atmosphere in its molecular form. Both photosystems I and II are involved in the process of non-cyclic photophosphorylation.

73. (A)
Plants, though they possess neither the sense organs nor the nervous system of animals, nevertheless respond to stimuli by means of tropisms. A tropism can be defined as a growth movement of an actively growing plant in response to a certain stimulus. This results in the differential growth or elongation of the plant toward or away from the stimulus. Tropisms are named after the kind of stimulus enticing them. Phototropism is a response to light; geotropism, a response to gravity; chemotropism, a response to a certain chemical; and thigotropism, a response to contact or touch.

Roots' downward growth is due in large part to the influence of gravity.

74. (A)
The most important functions of gibberellins are to lengthen stems and to stimulate pollen germination. They also play a role in inducing flower formation and in increasing the size of fruits in some species of plants.

Accelerating the rate of cell division and stimulating the elongation of cells fall under the control of another group of plant hormones, the cytokinins.

Inhibition of the growth of stems and leaves, and the promotion of shedding of both branches and leaves, are controlled by yet another plant hormone: abscisic acid. Abscisic acid, the key growth inhibitor in plants, provides a balance in growth by opposing the stimulating effects both of gibberellins and of cytokinins.

75. (D)
Acetylcholine and norepinephrine are two examples of neurotransmitters that affect gut motility. Paracrine

hormones differ from endocrine hormones in that they reach their target cells by diffusing through short intercellular spaces instead of through the circulatory system. Examples of paracrine hormones are somatostatin and vasoactive intestinal polypeptide (VIP).

76. (C)

Stomata are pores in plant stems and leaves and are not structures of a cell. The lipid membrane is common both to animals and plants, and chloroplasts, though structures in plant cells, are involved in the process of photosynthesis, not in the regulation of osmotic pressure.

The osmotic pressure of a plant cell is regulated by the cell wall. Consider the separate fates of an animal cell and a plant cell placed into very hypotonic solutions. The animal cell will take in water, causing it to swell, and if the original difference in osmotic pressure is great enough, the cell may take in more water and build up more pressure than its membrane can withstand, in which case the cell would burst (this is called lysis).

This lysis would not occur in a plant cell. A plant cell placed in a hypotonic medium would have water enter it, causing it to swell, however, an upper limit as to how much water can enter is imposed by the cell wall. As the cell swells, its plasma membrane exerts what is called turgor pressure on the cell wall. The wall exerts an equal and opposing pressure on the swollen membrane. Mature cell walls can be stretched only to a minute amount. When the pressure exerted by the cell wall is so great that further increase in cell size is not possible, water will cease to enter the cell. Thus, plant cells will only absorb a certain amount of water, even in an extremely dilute medium.

77. (E)

The femur is the large, upper leg bone which extends from the hip (pelvis) to the knee. The tibia and fibula are two smaller bones which extend from the knee to the ankle. The patella is the scientific name for the knee cap, and the humerus is the upper arm bone.

78. (C)

Procaryotic cells lack the internal membranous structure

characteristic of eucaryotic cells, and generally belong to the simple life forms, such as bacteria. Procaryotic cells lack endoplasmic reticulum, Golgi apparati, lysosomes, vacuoles, and a distinct nucleus. They are generally less organized and less advanced than eucaryotic cells.

79. (E)
Mitochondria are called the "powerhouse" of the cell because they are the major sites of ATP production in the cell. They contain the enzymes involved in both the Krebs (citric acid) cycle and the electron transport system, which work together to furnish 34 of the 36 ATPs produced by the complete oxidation of glucose. Another two ATPs are produced by glycolysis which occurs in the cytosol, making the total 36 ATPs per every oxidized molecule of glucose.

Pyruvate from glycolysis traverses both the outer and inner mitochondrial membranes easily because the membranes are totally permeable to pyruvate. Once inside the mitochondrion's matrix, pyruvate is acted upon by enzymes involved in the ATP-producing Krebs cycle. The large surface area of the greatly folded cristae facilitate the speed and efficiency of the reactions of the electron transport chain and of oxidative phosphorylation.

80. (D)
An amphoteric molecule can act either as an acid or a base, depending on the conditions of the reaction. Water is an example of an amphoteric molecule. Consider the following equations:

$$H_3O^+ \rightleftharpoons H^+ + H_2O \rightleftharpoons OH^- + 2H^+$$

Acid   Amphoteric   Base
         molecule

A given molecule acts as an acid when it donates a proton, thereby becoming a hydroxyl ion; or as a base when it accepts a proton.

81. (B)
The biosynthetic role of the citric acid cycle is exemplified by succinyl CoA which is the source of carbon atoms in porphyrins, α-ketoglutarate and oxaloacetate which are converted into amino acids.

82. (A)
Coenzyme A is one of the many carrier molecules in metabolism. It is named as Coenzyme A because it was found to be a cofactor required in many enzyme-catalyzed acetylations (A standing for acetylation). The terminal sulfhydryl group in CoA is the reactive site where acyl groups are linked to each other by a thioester bond ($R-\overset{\displaystyle O}{\overset{\displaystyle \|}{C}}\sim S\text{-CoA}$). The carrier molecule for phosphoryl group is ATP and that for methyl group is S-Adenosylmethionine.

83. (C)
Bacterial growth usually occurs by means of binary fission with one cell dividing into two and these two cells dividing into four, etc. The time interval required for each division (or for the entire population to double) is called the generation time.

To determine the generation time of a bacterial population, we must know the number of bacteria initially (No), the number of bacteria present at the end of a given time period (N), and the time period (t). We can determine the generation time by using some simple mathematical expressions. If we start with a single bacterium, the total population (N) after the nth generation is 2^n:

$$N = 1 \times 2^n.$$

Taking into account the fact that we usually start with more than one bacterium, we must modify the formula accordingly:

$$N = No \times 2^n.$$

The generation time G is simply equal to the time elapsed between No and N divided by the number of generations

$$G = \frac{t}{n}$$

Solving for n using logarithms, we find that

$$\log N = \log No + n \log 2$$

$$n = \frac{\log N - \log No}{\log 2}$$

$$n = \frac{\log N - \log No}{.301}$$

$$n = 3.3(\log N - \log No)$$

$$n = 3.3 \left(\log \frac{N}{No} \right)$$

$$n = 3.3 \left(\log \frac{20,000}{2} \right)$$

$$n = 3.3 (\log 10,000)$$

$$n = 3.3 \times 4$$

$$n = 13.2$$

$$G = \frac{t}{n}$$

$$G = \frac{132}{13.2}$$

$$G = 10$$

Therefore, generation time is equal to 10 minutes.

84. (A)

The steroid hormones consist of the glucocorticoids, the mineralcorticoids, and the sex hormones. They are secreted by the adrenal cortex. The two other major classes of hormones are the protein hormones and the fatty acid hormones. The protein hormones consist of five groups:

(1) amino acids

(2) small peptides - less than 30 amino acid groups

(3) large peptides - approximately 30 amino acid groups

(4) polypeptides - very long chains of amino acids (e.g. STH with 190 amino acid groups)

(5) glycoproteins - very high molecular weights. Included in this group are TSH, LH, FSH.

The fatty acid hormones have twenty carbon groups. The protein hormones and the fatty acid hormones are produced in various tissues throughout the body.

85. (D)

The first, second, and third order of neuroendocrine hormonal arrangement refer to the number of hormones involved in a particular regulation process. The first order, means that one hormone is used. An example is the regulation of the kidney by the hypothalamus, accomplished by the release of a single hormone, ADH. The second order means that two hormones are used. For example, a releasing hormone might be released from the pituitary, travel to another gland and stimulate it to release another hormone. Third order involves three hormones: for example, a releasing hormone stimulates a gland to release a hormone, which in turn stimulates another gland to release yet another hormone.

86. (D)
A kinase is an enzyme that transfers a phosphate from a donor to an acceptor. Typically, the donor is ATP - the energy-rich molecule manufactured in glycolysis and oxidative phosphorylation in mitochondria. A protein kinase is one which transfers the phosphate to a protein. Phosphate transfers occur in a vast number of critically important reactions.

87. (E)
Between the cavities of the atrium and ventricle in each half of the heart are the atrioventricular valves (AV valves), which permit blood to flow from the atrium to the ventricle but not from ventricle to atrium. The right and left AV valves are called, respectively, the tricuspid and the mitral valves.

88. (E)
The right side of the heart receives deoxygenated blood from the inferior and superior vena cava (veins). This deoxygenated blood is pumped by the right ventricle through the pulmonary artery to the lungs. There, the blood is oxygenated and is returned to the heart via the pulmonary vein. The pulmonary system is the opposite of the rest of the circulatory system: the pulmonary artery, unlike other arteries, carries deoxygenated blood, and the pulmonary vein, unlike other veins, carries oxygen-rich blood.

The pulmonary vein empties into the left atrium, which in turn pumps blood into the left ventricle. The left ventricle pumps oxygenated blood into the aorta, which branches into the many arteries which distribute blood to cells throughout the body.

89. (A)
Nondisjunction refers to the failure of chromosomes to separate after synapsis. Synapsis is a phase of meiosis in which homologous chromosomes wrap themselves around each other. When nondisjunction occurs, both chromosomes migrate into one of the sperm cells. If this sperm cell fertilizes an egg, the resulting zygote will have 47 chromosomes. The normal number of chromosomes for humans is 46.

90. (A)
Genes for hemophilia are carried on the X chromosome. In this example, the mother has no gene for the disease. The boy receives his X chromosome from the mother and the Y from the father. It is not possible for him to receive a gene for hemophilia.

91. (D)
External fertilization means that eggs are fertilized outside the female's body. It follows that if eggs are fertilized externally, then development of the embryo will take place externally also. Organisms in which fertilization is external must produce an enormous number of eggs. Since the egg cells (which are discharged from the body of the female) are unprotected, many of them are eaten or otherwise lost. Overproduction of eggs (producing more eggs than can be fertilized) is necessary in order to maintain the species number. External fertilization and external development of the young are the reproductive patterns present in most fish and amphibians.

92. (B)
Roan cattle result from crossing red cows with white bulls or vice versa. When roan cattle are interbred, the red and white strains segregate out as illustrated below.

		R	W	Roan bull
Roan cow	R	RR	RW	
	W	RW	WW	

RR = red
RW = roan
WW = white

The phenotype of the roan cattle is a result of the blending of genes for red and white color, neither of which is dominant over the other.

Independent assortment means that genes that are not linked on the same chromosome are inherited independently of each other. Dominance indicates a trait which is able to mask another; in this case, neither trait (red or white) is dominant. Natural selection is the condition in which only the most fit of a species survive.

93. (A)
Extracellular digestion means that digestion occurs outside of cells. All digestion in the paramecium is carried out within the single cell in the food vacuoles.

The alimentary canal of the earthworm consists of the mouth, pharynx, esophagus, crop, gizzard, intestine, and anus. These organs are arranged in a straight-line fashion. Digestion occurs inside of the intestine but outside of cells. Digestion therefore is extracellular.

Digestion in hydra is both extracellular and intracellular, occurring in both the digestive cavity and in cells that line the digestive cavity.

Digestion in the grasshopper is extracellular. The alimentary canal of the grasshopper consists of the esophagus, crop, gizzards, digestive glands, stomach, large intestine, small intestine, rectum, and anus. Digestion is completed in the stomach.

94. (E)
Calcium is important in blood clotting because it acts as cofactor for several enzymatic steps of the "cascade" by which many enzymes contribute to the eventual conversion of prothrombin to thrombin and fibrinogen to fibrin.

In transmission of nerve impulses, calcium serves as the link between membrane depolarization and neurotransmitter release. When an action potential in the pre-synaptic neuron reaches the end of the axon and depolarizes the terminal, small quantities of the neurotransmitter are released from the synaptic terminal into the synaptic cleft. This occurs because depolarization of the synaptic terminal causes an increase in the permeability of the terminal membrane to calcium ions. Calcium enters the presynaptic terminal during the action potentials and causes some of the vesicles to fuse with the cell membrane and liberate their contents into the synaptic cleft. Once released from the vesicles, the transmitter molecules diffuse across the cleft and bind to receptor sites on one membrane of the post synaptic neuron lying right under the synaptic terminal (subsynaptic membrane). The combination of the transmitter with the receptor site causes changes in the permeability of the subsynaptic membrane and thereby in the membrane potential of the postsynaptic cell.

Approximately 99 percent of the total body calcium is contained in bone, which consists primarily of a framework of organic molecules upon which calcium phosphate crystals are deposited. Contrary to popular opinion, bone is not an absolutely fixed, unchanging tissue but is constantly being remolded and, what is most important, is available for either the withdrawal or deposit of calcium from extracellular fluid.

In muscle contraction, calcium is important in determining when a muscle shall be in a state of contraction, and when it shall be in a state of relaxation. Relaxation is sustained by the inhibition of cross-bridge binding; this inhibition is due to two proteins, troponin and tropomyosin. In order for the cross-bridges to bind to actin (and thereby cause the muscle to contract), the tropomyosin molecules must be moved away from their blocking position. This occurs when calcium binds to a specific site on troponin. The binding produces a change in the shape of troponin such that it pulls the tropomyosin bound to it to one side, uncovering the cross-bridge binding sites on actin. Conversely, the removal of calcium from troponin reverses the process.

In summary, the availability of calcium ions to the troponin binding sites determines whether a muscle fiber is turned on or off.

Calcium does not play a role in ovulation.

95. (A)
In a reaction, molecules react only when they reach an adequate level of energy. This level of energy is called the activation energy for the reactions. When the temperature is raised, the molecules acquire greater energy, more molecules achieve activation energy and react, thus increasing (speeding up) the rate of reaction.

96. (C)
Increasing the temperature of the environment of a reaction speeds up the reaction, but after reaching a certain peak, the reaction rate tapers off and drops due to the

denaturation of enzyme at high temperature. There are other control mechanisms which keep reactions from becoming incontrollable. Two processes of control are induction and repression. Induction means that certain enzymes are synthesized only when substrate is available, through repression, the accumulation of product molecules turns off the synthesis of the enzyme responsible for forming the product. Both enzyme induction and repression are highly selective regulatory processes which act upon specific enzymes rather than affecting all of the enzymes in a cell.

The third factor determining the rate of an enzyme-mediated reaction is the catalytic activity of the individual enzyme in which the activity of certain enzymes is directly affected by the concentration of cofactors.

97. (E)
Smooth muscle is innervated by the autonomic nervous system, which is composed of the sympathetic and parasympathetic nervous systems. The autonomic nervous system is not normally under voluntary control. Skeletal muscle is ennervated by the somatic nervous system, and falls under voluntary control.

98. (B)
A reflex flexion is a fast reaction to a stimulus, such as the withdrawl of the foot when a painful stimulus is applied to the toe. When receptors in the toe are stimulated, the energy of the stimulus is transformed, via receptor potentials, into the electrochemical energy of action potentials in afferent neurons. The afferent fibers branch after entering the spinal cord, each branch terminating at a synaptic junction with another neuron. In the flexion reflex, the second neurons in the pathway are interneurons. Some of these interneurons carry information to the brain, others synapse upon the efferent neurons innervating flexor muscles. These muscles, when activated, cause flexion (bending) of the ankle and withdrawal of the foot from the stimulus.

Afferent neurons, in addition to stimulating neurons which trigger the withdrawal of the foot, serve to inhibit neurons which would oppose this action. This inhibition of antagonistic motor neurons is called reciprocal innervation. In this particular case, the motor neurons innervating the foot and leg extensor muscles (which straighten the leg and ankle) are inhibited.

Dual innervation occurs when glands, organs, or muscles are innervated by both the sympathetic and the parasympathetic nerve fibers of the autonomic nervous system.

99. (B)
The labyrinth of the inner ear has three semicircular canals, each consisting of a semicircular tube connected at both ends to the utriculus. Each canal lies in a plane perpendicular to the other two. At the base of each canal, where it leads into the utriculus, is a bulb-like enlargement (the ampulla) containing tufts of hair cells similar to those in the utriculus and sacculus, but lacking otoliths. These cells are stimulated by movements of the fluid (endolymph) in the canals. When a person's head is rotated, there is a lag in the movement of the endolymph in the canals.

Thus, the hair cells on the ampulla attached to the head rotate, in effect, in relation to the fluid. This movement of the hair cells with respect to the endolymph stimulates the former to send impulses to the cerebellum of the brain. There, these impulses are interpreted and a sensation of dizziness is felt.

100. (C)
A cross-section of the spinal cord reveals that the paired dorsal roots are the junctions where sensory fibers from the peripheral receptor areas enter, while the paired ventral roots exist where axons of the motor neurons leave the spinal cord and go to effectors, such as muscles. When the dorsal root of a spinal nerve is cut, the sensory fibers are severed and afferent impulses can no longer reach the spinal cord nor the brain. When the ventral root is severed, injury is inflicted on motor neurons. Hence, in the case of a dog in which both dorsal and ventral roots have been severed, the animal would be unable to feel the pain, and also would have lost the ability to withdraw its foot even if it could feel the pain.

101. (C)
In addition to hypoxia (lack of oxygen at the tissue level), decompression sickness may result from a rapid decrease in barometric pressure. In this event bubbles of nitrogen gas form in the blood and other tissue fluids, on the

condition that the barometric pressure drops below the total pressure of all gases dissolved in the body fluids. This can cause dizziness, paralysis, and unconsciousness.

Deep sea divers are greatly affected by the "bends". Divers descend to depths where the pressure may be three times as high as atmospheric pressure. Under high pressure, the solubility of gases (particularly nitrogen) in the tissue fluids increases. As divers rise rapidly to the surface of the water, the accompanying sharp drop in barometric pressure causes nitrogen to diffuse out of the blood as bubbles, resulting in decompression sickness.

102. (A)

An autosome is a chromosome that is not a sex chromosome. If two genes are located on the same chromosome, they will be inherited together.

Since an autosome is not a sex chromosome, the traits cannot be sex-linked. The fact that the two traits are on the same chromosome has nothing to do with whether they are dominant or recessive.

103. (D)

Striated muscle features a series of transverse bands forming a regular pattern along each muscle fiber. These bands are composed of thick myosin filaments and thin actin filaments. Both cardiac and skeletal muscle are of this type.

Smooth muscle, although it contains actin and myosin for contractile purposes, lacks the cross-striated banding pattern found in striated muscle.

104. (A)

Cross-bridges facilitate the sliding movement of myosin and actin filaments. They extend from thick filaments to thin filaments, and are arranged in a helix around the thick filament so that those extending from a single thick filament can make contact with each of the surrounding thin filaments. When a cross-bridge is activated, it moves in an arc parallel to the long axis of the thick filament, if, at this time, the cross-bridge is attached to a thin filament. This swiveling motion slides the thin filament toward the center of the A band, thereby producing shortening of the sarcomere.

105. (C)
A catalyst lowers the energy of activation of a reaction, thus facilitating transition from reactants to products. The catalyst does not, however, erase the energy barrier altogether, and the reactants and products start and arrive at the same energy levels as in the uncatalyzed reaction.

In graph C, the energy of activation is significantly lowered in the catalyzed reaction, while the reactants and products begin and end at the same levels in both the catalyzed and uncatalyzed reactions.

Graph B has the catalyzed and uncatalyzed reactions reversed.

In graph A, the catalyzed reaction shows an energy of activation which is lower than the final energy state of the products – this is impossible. Graph E shows the same thing – an energy of activation lower than the energy of the products – and also has the catalyzed reaction needing more energy than the uncatalyzed reaction. In graph D, the catalyzed and uncatalyzed reactions are reversed, and the energy levels of reactants and products before and after the reaction are not the same for both reactions.

106. (D)
The duodenum of the small intestine receives the secretions of two accessory organs, the liver and the pancreas. The liver, an organ of many functions, aids in digestion by secreting a slightly alkaline fat emulsifier known as bile. Bile is a complex substance containing salts, pigments, water, and a nucleoprotein. The bile salts are actually steroids, and are important in dissolving digested fats. The bile is stored in the gall bladder after it is produced in the liver. After a meal, the release of bile is brought about by the hormone cholecystokinin, which causes the gall bladder to contract. Bile reaches the duodenum via the bile duct, which is joined by highly alkaline fluids from the pancreatic duct just before it enters the intestine. The bile breaks down fats into minute droplets. As such, the fats can be digested by lipase (the fat-digesting enzyme). If the bile duct becomes obstructed, major digestive difficulties occur.

107. (E)
There are no centrosomes in plant cells, but spindle formation does occur during plant cell division. In plant cells, the spindle apparatus is synthesized directly by the

cytoplasm, whereas in animal cells, spindle formation occurs at the centrioles of the centrosome.

Plant and animal cell division also differ in the way new cells are formed after replication has taken place. Animal cells divide by a cleavage furrow - the outside cell membrane invaginates. Plant cells divide by the formation of a cell plate, on which a new cell wall eventually forms between the new daughter cells. This plate is synthesized with growth from the center outward.

108. (B)
Glycocholate is a major bile salt which is also derived from cholesterol. Sphingosine is the backbone of sphingomyelin, a phospholipid in membrane which is not derived from glycerol. Phosphatidate is a common intermediate in the synthesis of phosphoglycerides and triacylglycerols.

109. (C)
The membranes of the cells of small aquatic animals and plants are in direct contact with their aqueous environment, and therefore are able to exchange water and other important nutrients directly by diffusion. Transport systems are therefore unnecessary.

110. (D)
There are four blood types: A, B, AB, O. The basis of these bloodtypes is the presence or absence of certain antigens, called agglutinogens, on the surface of the erythrocytes. The erythrocytes of type A blood carry agglutinogen A; those of type B carry agglutinogen B. Type O has neither of these agglutinogens, while type AB has both A and B agglutinogens. These agglutinogens react with certain antibodies, called agglutinins, that may be present in the plasma. An agglutinin-agglutinogen reaction causes the cells to adhere to each other. This clumping, or agglutination, would then block the small blood vessels in the body, causing death.

Agglutination is the clumping of erythrocytes by agglutinins. The blood clotting mechanism, or coagulation, is not involved. Type A blood, which has the A agglutinogen, does not have the anti-A agglutinin in its plasma. But it does have the anti-B agglutinin. A person with type B blood has only anti-A agglutinin. A person with type O blood has both anti-A and anti-B; type AB has neither type of agglutinin. Basically, a person does not have the agglutinin in his plasma which would clump his own red blood cells. A summary of the antibodies and antigens that each blood type contains is outlined in the following table:

Blood Type	Agglutinogens (antigens)	Agglutinins (antibodies)	Able to accept blood type
A	A	anti-B	A, O
B	B	anti-A	B, O
AB	A and B	none	A, B, AB, O
O	none	anti-A and anti-B	O

Blood typing is a critical factor in blood transfusions. Transfusions are usually between people of the same blood type; however, one may use another bloodtype provided that it is compatible.

In this problem, we know that John can receive Carla's blood, but that Carla cannot receive John's. Choices (A) and (C) give blood types which fit this situation. In choice (B), John cannot receive Carla's blood, because his anti-A agglutinins would clump with Carla's A agglutinogens and agglutination would result.

111. (C)
The capillaries are extremely narrow, so much so that their diameter can accommodate only one red blood cell. They consequently have enormous surface area, and materials can readily diffuse through their thin walls. Materials are exchanged between the capillaries and the cells outside the bloodstream, and between the capillaries and the intercellular, tissue fluids. The main function of these microscopic blood vessels is to provide a means for exchange, via diffusion, between the bloodstream and the rest of the cells in the body.

112. (A)
FSH and LH are secreted by the anterior pituitary, while estrogen and progesterone are secreted by the ovaries.

318

113. (A)

Uracil is a base in the RNA molecule, not on the DNA molecule. Uracil replaces thymine. The other nitrogenous bases - adenine, guanine, and cytosine - are found in both RNA and DNA.

114. (B)

Nucleic acid contains the genetic information. When a virus infects a bacterium, it infuses its own nucleic acid into the bacterium, and the bacterium, using the genetic code of the virus, begins to produce hundreds of more viruses.

115. (B)

During transcription, free RNA nucleotides containing adenine will pair with any exposed thymine base in DNA; likewise, RNA nucleotides containing guanine, cytosine, and uracil will pair with exposed cytosines, guanine and adenine bases in DNA, respectively. In this way, the nucleotide sequence in DNA (the genetic code) determines the sequence of nucleotides along a molecule of messenger RNA.

116. (A)

STH - somatotropin - is the growth hormone which is species specific for humans. It is released from the anterior pituitary during periods of fasting and during sleep. It is released in spurts during sleep; in children we see the characteristic growth "spurts" due to this periodic release.

117. (C)

The disease can be transmitted to both sons and daughters. Both men and women have the X chromosome, and since this disease is X-chromosome dominant, both men and women will be afflicted if they inherit the X-chromosome carrying the disease.

118. (B)

When doing a problem such as this, it is helpful to list the order of dominance:

Order of Dominance

c^+ colored

c^{ch} chincilla

c^h himalayan

c albino

Then, draw a punnet square and compute the results:

$c^+c^h \times c^{ch}c$

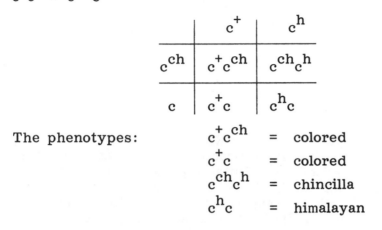

	c^+	c^h
c^{ch}	c^+c^{ch}	$c^{ch}c^h$
c	c^+c	c^hc

The phenotypes:

c^+c^{ch}	=	colored
c^+c	=	colored
$c^{ch}c^h$	=	chincilla
c^hc	=	himalayan

Thus the ratio is 2 colored:1 chincilla:1 himalayan.

119. (C)
DNA, or deoxyribonucleic acid, is in the form of a double helix having a deoxyribose sugar and phosphate backbone. The two helices are linked together by hydrogen bonds between nitrogenous bases bound to the sugar moiety of the backbone.

In DNA, the bases are adenine, guanine, cytosine, and thymine. RNA, or ribonucleic acid, differs from DNA in three important respects. First, the sugar in the sugar-phosphate backbone of RNA is ribose rather than deoxyribose. Ribose has hydroxyl groups on the number 2 and 3 carbons, whereas deoxyribose has a hydroxyl on the number 3 carbon only. Secondly, RNA has only a single sugar-phosphate backbone with attached single bases, and thus is similar to a single strand of DNA. Finally, the pyrimidine base uracil is found in RNA instead of the pyrimidine base thymine found in DNA.

120. (E)
The first cross tells you that tall is dominant to dwarf and that terminal is dominant to axial.

```
T  = tall
t  = dwarf
A  = terminal
a  = axial
```

Ta × tA → TtAa   F₁

TtAa × TTaa

	TA	Ta	tA	ta
Ta	TTAa tall terminal	TTaa tall axial	TtAa tall terminal	Ttaa tall axial
Ta	TTAa tall terminal	TTaa tall axial	TtAa tall terminal	Ttaa tall axial
Ta	TTAa tall terminal	TTaa tall axial	TtAa tall terminal	Ttaa tall axial
Ta	TTAa tall terminal	TTaa tall axial	TtAa tall terminal	Ttaa tall axial

Results: All tall; half terminal; half axial.

121.   (B)
```
R  = blond
r  = red
N  = tasting
n  = nontasting
```

Homozygous Blond Tasters: RRNN
Homozygous Blond Nontasters: RRnn

RRNN × RRnn = RRNn   F₁

RRNn × RRNn → F₂:

	RN	Rn	RN	Rn
RN	RRNN blond taster	RRNn blond taster	RRNN blond taster	RRNn blond taster
Rn	RRNn blond taster	RRnn blond nontaster	RRNn blond taster	RRnn blond nontaster
RN	RRNN blond taster	RRNn blond taster	RRNN blond taster	RRNn blond taster
Rn	RRNn blond taster	RRnn blond nontaster	RRNn blond taster	RRnn blond nontaster

RR ⟨ 1NN → RRNN
     2Nn → 2RRNn
     1nn → RRnn

¼ blond tasters
½ blond tasters
¼ blond nontasters
3/4 blond tasters:
1/4 blond nontasters

122. (D)

The affected daughter must be homozygous recessive; therefore each of the normal parents must be heterozygous.

Consider the possibilites for a normal son's genotype, as seen in a punnet square.

	N	n
N	NN	Nn
n	Nn	nn

N = normal
n = lobster claw

The son cannot be homozygous because he is normal. Of the three remaining possibilites, two contain the recessive allele, which would make him a carrier. Thus the odds that he is a carrier are 2/3, or 67%.

123. (A)

Circadian, annual, and monthly cycles are the most common. The most obvious circadian rhythm is sleep/wakefulness, which for many animals tends to parallel the 24 hours cycle of day (sunlight) and night (darkness). Annual cycles include migration, hibernation and mating. Monthly patterns include the ovulation cycle in many animals.

124. (E)

All of the statements are true regarding biological succession. If no disruptive factors interfere, most successions eventually reach a stage that is more stable than those that preceded it. The community of this stage is called the climax community. It has much less tendency than earlier successional communities to alter its environment in a manner injurious to itself. In fact, its more complex organization, larger organic structure, and more balanced metabolism enable it to control its own environment to such an extent that it can be self-perpetuating. Consequently, it may persist for long periods of time provided that major environmental factors remain essentially the same.

In general, the trend of most successions is toward a more complex ecosystem in which less energy is wasted and a greater biomass is supported.

125. (D)

The "ten percent rule" refers to the average ecological efficiency per trophic level. It is the amount of energy which is passed on from one level to the next. For example, consider a simple ecosystem in which there exists a field of clover, mice which eat the clover, rabbits which eat the mice, and wolves which eat the rabbits. For every 10,000 calories available from clover plants, only 1000 calories will be obtained for use by a mouse. When a rabbit eats the mouse, only 100 of the 1000 calories will be available to the rabbit. In the last trophic level, the wolf will obtain 10 percent of the 100 calories transferred to the rabbit. Thus, a mere 10 calories of the original 10,000 calories can be used for the metabolic processes of the wolf.

126. (B)
Sori are circular structures found on the underside of the leaves of many species of ferns. Each sorus contains a group of sporangia covered by a scale-like indusium.

127. (C)
The vascular system in monocotyledons occurs in scattered bundles and the vascular cambium is usually absent.

128. (D)
One of the distinguishing features of dicotyledons is the net-like arrangement of the leaf veins. In this arrangement there is usually one or more central veins from which several smaller veins branch off.

129. (E)
The red algae consists of about 300 species of seaweed belonging to the division Rhodophyta. All the species are aquatic and nearly all of them are marine.

130. (A)
Stoneworts are usually found at the bottom of sluggish streams, ponds and lakes. They derive their name from their ability to incorporate calcium salts into their cell walls.

131. (D) 132. (C) 133. (C) 134. (D)

131-134.

In the ABO blood group system, there are four types of blood determined genetically: A, B, AB, O. The type of blood is determined by the presence or absence of certain antigens (agglutinogens) on the surface of the erythrocytes. The erythrocytes of type A carry agglutinogen A; type B carry agglutinogen B; type AB has both A and B agglutinogens, while type O has no agglutinogens. These agglutinogens react with certain antibodies (agglutinins) that may be present in the plasma. This reaction leads to clumping of the erythrocytes by a process termed agglutination.

Group A blood, which contains agglutinogen A, has anti-B agglutinin in its plasma. Group B has anti-A agglutinin. Group O has both anti-A and anti-B agglutinins while group AB has neither type of agglutinin. It can be seen that a person does not have the agglutinin in his plasma which would agglutinate his own erythrocytes.

As group O blood has no agglutinogens, it can be given to any of the other blood groups without any serious effects. The agglutinins present in group O would produce some amount of agglutination but since they are diluted by the recipient's blood, agglutination is insignificantly low and is ignored. Hence group O is called the universal donor.

Group AB can receive from any other group as it has no agglutinins in its plasma to react with the donor's erythrocytes. It is therefore called the universal recipient.

135. (E)

Erythroblastosis fetalis is a condition in which the erythrocytes of a fetus have been agglutinated by the leakage of certain antibodies from the maternal circulation into the fetal circulation. The antibodies have been produced in response to the Rhesus antigen (Rhesus factor). Normally an individual's plasma will not have antibodies against the Rh antigen unless it has been sensitized to the antigen by previous exposure to it. Thus, while Rh-positive individuals have the Rh antigen, Rh-negative individuals have neither an antigen nor an antibody against the antigen.

The condition occurs when a Rh-negative mother marries a Rh-positive father. If the mother has a Rh-positive child and some blood from the fetus enters the maternal circulation during the birth process, its antigens will sensitize the mother, who will then produce antibodies. The antibodies are not produced rapidly enough and in enough amounts to affect the first child before birth. But if the mother has subsequent pregnancies, she will produce enough antibodies which can enter the blood of an Rh-positive fetus during the late stage of pregnancy and agglutinate its erythrocytes.

136. (A)

137. (E)

138. (C)

139. (D)

140. (A)

141. (C)

142. (B)

143. (B)

The pupil is a small opening through which light can enter the lens. The lens itself is transparent and elastic, and changes shape to focus on objects at different distances. The cornea is a protective, transparent, curved layer which covers the eye, and through which light passes into the eye. The region between the cornea and the lens is a cavity filled with a liquid called the aqueous humor. The larger chamber behind the lens is filled with a more viscous substance, the vitreous humor.

The amount of light entering the eye can be regulated by the contraction or expansion of the pupil. The pupil's size is controlled by a thin, muscular structure called the iris. Light, after passing through the cornea and pupil, is focused by the lens onto the retina. The retina, located at the rear of the eye, contains an abundance of sensory cells called rods and cones, which contain light-sensitive molecules called photopigments. Rods and cones function to perceive light and color.

The shape of the lens is controlled by small, fine muscles which extend from the tissue along the eye cavity to the lens itself. As these muscles contract and relax, the shape of the lens changes, allowing it to focus upon objects at various distances.

144. (B)

145. (C)

146. (A)

144-146.
The glomerulus is a tiny ball of blood capillaries surrounded by the Bowman's capsule. In each nephron, blood is filtered from the glomerulus to the interior space of the Bowman's capsule, where wastes are removed from the blood. Initial urine then flows through the nephric tubule, which is composed of the descending and ascending loop of Henle. The nephric tubule is highly convoluted and therefore extremely long and has enormous surface area, which is extremely important for the reabsorption of water. Water is in fact the most abundant of the substances that are filtered out from blood and later returned to it.

147 – 150.

147. (E)   148. (A)   149. (B)   150. (D)

An instinct is a pattern of unlearned, automatic behavior that is usually beneficial to the species. Insight learning is the ability to use reason in order to solve a problem at hand. It is most common among the primates, especially chimpanzees, monkeys, and man. A habit is a learned response. It is usually activity requiring little thought, thus, leaving the conscious free for more complicated thought processes. A tropism is a growth pattern such as the phototropism of a plant toward a light source. A taxis refers to an orientation pattern such as the photoaxis of bees following the sun in order to maintain a straight flight pattern.

151 – 155.

151. (E)   152. (C)   153. (A)   154. (D)   155. (B)

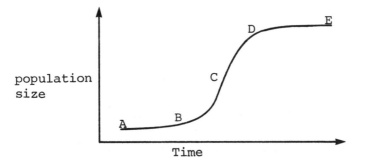

A = Lag phase of growth: that period of time during which the culture is first reproducing

B = Logarithmic phase of growth: that period of time during which there is a sufficient number of members reproducing exponentially.

C = Point of optimal yield: point of largest rule of increase within a given enviroment.

D = Stationary phase of growth: that period of time during which the birth rate is equal to the death rate.

E = Point of carrying capacity: point in time at which the largest number of organisms of a particular species that may be maintained in a given environment.

156. (A) 157. (B) 158. (D) 159. (C) 160. (E)

156–160.

Competition occurs when individuals or species vie for land or resources. When space and resources are limited, one individual or species will usually be defeated, and eliminated, by another.

Mimicry occurs because it is selectively favorable for one or more species. One species can avoid predators by mimicking the appearance of another species which is repulsive to predators.

Parasitism is a relationship in which one organism feeds off another organism. Most parasites live inside their hosts and suck up nutrients from various organs. The majority of parasitic relationships are harmful to the host, and some parasites can even be fatal to their hosts. Parasites would benefit were they to evolve into a form less damaging to their hosts. Killing their hosts can put the parasites at a great disadvantage.

Mutualism describes a relationship between two organisms in which both organisms benefit.

Allelopathy is the limiting of a population by the waste products which that population produces. In recent decades, much attention has been devoted to human allelopathy. Many environmentalists fear that the human race may poison itself to the point of extinction. Oil spills in the oceans, acid rain, deforestation and toxic wastes are only a few of the threats to the oxygen and food supply. It is not inconceivable that humans could destroy their environment to the point that it could no longer support them.

161. (D)

The nucleus is the control center of the cell. The nucleus contains the chromosomes, which bear the genes, the ultimate regulator of life.

The genes contain the information that specifies the precise nature of each protein synthesized by the cell. Control of protein synthesis is the key to controlling the activities and responses of the cell, since a tremendous array of important biological and biochemical processes are regulated by enzymatic proteins. By switching particular genes on and off, the cell controls not only the kinds of enzymes that it produces, but also the amounts. Fine control of enzymatic proteins is crucial to the proper functioning of the cell.

162. (A)
Mitochondria are membrane-bounded organelles concerned principally with the generation of energy to support the various forms of chemical and mechanical work carried out by the cell. They are distributed throughout the cytoplasm, and are most abundant in regions of the cell that consume large amounts of energy.

Mitochondria are enclosed by two membranes. The outer one is a continuous delimiting membrane. The inner membrane features many folds, called cristae, that extend into the interior of the mitochondrion. Inside the inner membrane are various enzymes that catalyze a number of different reactions, all of which lead to the production of ATP.

163. (B)
The endoplasmic reticulum is responsible for transporting certain molecules to specific areas within the cytoplasm. In addition to transporting lipids and proteins, the endoplasmic reticulum contains enzymes important in cellular metabolism.

The membranes of the endoplasmic reticulum (ER) form interconecting channels that can extend from the nuclear membrane to the plasma membrane. When the ER has ribosomes attached to its surface, it is called the rough endoplasmic reticulum (RER) and when there are no ribosomes attached, it is called the smooth endoplasmic reticulum (SER).

In cells actively engaged in protein synthesis, the protein or polypeptide chains are synthesized in the ribosomes and are then transported by the endoplasmic reticulum to other sites of the cell where they are needed.

164. (C)
The function of the Golgi apparatus is best understood in cells involved in protein synthesis and secretion. Protein synthesized on the rough endoplasmic reticulum is carried by vesicles to the Golgi bodies, where the protein is concentrated and condensed. The Golgi bodies release the protein in the form of secretory granules, which are then separated from the cytoplasm by a membrane that can fuse with the plasma membrane. When the secretory granule fuses with the plasma membrane, its content (protein in this case) is expelled from the cell, a process known as exocytosis.

The Golgi complex also has a synthetic role: it contains enzymes involved in the synthesis of complex carbohydrates and in the coupling of these carbohydrates to protein.

165. (E)
The plasma membrane consists primarily of lipid and protein. The lipid molecules are polar, with the two ends of each molecule having different electric properties. One end is hydrophobic, the other is hydrophilic. The lipid molecules arrange themselves in two layers in the plasma membrane so that the hydrophobic ends are near each other, and the hydrophilic ends face outside toward the water and are stabilized by water molecules. In this bilayer, individual lipid molecules can move laterally, so that the bilayer is actually fluid and flexible.

Protein molecules of the plasma membrane may be interspersed in various sites but embedded at different degrees in relationships to the bilayer. Some of them may be partially embedded in the lipid bilayer, some may be present only on the outer surfaces, others may be completely hidden in the interior, and still others may span the entire lipid bilayer from one surface to the other. The protein molecules, like the lipid molecules, are able to move laterally within the plane of the membrane.

The primary function of the membrane is to screen substances which would enter and leave the cell. The selectively permeable membrane allows the cell to regulate its internal environment.

The membrane is also important in cell adherence. The specificity of protein molecules on the membrane surface allows cells to recognize each other and to bind together through some interaction of the surface proteins. Communication between cells is important during cell division so that cells divide in an organized plane, rather than in random directions giving rise to an amorphous mass of cells as in cancer. Surface proteins play important roles in the immune response, hormonal communication, and conduction of impulses in nerve cells.

166-170.
166. (A) 167. (B) 168. (C) 169. (D) 170. (E)

Many of the answers for questions 166-170 are explained in the questions themselves, and should be self-explanatory.

Selective precipitation, also known as "salting out" involves the precipitation of proteins at high ionic strength; different proteins salt out at different ionic strengths.

Ion-exchange chromatography depends on the ionic attractions between protein molecules in solution and inert material (matrix) introduced to the solution.

Affinity chromatography purifies proteins by targeting specific proteins which are uniquely designed to combine with certain compounds, most notably certain other proteins.

Gel filtration relies on molecular weight. It involves a column of beads that acts as a molecular seive, allowing molecules only of a certain weight and size to pass through the column.

Finally, electrophoresis relies on both the electric charges of protein molecules and then molecular weight in order to separate them. Given the opportunity, proteins with an overall negative charge will move toward the positive pole, the anode, while positively charged proteins will move in the opposite direction - to the cathode. The distance that a particular protein migrates depends on its molecular weight and charge, which in turn depend on the pH of the medium. Since it is unlikely that two proteins will be present in a partially purified mixture with combinations of charge and size that cause them to migrate similarly, electrophoresis is a valuable technique to display the variety of proteins present in the mixture.

171. (C) 172. (B) 173. (A) 174. (D) 175. (E)

171-175.
The specific characteristics of the structures of the sarcomere are given in the questions themselves.

During contraction, the actin and myosin filaments slide together; however, the lengths of neither the actin nor the myosin filaments change during contraction. The H zone and the I band both decrease in width, as does the area between Z lines.

176. (B) 177. (A) 178. (C) 179. (E) 180. (D)

176 - 180
A resting muscle fiber contains all the ingredients necessary for cross-bridge activity - actin, myosin, ATP, and magnesium ions. Hence the question arises: Why are

muscles not in a continuous state of contraction? The reason is that in a resting, relaxed muscle fiber, the cross-bridges are unable to bind to actin and initiate the cross-bridge cycle that leads to contraction. This inhibition of cross-bridge binding is due to two proteins: troponin and tropomyosin. Tropomyosin molecules partially cover the myosin binding sites on the actin molecules, thereby preventing them from binding to the myosin cross-bridges. Each tropomyosin molecule is held in this blocking position by a molecule of troponin, itself bound to both tropomyosin and actin.

In order for the cross-bridges to bind to actin, the tropomyosin molecules must be moved away from their blocking position. This occurs when calcium binds to a specific site on troponin. Thus when calcium is present, the muscle contracts; when calcium is absent, the muscle relaxes.

The role of ATP in muscle contraction is essentially the same as in the rest of the body: it provides energy.

ATP performs two distinct roles in the cross-bridge cycle:

(A)   The energy released from the splitting of ATP provides the energy for cross-bridge movement;

(B)   the binding (not splitting) of ATP to myosin breaks the link between actin and myosin at the end of a cross-bridge cycle, allowing it to be repeated. Thus both contraction and relaxation of muscles requires energy derived from ATP.

Myoglobin is a protein similar to hemoglobin found in red blood cells. It binds oxygen for diffusion into muscle in high oxidative fibers – that is, fibers which rely primarily on oxidative phosphorylation rather than glycolysis for ATP. Oxidative phosphorylation requires oxygen (hence the importance of myoglobin) whereas glycolysis does not require oxygen. Fast twitch muscles have a high capacity for glycolysis, and hence are surrounded by few capillaries and contain little myoglobin.

181. (E) 182. (B) 183. (A) 184. (E) 185. (D) 186. (C)
181-186.
The heart pumps blood via controlled, coordinated contractions of its parts. The left and right atria contract simultaneously, filling the ventricles with blood. Then the ventricles contract, sending blood to the lungs via the right ventricle, and through the aorta to the rest of the body via the left ventricle. The period of ventricular contraction

is called systole; it is followed by a period of ventricular relaxation called diastole.

Cardiac output is a measure of heart productivity. It is defined as the amount of blood pumped by each ventricle per minute (note that this is not the same as the amount of blood pumped by both ventricles per minute). The stroke volume is the amount of blood pumped by each ventricle per heartbeat.

Cardiac output can be determined by the following equation:

$$\text{cardiac output} = \text{heart rate} \times \text{stroke volume}$$

$$\text{L/min} \qquad \text{beats/min} \qquad \text{L/beat}$$

The end-diastolic volume and end-systolic volume are the quantities of blood in the ventricles at different stages of contraction and relaxation. During diastole, blood flows from the atrium (which is receiving blood from the pulmonary artery) to the ventricle. Actually this filling of the ventricle before atrial contraction accounts for approximately 80 percent of the total ventricular volume. The rest of the blood flows into the ventricle when the atrium contracts. The amount of blood in the ventricle just prior to systole is called the end-diastolic volume. When the ventricle contracts, it pumps blood into the aorta (left ventricle) or the pulmonary artery (right ventricle). However, the ventricles do not empty completely during systole, and the amount of blood remaining after ejection is called the end-systolic volume.

187. (E) 188. (B) 189. (A) 190. (C) 191. (D)

187-191.
Tetanus is the state of greatest possible tension that a muscle can develop. It is a response to summation, which is the increased contraction of a muscle in response to action potentials occurring in rapid succession. The greater the frequency of stimulation, the greater is the intensity of the mechanical response (summation) until a frequency is reached beyond which the response no longer increases. This is the greatest tension the muscle can develop and is generally about three to four times greater than the isometric twitch tension produced by a single stimulus.

Recruitment is an increase in the number of motor neurons that are actively discharging action potentials to a given muscle. The more neurons firing, the greater the speed and tension with which the muscle contracts. Thus, a muscle will jerk a hand away more violently from a burning hot surface than from a mildly hot surface, because more motor neurons have fired.

Hypertrophy is the increase in size of muscles due to repeated neural stimulation. In addition to the increase in size, the chemical composition of the muscles may be altered as well. Action potentials in nerve fibers appear to release chemical substances which influence the biochemical activities of the muscle fiber. The identity of these tropic agents is unknown.

Pacemaker potentials occur primarily in smooth muscle. Some types of smooth muscle fibers generate action potentials spontaneously in the absence of any neural or hormonal input. These muscle fibers contain membranes which are unstable and do not maintain a constant potential, but gradually depolarize until they reach the threshold potential at which point an action potential is initiated. The spontaneous depolarization to threshold is known as the pacemaker potential.

192. (D)
It is the increased secretion of estrogen by the placenta that causes the estrogen concentration to increase.

Since the placenta eventually is able to secrete enough progesterone and estrogen to maintain pregnancy, the placenta, after a few months, secretes little gonadtrophic hormone, as the maintenance of the corpus luteum (the corpus luteum secretes progesterone also) is no longer necessary.

The ovaries do secrete estrogen, but they are not responsible for the rise in estrogen concentration. The corpus luteum does not secrete pregnanediol. The placenta increases its rate of steroid (such as estrogen) secretion, and does not decrease it.

193. (E)
After six months, enough estrogen would be produced by the placenta that removal of the ovaries would not affect the estrogen level of the urine. Pregnanediol is not secreted by the ovaries, and neither is gonadotrohic hormone.

194. (C)
Human chorionic gonadotrophin (hCG) is secreted by

trophoblast cells during the first trimester of pregnancy. After that time the placenta becomes the major sex hormone-producing gland, secreting increasing amounts of steroids.

195. (D)
hCG is identical in effect to LH and is therefore able to maintain the corpus luteum. Thus, the secretion of estradiol and progesterone is maintained and menstruation is normally prevented.

196. (A)
The control of any experiment should experience all of the same conditions that are being experienced by the experimental groups, except for the variable being tested for. In this experiment, the rate of conversion of succinic acid by succinic dehydrogenase was measured. This rate is dependent upon two variables: the concentration of succinic acid and of succinic dehydrogenase (which is supplied by the heart homogenate). However, since it is the action of the enzyme that is being demonstrated, it is not necessary to vary the amounts of succinic acid used, which means that (B) and (D), as well as (C) and (E), are incorrect.

197. (A)
The dehydrogenation (oxidation) of succinic acid is part of an oxidation reaction. The enzyme contributed by the heart homogenate therefore is oxidative. The in vivo agent that is reduced is FAD.

198. (E)
Methylene blue is the oxidizing agent in the in vitro reaction and is therefore reduced in the process.

199. (B)
Malonic acid is a competitive inhibitor of succinic acid. It binds to the same site of the enzyme as succinic acid.

Therefore, it will decrease the rate of reaction between succinic acid and the dehydrogenase enzyme.

200.   (B)
Each FAD yields 2 ATP and each NAD yields 3 ATP molecules in the subsequent process of electron transport. That leaves a total of 5 ATP molecules produced for each Acetyl-CoA. A sixteen-carbon fatty acid yields $\left(\dfrac{n}{2}\right) - 1$ Acetyl-CoA or 7. $7 \times 5 = 35$.

201.   (C)
We already know from question 200 that 35 ATP molecules have been produced. Each molecule of Acetyl-CoA may enter the Krebs cycle and yield an additional 12 molecules of ATP. $7 \times 12 = 84$. $84 + 35 = 119$.

202.   (E)
Hypoglycemia – low blood sugar – depletes the cell's supply of carbohydrates, its prime source of energy. Lipid metabolism ensues. Lactic acid buildup can result from muscle development and both result in the conversion of lactic acid to pyruvic acid in the liver. Ketosis and acidosis result from the buildup of Acetyl-CoA as a result of lipolysis.

203.   (C)
Although this is a form of fatty-acid metabolism, it is not a cycle. It is the oxidation of the β carbon of the fatty acid. The citric acid cycle is more commonly known as the Krebs cycle. Gluconeogenesis is the formation of glucose from non-carbohydrate molecules such as amino acids and lactic acid. Glucogenesis is the formation of glycogen from glucose.

204.   (A)
We see that even the final filtrate causes disease in the plants. Therefore, the pathogen must be smaller than .4 microns. Of the choices given, only viruses and

bacteriophages are this small. Bacteriophages are bacteria-specific viruses. This question asks about plant viruses so that of the two, virus is the more correct.

205. (D)
The process described is a form of dilution. If the toxin were present in the second filtrate, then it would be assumed that the toxin-producing agent had been filtered out and, therefore, no new toxin is being released. This limiting amount of toxin would show decreased activity with increased dilution. However, the degree of infection remains constant with each dilution. This can only mean that a pathogenic agent is present and that its activity remains constant through each dilution so that ensuing second residues remain uninfectious.

206. (C)
We would normally expect growth on fully nutritive media from bacteria or fungi. However, virus particles require not nutrient media, but genetic vectors, i.e. bacteria, for their growth.

207. (B)
These rats do not have functional β-cells. β-cells are found in the isles of Langerhans and produce insulin. Insulin is needed in the transport of glucose across cellular membranes. Without it, the cell starves. This leaves the animal hungry. Thirst ensues due to the resulting dehydration of the cell. Dehydration is caused by the imbalance of cellular electrolytes and salts due to the absence of glucose.

208. (E)
Celery and lettuce cannot be digested by the mammalian digestive system. They are composed of the sugar cellubiose, not glucose. Mammals do not have enzymes which can digest this sugar, thus the removal or alteration of the pancreas does not alter its ability to digest these vegetables.

209.   (C)

Tying off the pancreatic duct causes β-cells   to atrophy, ending the production of digestive enzymes.

210.   (E)

Diabetes mellitus is a disease in which the body is unable to utilize glucose as an energy source. This is due to the absence of insulin, which is produced by the   β-cells   of the pancreas.   Diabetes insipidus is a disease of the posterior pituitary and its inability to release antidiuretic hormone.

GRE

BIOLOGY TEST

MODEL TEST IV

THE GRADUATE RECORD EXAMINATION

BIOLOGY TEST

ANSWER SHEET

1. Ⓐ Ⓑ Ⓒ Ⓓ Ⓔ
2. Ⓐ Ⓑ Ⓒ Ⓓ Ⓔ
3. Ⓐ Ⓑ Ⓒ Ⓓ Ⓔ
4. Ⓐ Ⓑ Ⓒ Ⓓ Ⓔ
5. Ⓐ Ⓑ Ⓒ Ⓓ Ⓔ
6. Ⓐ Ⓑ Ⓒ Ⓓ Ⓔ
7. Ⓐ Ⓑ Ⓒ Ⓓ Ⓔ
8. Ⓐ Ⓑ Ⓒ Ⓓ Ⓔ
9. Ⓐ Ⓑ Ⓒ Ⓓ Ⓔ
10. Ⓐ Ⓑ Ⓒ Ⓓ Ⓔ
11. Ⓐ Ⓑ Ⓒ Ⓓ Ⓔ
12. Ⓐ Ⓑ Ⓒ Ⓓ Ⓔ
13. Ⓐ Ⓑ Ⓒ Ⓓ Ⓔ
14. Ⓐ Ⓑ Ⓒ Ⓓ Ⓔ
15. Ⓐ Ⓑ Ⓒ Ⓓ Ⓔ
16. Ⓐ Ⓑ Ⓒ Ⓓ Ⓔ
17. Ⓐ Ⓑ Ⓒ Ⓓ Ⓔ
18. Ⓐ Ⓑ Ⓒ Ⓓ Ⓔ
19. Ⓐ Ⓑ Ⓒ Ⓓ Ⓔ
20. Ⓐ Ⓑ Ⓒ Ⓓ Ⓔ
21. Ⓐ Ⓑ Ⓒ Ⓓ Ⓔ
22. Ⓐ Ⓑ Ⓒ Ⓓ Ⓔ
23. Ⓐ Ⓑ Ⓒ Ⓓ Ⓔ
24. Ⓐ Ⓑ Ⓒ Ⓓ Ⓔ

25. Ⓐ Ⓑ Ⓒ Ⓓ Ⓔ
26. Ⓐ Ⓑ Ⓒ Ⓓ Ⓔ
27. Ⓐ Ⓑ Ⓒ Ⓓ Ⓔ
28. Ⓐ Ⓑ Ⓒ Ⓓ Ⓔ
29. Ⓐ Ⓑ Ⓒ Ⓓ Ⓔ
30. Ⓐ Ⓑ Ⓒ Ⓓ Ⓔ
31. Ⓐ Ⓑ Ⓒ Ⓓ Ⓔ
32. Ⓐ Ⓑ Ⓒ Ⓓ Ⓔ
33. Ⓐ Ⓑ Ⓒ Ⓓ Ⓔ
34. Ⓐ Ⓑ Ⓒ Ⓓ Ⓔ
35. Ⓐ Ⓑ Ⓒ Ⓓ Ⓔ
36. Ⓐ Ⓑ Ⓒ Ⓓ Ⓔ
37. Ⓐ Ⓑ Ⓒ Ⓓ Ⓔ
38. Ⓐ Ⓑ Ⓒ Ⓓ Ⓔ
39. Ⓐ Ⓑ Ⓒ Ⓓ Ⓔ
40. Ⓐ Ⓑ Ⓒ Ⓓ Ⓔ
41. Ⓐ Ⓑ Ⓒ Ⓓ Ⓔ
42. Ⓐ Ⓑ Ⓒ Ⓓ Ⓔ
43. Ⓐ Ⓑ Ⓒ Ⓓ Ⓔ
44. Ⓐ Ⓑ Ⓒ Ⓓ Ⓔ
45. Ⓐ Ⓑ Ⓒ Ⓓ Ⓔ
46. Ⓐ Ⓑ Ⓒ Ⓓ Ⓔ
47. Ⓐ Ⓑ Ⓒ Ⓓ Ⓔ
48. Ⓐ Ⓑ Ⓒ Ⓓ Ⓔ

49. Ⓐ Ⓑ Ⓒ Ⓓ Ⓔ
50. Ⓐ Ⓑ Ⓒ Ⓓ Ⓔ
51. Ⓐ Ⓑ Ⓒ Ⓓ Ⓔ
52. Ⓐ Ⓑ Ⓒ Ⓓ Ⓔ
53. Ⓐ Ⓑ Ⓒ Ⓓ Ⓔ
54. Ⓐ Ⓑ Ⓒ Ⓓ Ⓔ
55. Ⓐ Ⓑ Ⓒ Ⓓ Ⓔ
56. Ⓐ Ⓑ Ⓒ Ⓓ Ⓔ
57. Ⓐ Ⓑ Ⓒ Ⓓ Ⓔ
58. Ⓐ Ⓑ Ⓒ Ⓓ Ⓔ
59. Ⓐ Ⓑ Ⓒ Ⓓ Ⓔ
60. Ⓐ Ⓑ Ⓒ Ⓓ Ⓔ
61. Ⓐ Ⓑ Ⓒ Ⓓ Ⓔ
62. Ⓐ Ⓑ Ⓒ Ⓓ Ⓔ
63. Ⓐ Ⓑ Ⓒ Ⓓ Ⓔ
64. Ⓐ Ⓑ Ⓒ Ⓓ Ⓔ
65. Ⓐ Ⓑ Ⓒ Ⓓ Ⓔ
66. Ⓐ Ⓑ Ⓒ Ⓓ Ⓔ
67. Ⓐ Ⓑ Ⓒ Ⓓ Ⓔ
68. Ⓐ Ⓑ Ⓒ Ⓓ Ⓔ
69. Ⓐ Ⓑ Ⓒ Ⓓ Ⓔ
70. Ⓐ Ⓑ Ⓒ Ⓓ Ⓔ
71. Ⓐ Ⓑ Ⓒ Ⓓ Ⓔ
72. Ⓐ Ⓑ Ⓒ Ⓓ Ⓔ

73. Ⓐ Ⓑ Ⓒ Ⓓ Ⓔ	97. Ⓐ Ⓑ Ⓒ Ⓓ Ⓔ	121. Ⓐ Ⓑ Ⓒ Ⓓ Ⓔ
74. Ⓐ Ⓑ Ⓒ Ⓓ Ⓔ	98. Ⓐ Ⓑ Ⓒ Ⓓ Ⓔ	122. Ⓐ Ⓑ Ⓒ Ⓓ Ⓔ
75. Ⓐ Ⓑ Ⓒ Ⓓ Ⓔ	99. Ⓐ Ⓑ Ⓒ Ⓓ Ⓔ	123. Ⓐ Ⓑ Ⓒ Ⓓ Ⓔ
76. Ⓐ Ⓑ Ⓒ Ⓓ Ⓔ	100. Ⓐ Ⓑ Ⓒ Ⓓ Ⓔ	124. Ⓐ Ⓑ Ⓒ Ⓓ Ⓔ
77. Ⓐ Ⓑ Ⓒ Ⓓ Ⓔ	101. Ⓐ Ⓑ Ⓒ Ⓓ Ⓔ	125. Ⓐ Ⓑ Ⓒ Ⓓ Ⓔ
78. Ⓐ Ⓑ Ⓒ Ⓓ Ⓔ	102. Ⓐ Ⓑ Ⓒ Ⓓ Ⓔ	126. Ⓐ Ⓑ Ⓒ Ⓓ Ⓔ
79. Ⓐ Ⓑ Ⓒ Ⓓ Ⓔ	103. Ⓐ Ⓑ Ⓒ Ⓓ Ⓔ	127. Ⓐ Ⓑ Ⓒ Ⓓ Ⓔ
80. Ⓐ Ⓑ Ⓒ Ⓓ Ⓔ	104. Ⓐ Ⓑ Ⓒ Ⓓ Ⓔ	128. Ⓐ Ⓑ Ⓒ Ⓓ Ⓔ
81. Ⓐ Ⓑ Ⓒ Ⓓ Ⓔ	105. Ⓐ Ⓑ Ⓒ Ⓓ Ⓔ	129. Ⓐ Ⓑ Ⓒ Ⓓ Ⓔ
82. Ⓐ Ⓑ Ⓒ Ⓓ Ⓔ	106. Ⓐ Ⓑ Ⓒ Ⓓ Ⓔ	130. Ⓐ Ⓑ Ⓒ Ⓓ Ⓔ
83. Ⓐ Ⓑ Ⓒ Ⓓ Ⓔ	107. Ⓐ Ⓑ Ⓒ Ⓓ Ⓔ	131. Ⓐ Ⓑ Ⓒ Ⓓ Ⓔ
84. Ⓐ Ⓑ Ⓒ Ⓓ Ⓔ	108. Ⓐ Ⓑ Ⓒ Ⓓ Ⓔ	132. Ⓐ Ⓑ Ⓒ Ⓓ Ⓔ
85. Ⓐ Ⓑ Ⓒ Ⓓ Ⓔ	109. Ⓐ Ⓑ Ⓒ Ⓓ Ⓔ	133. Ⓐ Ⓑ Ⓒ Ⓓ Ⓔ
86. Ⓐ Ⓑ Ⓒ Ⓓ Ⓔ	110. Ⓐ Ⓑ Ⓒ Ⓓ Ⓔ	134. Ⓐ Ⓑ Ⓒ Ⓓ Ⓔ
87. Ⓐ Ⓑ Ⓒ Ⓓ Ⓔ	111. Ⓐ Ⓑ Ⓒ Ⓓ Ⓔ	135. Ⓐ Ⓑ Ⓒ Ⓓ Ⓔ
88. Ⓐ Ⓑ Ⓒ Ⓓ Ⓔ	112. Ⓐ Ⓑ Ⓒ Ⓓ Ⓔ	136. Ⓐ Ⓑ Ⓒ Ⓓ Ⓔ
89. Ⓐ Ⓑ Ⓒ Ⓓ Ⓔ	113. Ⓐ Ⓑ Ⓒ Ⓓ Ⓔ	137. Ⓐ Ⓑ Ⓒ Ⓓ Ⓔ
90. Ⓐ Ⓑ Ⓒ Ⓓ Ⓔ	114. Ⓐ Ⓑ Ⓒ Ⓓ Ⓔ	138. Ⓐ Ⓑ Ⓒ Ⓓ Ⓔ
91. Ⓐ Ⓑ Ⓒ Ⓓ Ⓔ	115. Ⓐ Ⓑ Ⓒ Ⓓ Ⓔ	139. Ⓐ Ⓑ Ⓒ Ⓓ Ⓔ
92. Ⓐ Ⓑ Ⓒ Ⓓ Ⓔ	116. Ⓐ Ⓑ Ⓒ Ⓓ Ⓔ	140. Ⓐ Ⓑ Ⓒ Ⓓ Ⓔ
93. Ⓐ Ⓑ Ⓒ Ⓓ Ⓔ	117. Ⓐ Ⓑ Ⓒ Ⓓ Ⓔ	141. Ⓐ Ⓑ Ⓒ Ⓓ Ⓔ
94. Ⓐ Ⓑ Ⓒ Ⓓ Ⓔ	118. Ⓐ Ⓑ Ⓒ Ⓓ Ⓔ	142. Ⓐ Ⓑ Ⓒ Ⓓ Ⓔ
95. Ⓐ Ⓑ Ⓒ Ⓓ Ⓔ	119. Ⓐ Ⓑ Ⓒ Ⓓ Ⓔ	143. Ⓐ Ⓑ Ⓒ Ⓓ Ⓔ
96. Ⓐ Ⓑ Ⓒ Ⓓ Ⓔ	120. Ⓐ Ⓑ Ⓒ Ⓓ Ⓔ	144. Ⓐ Ⓑ Ⓒ Ⓓ Ⓔ

145. Ⓐ Ⓑ Ⓒ Ⓓ Ⓔ	167. Ⓐ Ⓑ Ⓒ Ⓓ Ⓔ	189. Ⓐ Ⓑ Ⓒ Ⓓ Ⓔ
146. Ⓐ Ⓑ Ⓒ Ⓓ Ⓔ	168. Ⓐ Ⓑ Ⓒ Ⓓ Ⓔ	190. Ⓐ Ⓑ Ⓒ Ⓓ Ⓔ
147. Ⓐ Ⓑ Ⓒ Ⓓ Ⓔ	169. Ⓐ Ⓑ Ⓒ Ⓓ Ⓔ	191. Ⓐ Ⓑ Ⓒ Ⓓ Ⓔ
148. Ⓐ Ⓑ Ⓒ Ⓓ Ⓔ	170. Ⓐ Ⓑ Ⓒ Ⓓ Ⓔ	192. Ⓐ Ⓑ Ⓒ Ⓓ Ⓔ
149. Ⓐ Ⓑ Ⓒ Ⓓ Ⓔ	171. Ⓐ Ⓑ Ⓒ Ⓓ Ⓔ	193. Ⓐ Ⓑ Ⓒ Ⓓ Ⓔ
150. Ⓐ Ⓑ Ⓒ Ⓓ Ⓔ	172. Ⓐ Ⓑ Ⓒ Ⓓ Ⓔ	194. Ⓐ Ⓑ Ⓒ Ⓓ Ⓔ
151. Ⓐ Ⓑ Ⓒ Ⓓ Ⓔ	173. Ⓐ Ⓑ Ⓒ Ⓓ Ⓔ	195. Ⓐ Ⓑ Ⓒ Ⓓ Ⓔ
152. Ⓐ Ⓑ Ⓒ Ⓓ Ⓔ	174. Ⓐ Ⓑ Ⓒ Ⓓ Ⓔ	196. Ⓐ Ⓑ Ⓒ Ⓓ Ⓔ
153. Ⓐ Ⓑ Ⓒ Ⓓ Ⓔ	175. Ⓐ Ⓑ Ⓒ Ⓓ Ⓔ	197. Ⓐ Ⓑ Ⓒ Ⓓ Ⓔ
154. Ⓐ Ⓑ Ⓒ Ⓓ Ⓔ	176. Ⓐ Ⓑ Ⓒ Ⓓ Ⓔ	198. Ⓐ Ⓑ Ⓒ Ⓓ Ⓔ
155. Ⓐ Ⓑ Ⓒ Ⓓ Ⓔ	177. Ⓐ Ⓑ Ⓒ Ⓓ Ⓔ	199. Ⓐ Ⓑ Ⓒ Ⓓ Ⓔ
156. Ⓐ Ⓑ Ⓒ Ⓓ Ⓔ	178. Ⓐ Ⓑ Ⓒ Ⓓ Ⓔ	200. Ⓐ Ⓑ Ⓒ Ⓓ Ⓔ
157. Ⓐ Ⓑ Ⓒ Ⓓ Ⓔ	179. Ⓐ Ⓑ Ⓒ Ⓓ Ⓔ	201. Ⓐ Ⓑ Ⓒ Ⓓ Ⓔ
158. Ⓐ Ⓑ Ⓒ Ⓓ Ⓔ	180. Ⓐ Ⓑ Ⓒ Ⓓ Ⓔ	202. Ⓐ Ⓑ Ⓒ Ⓓ Ⓔ
159. Ⓐ Ⓑ Ⓒ Ⓓ Ⓔ	181. Ⓐ Ⓑ Ⓒ Ⓓ Ⓔ	203. Ⓐ Ⓑ Ⓒ Ⓓ Ⓔ
160. Ⓐ Ⓑ Ⓒ Ⓓ Ⓔ	182. Ⓐ Ⓑ Ⓒ Ⓓ Ⓔ	204. Ⓐ Ⓑ Ⓒ Ⓓ Ⓔ
161. Ⓐ Ⓑ Ⓒ Ⓓ Ⓔ	183. Ⓐ Ⓑ Ⓒ Ⓓ Ⓔ	205. Ⓐ Ⓑ Ⓒ Ⓓ Ⓔ
162. Ⓐ Ⓑ Ⓒ Ⓓ Ⓔ	184. Ⓐ Ⓑ Ⓒ Ⓓ Ⓔ	206. Ⓐ Ⓑ Ⓒ Ⓓ Ⓔ
163. Ⓐ Ⓑ Ⓒ Ⓓ Ⓔ	185. Ⓐ Ⓑ Ⓒ Ⓓ Ⓔ	207. Ⓐ Ⓑ Ⓒ Ⓓ Ⓔ
164. Ⓐ Ⓑ Ⓒ Ⓓ Ⓔ	186. Ⓐ Ⓑ Ⓒ Ⓓ Ⓔ	208. Ⓐ Ⓑ Ⓒ Ⓓ Ⓔ
165. Ⓐ Ⓑ Ⓒ Ⓓ Ⓔ	187. Ⓐ Ⓑ Ⓒ Ⓓ Ⓔ	209. Ⓐ Ⓑ Ⓒ Ⓓ Ⓔ
166. Ⓐ Ⓑ Ⓒ Ⓓ Ⓔ	188. Ⓐ Ⓑ Ⓒ Ⓓ Ⓔ	210. Ⓐ Ⓑ Ⓒ Ⓓ Ⓔ

THE GRE BIOLOGY TEST

MODEL TEST IV

Time: 170 Minutes
        210 Questions

DIRECTIONS: *Choose the best answer for each question and mark the letter of your selection on the corresponding answer sheet.*

1. The SPONCH element which forms compounds that are a basic subunit of all biological membranes is

    (A)   carbon   (D)   phosphorous

    (B)   oxygen   (E)   sulphur

    (C)   hydrogen

2. In the gastrointestinal tract sucrose is broken down by

    (A)   dehydrolysis   (D)   dehydration linkage

    (B)   hydrophobic cleavage   (E)   dephosphorylation

    (C)   hydrolytic cleavage

3. Amylopectin is a large molecule which consists of:

    (A)   1-2 glucose chains

    (B)   1-2 glucose chains cross linked with 1-4 and 1-6 linkages

(C) 1-4 glucose chains

(D) 1-4 glucose chains cross linked with 1-6 linkages

(E) 1-4 glucose chains cross linked with 1-6 and 1-3 linkages

4. The R group of amino acids is always attached to a

(A) nitrogen atom   (D) hydrogen atom

(B) carbon atom   (E) None of the above

(C) oxygen atom

5. Dimers are formed at the _____ stage of a protein.

(A) primary   (D) conjugated

(B) secondary   (E) quaternary

(C) tertiary

6. As a cell increases in size, the volume is

(A) constant   (D) stagnant

(B) doubled   (E) None of the above

(C) growing faster than
     the surface area

7. Plants use cytoplasmic or cell streaming to

(A) regulate the surface-volume ratio

(B) support the roots

(C) control intracellular activities

(D) reduce the cytoplasm

(E) All of the above

8. Danielli, who specialized in lipid biochemistry, discovered

(A)   simple diffusion   (D)   active transport

(B)   membranous structure   (E)   facilitated diffusion

(C)   random movement

9.   It has been recently discovered that O_2 moved from the environment into cells and blood not solely by diffusion but by utilizing a heme protein called

(A)   cytochrome P 450   (D)   cytochrome R 240

(B)   cytochrome P 370   (E)   None of the above

(C)   cytochrome Q 90

10.   The ____ microscope has the distinct advantage of producing the illusion of three-dimensional images with unusually great depth of field.

(A)   Light   (D)   Telescopic electron

(B)   Transmission electron (E)   All of the above

(C)   Scanning electron

11.   A produce man sprayed his lettuce with a solution of salt in which the number of salt molecules equaled that in the cytoplasm of the lettuce cells. The spray would be _____ .

(A)   hypotonic   (D)   isokinetic

(B)   isotonic   (E)   supertonic

(C)   hypertonic

12.   One function not carried out by the smooth endoplasmic reticulum is

(A)   the manufacture of   (D)   the synthesis of lipids
      proteins
                             (E)   the storage of non-protein
(B)   the synthesis of   products
      carbohydrates

(C)   the synthesis of
      steroid hormones

13. The eucaryotic cell would quickly self-destruct if powerful hydrolytic enzymes were released from

(A) peroxisomes

(D) lysosomes

(B) endoplasmic reticulum

(E) plastids

(C) golgi bodies

14. Like chloroplasts, mitochondria

(A) are enclosed in their own double membranes

(B) have their own circular DNA

(C) have their own ribosomes

(D) have their own machinery of protein synthesis

(E) All of the above

15. The nine plus two arrangement of microtubules can be seen in

(A) the rods and cones of the retina

(B) the olfactory fibers of the nasal epithelium

(C) the sensory hairs of the cochlea and semicircular canals

(D) None of the above

(E) All of the above

16. NADP is an important soluble electron carrier of the cell which is used more in

(A) oxidative respiration

(D) aerobic respiration

(B) photosynthesis

(E) non-oxidation-reduction reactions

(C) glycolysis

17. In the photolysis of water

I. Two electrons enter the electron transport system.

II. Photosystems are utilized.

III. Two hydrogen ions are released into the atmosphere.

IV. NAD is the final electron acceptor.

(A) I, II and III are correct

(B) I and II are correct

(C) IV is correct

(D) II and IV are correct

(E) None of the above are correct

18. In the chloroplasts it has been found that light-independent reactions actually occur in

(A) the thylakoid   (D) the lamellae

(B) the grana   (E) the quantasomes

(C) the stroma

19. Pyruvate, the final product of glycolysis which contains a considerable amount of potential energy, has three fates. One of these is its oxidation by NAD and the removal of CO_2 to produce

(A) ethyl alcohol   (D) ATP

(B) lactic acid   (E) AMP

(C) acetyl CoA

20. The electron transport system of chloroplasts and mitochondria, even though similar in their use of electron carriers, are different in that

I. in electron transport system of chloroplasts CFI particles project outward of the thylakoid whereas in the electron transport system of mitochondria stalked FI particles project inwardly

II. the Z scheme of photosynthesis began with the splitting of water into hydrogen ions, electrons, and molecular oxygen while in the mitochondrial transport system, the flow ended with the combining of hydrogen ions, electrons, and molecular oxygen into water

III.   the photosynthetic electron transport chain started with
       NADP and the mitochondrial chain ended with NADP

IV.   chloroplasts turn energy into carbon compounds and
       oxygen while mitochondria turn carbon compounds and
       oxygen into energy

(A) III only   (D) I, II and IV

(B) I, II and III   (E) I and II

(C) III and IV

21. Mitotic problems might very well be encountered if animal cells
     have lost their

(A) cilia   (D) ribosomes

(B) centrioles   (E) None of the above

(C) plastids

22. Of the following which is absent in both procaryotes and higher
     plants?

(A) Microtubules in 9 + 2 arrangement

(B) Centrioles

(C) Protein cytoskeleton

(D) All of the above are absent

(E) None of the above are absent

23. The intracellular transport system of cells is greatly increased
     by

(A) porous cell walls   (D) cell surface proteins

(B) an extensive endoplas- (E) None of the above
     mic reticulum

(C) gap junctions

24. In sustained muscular activity, the ATP reserves are depleted

in the first few minutes. To replenish the supply the reservoir tapped is

(A)   lactic acid

(D)   PGAL

(B)   creatine phosphate

(E)   All of the above

(C)   AMP

25.   Specific codons of mRNA have been experimentally found to be recognized by:

(A)   the -CCA stem

(D)   D loop

(B)   the anticodon loop

(E)   None of the above

(C)   T loop

26.   "Stop" codons which specify the end of a protein are

   I.   UAA

   II.   UAG

   III.   UUA

   IV.   UGA

(A)   I only

(D)   I, II and IV only

(B)   I, II and III only

(E)   I, II, III and IV

(C)   I and IV only

27.   Initiator tRNA with an anticodon recognizes and pairs with the initiation codon

(A)   5'-CAU-3'.

(D)   5'-UUA-3'.

(B)   5'-AUG-3'

(E)   5'-TAC-3'

(C)   5'-CUA-3'

28.   A particular group of cells and their descendants are doomed to perish. These cells would most likely be

(A)   gametic cells

(B) somatic cells   (D) None of the above

(C) sex cells   (E) All of the above

29. A baby born in the hospital has a slim chance of survival; he
     was found to have three copies of chromosome 21. Should he
     grow to adulthood his abnormality would be known as

     (A) Parkinson's disease   (D) Tsutsugamushi disease

     (B) Sylvan plague   (E) Down's syndrome

     (C) Bang's disease

30. Translocation is a chromosomal mutation which today is largely
     responsible for the introduction of various types of diseases
     in the world. In this type of mutation

     (A) a segment of the chromosome is deleted

     (B) a segment of the chromosome is represented twice and is
          expressed

     (C) a segment of the chromosome is transferred to another
          homologous chromosome

     (D) a segment of one chromosome is transferred to another
          non-homologous chromosome

     (E) a segment is removed and reinserted at another point on
          the same chromosome

31. Which of the following groups would be considered a society?

     (A) A swarm of flies attracted to a rotting fruit

     (B) A collection of different insects in a jar

     (C) A group of male crickets attracted to the same female

     (D) A collection of birds in a zoo

     (E) A pack of dogs chasing their prey in relays

32. The nuclei were first separated from the cytoplasm of cells
     and nucleic acid isolated from them by:

(A)   Friedrick Miescher.   (D)   Erwin Chargaff

(B)   Avery, MacLeod and   (E)   Phoebus A. Levene
        McCarty

(C)   Frederick Griffith

33.   A cell which is placed in a solution of dye was found after a
      while to be more concentrated intracellularly than extracellu-
      larly. When a metabolic inhibitor was added to the solution,
      the dye was discovered to equilibrate across the cell membrane
      until the intracellular and extracellular concentrations were
      equal. It is highly probable that the metabolic inhibitor

      (A)   inhibited protein   (D)   inhibited ATP produc-
              synthesis   tion

      (B)   delayed chromosomal   (E)   accelerated meiotic pro-
              replication   cesses

      (C)   accelerated aerobic
              respiration

34.   The term cephalochordata refers to a

      (A)   genus   (D)   group

      (B)   phyla   (E)   class

      (C)   subphyla

35.   Aves belong to a

      (A)   class   (D)   genus

      (B)   phyla   (E)   group

      (C)   subphyla

36.   The phylogenetic tree of humans is postulated to have begun
      with

      (A)   homo erectus   (D)   homo sapiens sapiens

      (B)   ramapithecus   (E)   homo sapiens

      (C)   homo habilis

37. From the fossils discovered by Louis Leakey of the homo habilis we can say that homo habilis did not

    (A) have a greater cranial capacity than Australopithecus

    (B) seem closer to human form

    (C) belong to the genus homo

    (D) possess strong jaws and teeth

    (E) appear closer to the human line than Australopithecus

38. The formula presented below is the formula of the _____.

$$6Ru - P + 6CO_2 + 12NADPH_2 + 18\ ATP \rightarrow$$

$$C_6H_{12}O_6 + 18\ ADP + 18Pi + 6Ru - P + 6H_2O$$

    (A) photosynthetic cycle   (D) Calvin cycle

    (B) respiration cycle   (E) Krebs cycle

    (C) fermentation

39. Which of the following chemical transmitters may be chemically similar to heroin and may serve as the body's own natural pain killer?

    (A) The endorphins   (D) Acetyl choline

    (B) Serotonin (5-HT)   (E) None of the above

    (C) Norepinephrine

40. The constriction of the blood vessels of the skin in response to a cold is a good illustration of a reflex arc. As cold temperature strikes the receptors and sensory neurons near the blood vessels which of the following statements would most likely be false?

    (A) Initial sensory impulses will be carried by the afferent nervous system to the spinal cord.

    (B) This sensory neuron may then synapse with an association neuron.

    (C) The association neuron may then synapse with a neuron of the somatic efferent nervous system.

(D)   The stimulated motor neuron then stimulates the smooth
      muscles of the blood vessel to automatically constrict its
      diameter.

(E)   All of the above are correct.

41.   In myelinated axons and dendrites, the reversed potential
      leaps from one node of Ranvier to the next by a process
      called

(A)   hydrophobic induction. (D)   electrophillic conduction.

(B)   saltatory conduction.   (E)   constriction and dilation.

(C)   electron transport.

Questions 42-45 concern the diagram.

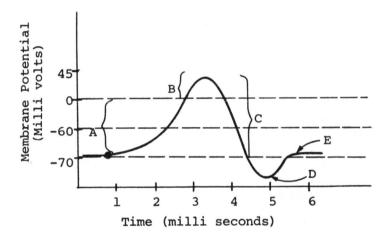

42.   Part A of the curve represents a condition called

(A)   depolarization   (D)   destabilization

(B)   reversed potential   (E)   restabilization

(C)   hyperpolarization

43.   If a stimulus of -66 millivolts was applied:

(A)   depolarization will be effected

(B) Na^+ ions will flow out of the system

(C) an action potential will not be created

(D) hyperpolarization would be effected

(E) None of the above will happen

44. Part C of the curve is due to the

(A) outflow of Na^+

(B) outflow of K^+

(C) influx of Na^+

(D) influx of K^+

(E) outflow of Na^+ and influx of K^+

45. Parts A and B of the curve are due to the

(A) outflow of Na^+

(B) influx of Na^+

(C) outflow of K^+

(D) influx of K^+

(E) All of the above

46. Of the following, which represents the most logical evolutionary advantage of multicellularity?

(A) Multicellular organisms are able to become quite large.

(B) Multicellular organisms are more efficient than single-celled.

(C) Multicellular organisms can cope differently with the environment and its resources.

(D) Multicellular organisms are better equipped to survive than single-celled.

(E) Of the above none are logical evolutionary advantages.

47. Which of the following scientists contributed to the cell theory?

(A) Mathias Schleiden

(B) Theodore Schwann

(C) Robert Hooke

(D) Rudolph Virchow

(E) All of the above

48. Geneticists Watson and Crick are known to have postulated that because of the arrangement and nature of the nitrogenous bases, DNA replication is

    (A)   conservative   (D)   directional

    (B)   semi-conservative   (E)   None of the above

    (C)   dispersive

49. As each electron completes its movement back and forth across the mitochondrial membrane, starting at the $FADH_2$ level, the chemiosmotic differential is enriched by:

    (A)   one H^+ (D)   four H^+

    (B)   two H^+ (E)   None of the above

    (C)   three H^+

50. A human embryo in its mother's womb is now developing form. This process is called

    (A)   catharsis   (D)   morphogenesis

    (B)   parthenogenesis   (E)   differentiation

    (C)   ontogeny

51. Stability in the helical structure of DNA is maintained by

    (A)   carbon bonding   (D)   ATP

    (B)   phosphate bonding   (E)   None of the above

    (C)   hydrogen bonding

52. The nervous system operates like a well-oiled machine. When a depolarizing wave reaches the synaptic knobs the order in which the impulse is transmitted is

    (A) Ca^{2+} ions are admitted from surrounding fluid, micro-tubular system is activated, microtubules attach to vesicles, neurotransmitters are released

    (B)   microtubular system is activated, microtubles attach to vesicles, Ca^{2+} ions are admitted from surrounding fluid, neurotransmitters are released.

(C)   microtubular system is activated, Ca^{2+} ions are released into surrounding fluid, microtubules attach to vesicles, neurotransmitters are released.

(D) Ca^{2+} ions are released into surrounding fluid, microtubular system is activated, microtubules attach to vesicles, neurotransmitters are released.

(E)   None of the above

53.   This neurotransmitter is found at neuromuscular junctions in the brain and at junctions in the internal organs.

(A)   Norepinephrine   (D)   Acetylcholine

(B)   Serotonin   (E)   Epinephrine

(C)   Dopamine

54.   You are startled by a large bear which suddenly rushes into the room where you are quietly reading. Your heart rate increases, your pupils dilate and blood is shunted away from the digestive tract and peripheral vessels. Changes such as these are directed by

(A)   the parasympathetic system of the autonomic nervous system

(B)   the parasympathetic system of the somatic nervous system

(C)   the sympathetic system of the autonomic nervous system

(D)   the sympathetic system of the somatic nervous system

(E)   All of the above

55.   A cat which has been stimulated in a certain area of the brain displays unusual hunger and spits at those she is fond of in anger. The part of the brain stimulated would most likely be the

(A)   corpus callosum

(B)   hypothalamus

(C)   olfactory bulb

(D)   pituitary

(E)   thalamus

56.   When visual purple absorbs a photon of light, the following
      series of events occurs.

      (A)   opsin breaks away → trans retinal → cis retinal

      (B)   cis retinal → trans retinal

      (C)   opsin → cis retinal → trans retinal

      (D)   trans retinal → cis retinal → opsin breaks away

      (E)   cis retinal → trans retinal → opsin breaks away

57.   Light receptors are sensitive to this part of the spectrum.

      (A)   430 - 750 nm   (D)   600 - 850 nm

      (B)   430 - 500 nm   (E)   None of the above

      (C)   450 - 900 nm

58.   The sensory hairs of the hair cells are modified cilia which
      push against the

      (A)   oval window   (D)   malleus

      (B)   tectorial membrane   (E)   eustachian tube

      (C)   stapes

59.   The hormone that increases glucose, protein and fat metabolism
      and reduces inflammation is

      (A)   cortisol   (D)   calcitonin

      (B)   oxytocin   (E)   aldosterone

      (C)   antidiurectic hormone

60. A population of deer in an overcrowded situation were noticed to react strangely. They were provided with all the food and water they needed and seemed to be healthy and grew in number. But then they apparently experience "physiological shock" and stopped reproducing. The most likely cause of this was

(A) high GH levels and adrenal steroid levels

(B) high LH levels and adrenal levels

(C) low ACTH levels

(D) high ACTH levels and adrenal steroid levels

(E) None of the above

61. The order of the contraction of the heart is

(A) A-V node → atria → bundle of His → ventricles

(B) S-A node → A-V node → bundle of His → ventricles

(C) atria → S-A node → A-V node → ventricles → bundle of His

(D) S-A node → atria → A-V node → bundle of His → ventricles

(E) None of the above

62. The order of the clotting process is

(A) prothrombin → thrombin → fibrinogen → fibrin → clot

(B) fibrinogen → fibrin → prothrombin → thrombin → clot

(C) fibrin → fibrinogen → thrombin → prothrombin → clot

(D) thrombin → prothrombin → fibrin → fibrinogen → clot

(E) None of the above orders are correct

63. The process of fertilization is considered to be complete when

(A) the second division cycle of meiosis has ended

(B) the disjunction of maternal and paternal chromosomes occur

(C)   the sperm and ovum pronuclei fuse

(D)   the second polar body forms

(E)   All of the above are correct

64.   During muscular contraction the protein that also serves as an enzyme that breaks down $ATP \rightarrow ADP + P_i$ is

(A)   tropomyosin   (D)   troponin

(B)   fibrinogen   (E)   regulator protein

(C)   actomyosin

65.   The continued upward growth of the plant even though lateral branches are being steadily produced is known as

(A)   positive phototropism   (D)   negative geotropism

(B)   negative phototropism   (E)   auxotrophy

(C)   positive geotropism

66.   A genetic engineering process usually done in the laboratory which involves the transfer of genetic material from one bacterium to another using a bacteriophage as the carrier is

(A)   transduction

(B)   transcription

(C)   plasmid transfer

(D)   transformation

(E)   None of the above

67.   Studies of genes and gene frequencies have shown that individuals do not evolve but rather population. This evolution essentially involves a change in the frequencies of:

(A)   genes   (D)   recombinance

(B)   alleles   (E)   All of the above

(C)   mitotic divisions

68. Chemotrophic autotrophs are organisms which derive their energy from

(A) sunlight

(B) light energy systems

(C) organic substances

(D) inorganic substances

(E) None of the above

69. According to the trophic levels of ecosystems, herbivores are

(A) primary consumers

(B) primary producers

(C) secondary consumers

(D) tertiary consumers

(E) quaternary producers

70. A male European robin in breeding condition will attack a tuft of red feathers placed in his territory. Since red feathers are usually on the breast of his competitor, it is to his reproductive advantage to behave aggressively at the sight of them. This is an illustration of

(A) instinctive behavior

(B) operants

(C) releasers

(D) fixed action pattern

(E) appetitive behavior

71. Ligation of the Islets of Langerhans of the pancreas will deprive the circulatory system of

(A) insulin

(B) trypsin

(C) serotonin

(D) bile

(E) pepsin

72. The interaction of dominant genes S and P results in black fur color in dogs; if either is absent the phenotype is grey. If a dog with black fur color is self-fertilized which of the following is LEAST likely to be expressed in the progeny?

(A) 5 black to 5 grey

(B) all grey fur colored

(C)   all black fur colored

(D)   7 black to 3 grey

(E)   3 black to 1 grey

73.   Which of the following is not an example of instinctive behavior?

(A)   The pattern of breeding behavior in male sickleback fish.

(B)   A dog's salivating at the sight of food .

(C)   A human's heartbeat accelerating at the sight of a charging mad bull .

(D)   The flight of birds to warm places during winter .

(E)   All are examples of instinctive behavior.

74.   Fertilization of the egg cell in flowering plants takes place in the

(A)   micropyle

(B)   ovary

(C)   embryo sac

(D)   endosperm mother cell

(E)   pollen tube

75.   It has been found that the hair rises under eery conditions. This is characteristic of

(A)   skeletal muscle

(B)   striated muscle

(C)   cardiac muscle

(E)   voluntary muscle

(E)   smooth muscle

76.   Proteins are finally digested in the duodenum after a long process by

(A)   amylase

(B)   enterokinase

(C)   pepsin

(D)   trypsin

(E)   carboxypeptidase

77.   The Carotid reflex, which is located just above the bifurcation of carotid arteries, synapses with the

(A)   cardio-inhibitory center of the medulla of the parasympathetic system

(B)   cardio-acceleratory center of the sympathetic system

(C)   cardio-inhibitory center of the medulla of the sympathetic system

(D)   cardio-acceleratory center of the medulla of the parasympathetic system

(E)   None of the above are correct

78.   Of the following, which does not have a closed circulatory system?

(A)   Mammals   (D)   Annelids

(B)   Fish   (E)   Earthworms

(C)   Mollusks

79.   A diver at about 100m breathes compressed air with a total pressure of about 10 atmospheres; the partial pressure of O_2 is 2.1 atmospheres.   Which of these statements is correct?

(A)   The partial pressure is too low for normal physiological processes.

(B)   The partial pressure is conducive for normal physiological processes.

(C)   The partial pressure is too high for normal physiological processes.

(D)   The partial pressure agrees with what is required for normal living.

(E)   For prolonged activity of such depths the diver may breathe a mixture of gases that is 98% O_2 and 2% helium.

80. In the capillary network, blood entering the alvioli is

(A) rich in CO_2 and nearly depleted of O_2

(B) poor in CO_2 and rich in O_2

(C) changed into carbaminohemoglobin

(D) highly oxygenated

(E) None of the above are correct

81. In terms of respiration and gas exchange, the vertebrate with the most advanced respiratory system is the

(A) mammal   (D) fish

(B) bird   (E) bat

(C) frog

82. Blood from the right and left branches of the pulmonary arteries enters the capillaries surrounding the alveoli of the lungs and gases are exchanged. The capillaries then rejoin to form the pulmonary veins which return the blood to the left atrium. The blood then passes through the valve called the

(A) bicuspid valve   (D) pulmonary aortic valve

(B) tricuspid valve   (E) aortic semilunar valve

(C) pulmonary semilunar valve

83. Vasoconstriction and vasodilation of the arterioles, which is augmented by hormonal influence, is partly under the control of

(A) the parasympathetic fibers of the autonomic nervous system

(B) the sympathetic fibers of the autonomic nervous system

(C) the parasympathetic fibers of the peripheral nervous system

(D) the sympathetic fibers of the peripheral nervous system

(E) None of the above

84. An unusually nervous, skinny and hyperactive individual who suffers from insomnia is most likely to be afflicted with

(A) hypoglycemia

(D) hypothyroidism

(B) hyperglycemia

(E) schizophrenia

(C) hyperthyroidism

85. A mother hears her baby crying; thinking that he is hungry she rushes to him and tries to feed him but this only makes him cry more. This behavior is an example of

(A) negative feedback

(D) positive feedback

(B) positive conditioning

(E) negative communication

(C) operant behavior

86. The walls of the collecting ducts and distal convoluted tubule are made permeable to water for the animal that wants to conserve water by

(A) aldosterone

(D) antidiurectic hormone (ADH)

(B) parathormone

(E) calcitonin

(C) luteinizing hormone (LH)

87. Altruism is best expressed in which of the following behaviors?

(A) A dog rushing in front of a speeding train to save a chicken

(B) A mother rushing to the defense of her child

(C) A pregnant, swollen woman

(D) None of the above

(E) All of the above

88. The best technique for separating amino acids in hydrolysate is

(A) centrifugation

(B) chromatography

(C) spectrophotometry

(D) hydrolysis

(E) ligation

89. A mutation in a certain group of animals was traced and was found to have been caused by the addition of a single nucleotide A. Instead of reading

CAT CAT CAT CAT CAT CAT the
$\uparrow$
A

nucleotide sequence now read

(A) CAT CAT CAT CAT CAT CAT A

(B) CAT CAT CAT A CAT CAT CAT

(C) CAT CAA TAC CAT CAT TCA T

(D) CAT CAA TCA TCA TCA TCA T

(E) CAT CAA TCA CAT CAT CAT

90. All of the following are territorial behavior except

(A) a male fish swimming on guard

(B) the singing and isolation of a yellowhammer in early spring

(C) aggressive behavior of birds towards neighbors

(D) mutual avoidance of lions of each other

(E) the control of the predator over the prey

Questions 91 - 92 concern the population curve below.

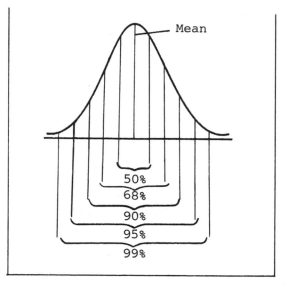

91. According to the curve most of the population seem to center

(A) around the mean   (D) between 68 and 90%

(B) along the ends   (E) All the above are correct

(C) in the 68% zone

92. The curve approximates

(A) the generalized distribution

(B) the difference between members of the population

(C) the idealized distribution

(D) the normal distribution

(E) All the above are correct

Questions 93 - 95 concern the curve which shows the effect of substrate concentration .

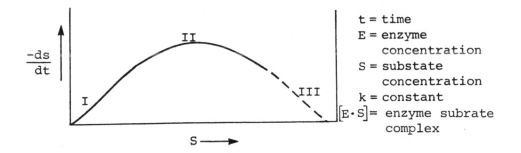

t = time
E = enzyme concentration
S = substate concentration
k = constant
[E·S] = enzyme subrate complex

If the concentration of enzyme is held constant and the substrate concentration is varied, plotting the reaction velocity against S gives these results.

93. The first portion of the curve is a straight line whose characteristics may be expressed as

(A) $- \dfrac{dS}{dt} = k \cdot E$

(B) $-dS = \dfrac{k \cdot E}{dt}$

(C) $- \dfrac{dS}{dt} = k[E \cdot S]$

(D) $- \dfrac{dS}{dt}$

(E)   None of the above formulas are correct

94. The third stage of the reaction (III) occurs

(A)   when enzyme activity is inhibited at high concentrations of substrate

(B)   when with increasing substrate concentration a plateau is reached

(C)   when the reaction velocity remains unaffected

(D)   None of the above are true of the third stage

(E)   All of the above are true of the third stage

95. The second stage indicates that

(A)   the enzyme and substrate are incompatible

(B)   a plateau is reached

(C)   all the resources of the enzyme have been utilized.

(D)   the substrate has no affinity for the enzyme

(E)   None of the above is indicative of this stage

Questions 96 - 98   concerns the below curve of excretion of sex
hormones during pregnancy.

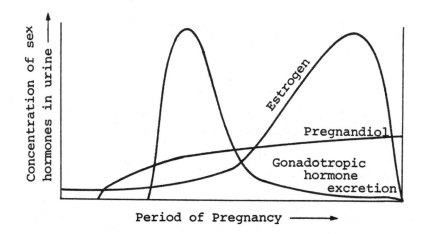

During pregnancy the sex hormones are secreted in the urine
of the mother.

96.   The appearance of pregnandiol in the urine indicates the
       presence of

(A)   parathermone   (D)   oxytocin

(B)   prolactin   (E)   follicle-stimulating hor-
                                    mones (FSH)
(C)   progesterone

97.   The substance secreted by the placenta that is similar to LH
       (luteinizing hormone) of the pituitary is

(A)   chorionic gonadotropin

(B)   pregnandiol

(C)   estrogen

(D)   androgen

(E)   progesterone

98.   The hormone produced in large amounts during the first two
       weeks of pregnancy and which serves as a basis for pregnancy
       tests is

       (A)   gonadotropic hormone   (D)   None is used for preg-
                                             nancy tests
       (B)   estrogen
                                      (E)   All are present in large
       (C)   pregnandiol   amounts

Questions 99 - 100

       Each gas in a mixture of gases exerts a partial pressure equal
to the pressure it would exert if it existed alone. The total pressure
of the mixture is dependent on the partial pressure of each gas.

99.   At sea level (760 mm Hg) oxygen makes up 20.93% by volume
       of the air. Its partial pressure in the inspired air is there-
       fore

       (A)   15906.80 mmHg   (D)   39.200 mmHg

       (B)   176.5 mmHg   (E)   None of the above is
                                             correct
       (C)   159.1 mmHg

100.   If we were however at an altitude of 18,000 ft. where the at-
       mospheric pressure is 380 mmHg, the partial pressure of oxy-
       gen would be

       (A)   76.7 mm   (D)   7953.4 mm

       (B)   78.55 mm   (E)   3976.7 mm

       (C)   79.5 mm

101 - 103:   The figure below shows the wave form of a normal ECG.
             Select the answer which best fits the question.

371

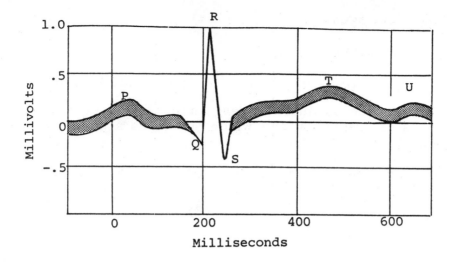

101. The P wave is indicative of

   (A) the initial repolarization of the SA node

   (B) the initial depolarization of the SA node

   (C) depolarization of the AV node

   (D) depolarization of the ventricles

   (E) repolarization of the myocardial tissue

102. The QRS complex of the ECG is

   (A) ventricular systole

   (B) gradual repolarization of the papillary muscles

   (C) depolarization of the ventricles preceding ventricular contraction

   (D) depolarization of the AV node

   (E) atrial contraction

103. If the R and S peaks were reversed the function affected would be

   (A) depolarization of the SA node

   (B) atrial contraction

(C)   depolarization of the AV node

(D)   ventricular diastole

(E)   depolarization of the myocardial tissue

104.   The tissue that lines the inside of the blood vessels is known as

(A)   epithelial tissue   (D)   hyaline cartilage

(B)   connective tissue   (E)   reticular tissue

(C)   adipose tissue

105.   A bacterium which can grow on the most minimal medium and which synthesizes all the essential organic compounds it needs is a/an

(A)   auxotroph   (D)   chemio-organotroph

(B)   heterotroph   (E)   prototroph

(C)   organotroph

106.   The enzyme hyaluronidase which is very active in the process of fertilization

(A)   causes the formation of the vitelline membrane

(B)   fuses the follicle cells to prevent the attachment of other sperms to the ovary

(C)   breaks down the corona radiata

(D)   causes the penetration of the zona pellucida by microvilli

(E)   allows the acrosomal membrane to fuse with the egg membrane

107.   The orienting of a grayling butterfly towards the sun thus causing the pursuing predator to be partly blinded is an example of

(A)   kinesis

(B)   tropism

(C)   taxis

(D)   instinctive behavior

(E)   conditioning

108.   Gibberellins play a very important role in

(A)   development of the apical meristem

(B)   dissociation of α amylose so that starch hydrolysis can be inhibited

(C)   plant growth retardation

(D)   polar transportation in plant cells

(E)   directional regulatory influences within the plant environment

109.   A group of cells were treated with radioactive compounds, deoxythymidine, deoxycytidine, glycine, glucose and creatine. Upon examination the compound that appeared predominantly in the nucleus was found to be

(A)   glucose and creatine

(B)   glucose and glycine

(C)   deoxythymidine and deoxycytidine

(D)   deoxythymidine and glycine

(E)   deoxycytidine and creatine

110.   The material responsible for conferring resistance to antibiotics is the

(A) F^- factor

(B)   Hfr factor

(C)   Col plasmid

(D)   R plasmid

(E)   F' factor

Questions 111 - 113 are based on the information and diagrams below.

Diagram 1 shows the area originally occupied by species A. A barrier arose splitting species A into two separate populations (diagram 2). From the two separted populations two new species, C and B, evolved (diagram 3).

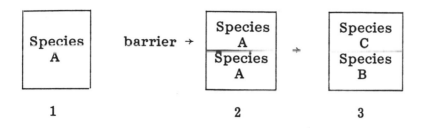

1   2   3

111.   This situation has distinct similarity to the evolutionary process as influenced by

        (A)   homologous structures.

        (B)   geographic isolation

        (C)   artificial selection

        (D)   migration

        (E)   genetic drift

112.   The two new sets of species, B and C, most likely appeared because

        (A)   organisms of species A migrated to a new environment

        (B)   sexual reproduction was impossible because of the barrier which prevented interbreeding

        (C)   organisms of one region of barricade underwent mutation

        (D)   there was tremendous competition between the two new species for food

        (E)   organisms were in competition for the same environmental niche

113.   Based on the evolution of the two new species, B and C, it can now be expected that species B and C will

        (A)   be most likely unrelated to each other

        (B)   be capable of interbreeding within species

        (C)   be incapable of interbreeding

        (D)   be affected by natural selection

        (E)   undergo no further evolutionary changes

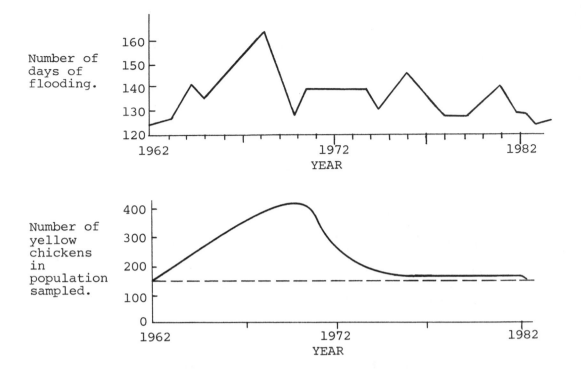

Number of
days of
flooding.

YEAR

Number of
yellow
chickens
in
population
sampled.

YEAR

Graph 1 shows the number of days of flooding from 1962–1982 and graph 2 shows the number of yellow chickens in the population sampled during the same period.

114.   Which statement is supported by the data in the graph?

(A)   The count of yellow chickens was greatest during the years of longest flooding.

(B)   The count of albino chickens was greatest during the years of longest flooding.

(C)   The actual number of yellow and albino chickens was greatest during years of least flooding.

(D)   The actual number of albino chickens was least during the years of least flooding.

(E)   It is impossible to tell when the count of the chickens rose or fell.

115.   The maximum number of yellow chickens appeared

(A)   after the maximum days of flooding

(B)   before the maximum days of flooding

(C)   coincidentally with the maximum days of flooding

(D)   at the minimum days of flooding

(E)   The graph does not supply enough days for this to be determined

Questions 116 – 118 are based on the graph below. Studies of carbon dioxide content in the atmosphere at the Mauna Loa Observatory.

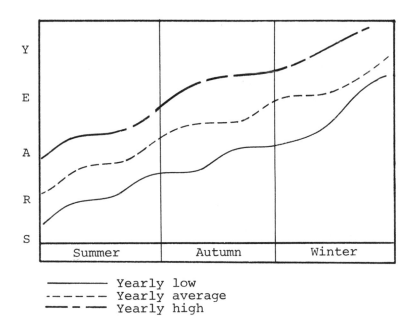

———————— Yearly low
· ― ― ― ― Yearly average
―― ― ―― Yearly high

116.   The low CO_2 content in the atmosphere during summer can be attributed to

(A)   increased replication of DNA molecules in animal cells

(B)   plants undergoing adaptive forms to a warmer environment

(C)   the entering of CO_2 into the photosynthetic cycle from the atmosphere

(D)   sudden increases in populations of autotrophs

(E)   carbon, which is present in very small amounts in the atmosphere becoming a rare gas

117. Observable patterns during winter months show that the CO_2 present in the atmosphere rises. This increase in available CO_2 at this time is due in part to

  (A) the decreasing rate of fossil fuel expenditure by humans

  (B) the fact that new aerobic plant mutants are successful in the adaptation to winter and the frost

  (C) the migration of ciliates to the deep waters of the ocean

  (D) the fact that many of the plants are seasonal

  (E) All of the above are valid reasons for the increase in available CO_2 during winter months

118. On the basis of the diagram one can be justified in assuming that

  (A) carbon dixoide while being essential for plant growth, has no real function in animal growth

  (B) the seasonally highs and lows reflect variation in carbon fixed during photosynthesis

  (C) the concentrations of carbon dioxide and oxygen are relatively uniform in the atmosphere

  (D) None of the above

  (E) All of the above

119. Inversion, a class of structural variation is a type of chromosomal aberration in which

  (A) an extra chromosomal segment is included in the haploid pairing of chromosomes

  (B) a segment of chromosome is removed from one location and is placed at another site on the chromosome

  (C) chromosomal segments are placed along the chromosome thus causing the formation of new alleles

  (D) a segment of a chromosome is turned around 180° and is reinserted into the chromosome

  (E) there is a variation in the chromosome numbers which ranges from the addition or loss of one or more chromosomes to the addition or loss of one or more pairs of haploid sets of chromosomes

120. On a small island off the coast of China a large number of people are polydactyl (having more than five fingers or toes). Which factor most likely contributed to this phenomenon?

(A) Overcrowding

(B) Overproduction

(C) Variation

(D) Natural selection

(E) Isolation

121. The principle that stated that a population remained genetically stable in succeeding generations was the

(A) Hardy-Weinberg principle

(B) Darwinian principle

(C) Mendelian principle

(D) Avery, McLeod and McCarty principle

(E) Levine principle

122. A eutrophic aquatic system is one where

(A) plankton is at a very low density

(B) phosphate and sodium detergents are used

(C) there is low productivity

(D) plant and animal life die due to inavailability of nutrients

(E) the biomass is steadily increasing

DIRECTIONS: *For each group of questions below, match the numbered word, phrase, or sentence to the most closely related lettered heading and mark the letter of your selection on the corresponding answer sheet. A lettered heading may be chosen as the answer once, more than once, or not at all for the question in each group.*

Questions 123-127

(A) Gap$_1$ stage

(B)   Prophase   (D)   S stage

(C)   Cytokinesis   (E)   Anaphase

123.   At this stage DNA replication and the formation of new chromo-
       somes are observed. It is the stage at which the cell incorp-
       orates radioactive thymine.

124.   The chromosomes were observed to condense at this stage and
       the nucleolus to have disappeared.

125.   Chromosomes move rapidly apart and the fibers connected to
       their centromeres shorten.

126.   The constricting ring has closed down on the fibrous remains
       of the spindle.

127.   This is generally a very active period. It is the time when
       the cell synthesizes the enzymes and structural proteins neces-
       sary for cell growth.

128-132 Select the letter of the terms that correspond to the statement.

       (A)   Pia mater   (D)   Meissner's corpuscle

       (B)   Joint kinesthetic   (E)   Golgi tendon organ

       (C)   Baroreceptor

128.   This receptor detects the relative position of bones.

129.   This receptor detects the degree of stretch and contraction of
       a muscle.

130.   This meninx is closest to the surface of the brain.

131. This receptor detects changes in blood pressure.

132. This receptor detects light or fine touch.

Questions 133 - 136

    (A)   Tropical rain forest   (D)   Temperate decidious forest

    (B)   Desert   (E)   Prairie

    (C)   Tundra

133. Rich vegetation covered by a carpet of herbs

134. Small rapid-growing annual herbs with seeds that germinate only when there is heavy rain

135. Soil drainage, decomposition and activities of soil animals impeded

136. Creepers and epiphytes are dominant in this biome

Questions 137 - 141

    (A)   Parasitism   (D)   Mimicry

    (B)   Commensalism   (E)   Allelopathy

    (C)   Mutualism

137. Bacteria fail to reproduce successfully when their colony is seeded with fungi

138. The appearance of the Oncidium orchid like a male bee

139.   The extinction of the American chestnut tree from the Appala-
        chian forests due to the introduction of the sac fungus from
        China

140.   The relationship between a termite and the microorganisms in
        its digestive tract

141.   The growth of epiphytes on the branches of large trees

Questions 142 - 146

        (A)   Vitamin B_1 (thiamine)

        (B)   Vitamin B_6 (pyridoxal)

        (C)   Vitamin A

        (D)   Vitamin D (calciferol)

        (E)   Folic acid

142.   Is involved in calcium absorption and metabolism. Its defic-
        iency causes skeletal deformity

143.   Is important in the prevention of the disease xerophthalmia
        which causes keratinization of tissues of the eye that can
        lead to permanent blindness

144.   Component of a co-enzyme involved in reactions transferring
        amino acids from one compound to another

145.   Component of a co-enzyme that catalyzes the oxidation of pyru-
        vic acid

146.   Is important in the synthesis of some of the nucelotides and is
        a very important component of cell division

Questions 147 - 151

(A) Colchicine   (D) Cycloheximide

(B) Actinomyosin D   (E) Puromycin

(C) Cytochalasin B

147. Reagent that alters the properties of actin and which is used to reveal cellular processes in which microfilaments may be involved

148. Causes the breakdown of microtubules and prevents assembly of new ones

149. Inhibits ribonucleic acid synthesis

150. Causes the blockage of long term memory by affecting protein synthesis

151. Interferes with protein synthesis by the cytoplasmic ribosomes of eukaryotic cells but not by the cytoplasmic ribosomes of prokaryotic cells.

Questions 152 - 156

(A) Endoplasmic reticulum   (D) Cell membrane

(B) Golgi apparatus or dictyosome   (E) Nuclear membrane

(C) Cell wall

152. Keeps the cell cytoplasm stable by being selectively permeable to intracellular as well as extracellular substances and particles

153. An internal membrane that is continuous with the outer membrane of the nucleus

383

154. Peculiar organelle which shows up when treated with silver salts as a flattened, membranous sac lying close to the nucleus

155. A double membrane which is made up of lipids and associated proteins and carbohydrates

156. Semirigid, extracellular encasement which is considered a dead structure

Questions 157 - 161

(A) Apical meristem   (D) Vascular cambium

(B) Cortex   (E) Parenchyma cells

(C) Phloem

157. Surrounds the large leaf veins and contains most of the chloroplasts of leaves

158. This is a thin layer of undifferentiated cells

159. This structure is made up of parenchyma and collenchyma

160. Tissue consisting of sieve-tube members, sieve-tube and companion cells

161. These cells divide actively and are of the growing tip of a root or shoot

Questions 162 - 166

(A) Echinodermata

(B) Annelida

(C)   Mollusca

(D)   Chordata

(E)   Coelenterata

162.   Gastrovascular cavity with simple opening

163.   Segmented body

164.   Muscular foot containing sensory and motor systems

165.   Spiny, crusted covering which is an endoskeleton

166.   Dorsal, hollow turgid rod which serves as a skeletal support
       and has pharyngeal gill slits

Questions 167 – 171

(A)   Bryophyta   (D)   Pterophyta

(B)   Spermophyta   (E)   Chlorophyta

(C)   Angiosperms

167.   Is divided into monocots and diocots

168.   Advanced vascular plants, all produce seeds and pollen.

169.   Spore formers which alternate generations

170.   Contains chlorophylls a and b and has little cell differentiation

171.   Contains chlorophylls a and b, are all multicellular with considerable cell specialization

Questions 172 – 176

    (A)   Edward Jenner

    (B)   Louis Pasteur

    (C)   Sir Alexander Fleming

    (D)   J. Watson

    (E)   John Tyndall

172.   Was the first to successfully vaccinate against smallpox

173.   Finally disproved the theory of spontaneous generation

174.   Discovered penicillin which opened the era of antibiotics

175.   Proved that dust carries germs

176.   Isolated the germ responsible for chicken cholera

Questions 177 – 181

    (A)   Density-independent mortality

    (B)   Abiotic control

    (C)   Age profiles

    (D)   Carrying capacity

    (E)   Programmed death

177. Birds perish if they begin their migration in the spring and are caught by a late cold spell.

178. The population of India shows sign of growth out of control and exponential increase.

179. Brazilian fish living in temporary ponds which exist only during rainy seasons.

180. During severe drought many plants are killed by the parching sun. Similarly in an area saturated with DDT most insects die.

181. The population's rate of increase is reduced when an essential commodity in the environment comes into short supply and environmental resistance is encountered.

Questions 182 - 186

    (A)   Erythrocytes   (D)   Lymphocytes

    (B)   Neutrophils   (E)   Basophils

    (C)   Eosinophils

182. Phagocytic cells concerned primarily with local infections

183. Protein substance in blood that inactivates foreign protein

184. Leukocyte that probably produces heparin (an anticoagulant), histamine and serotonin

185. Cells that transport O_2 and CO_2

186. Phagocytic cells concerned, primarily, with generalized infections

Questions 187 – 189

(A) Altruism

(D) Orientation

(B) Oligotrophy

(E) Consolidation

(C) Communication

187. The curling back of the lips of a dog to expose his teeth

188. The storage of memory in two ways; short term and long term

189. Sparse littoral vegetation

DIRECTIONS: *The following groups of questions are based on laboratory or experimental situations. Choose the best answer for each question and mark the letter of your selection on the corresponding answer sheet.*

Questions 190 and 191

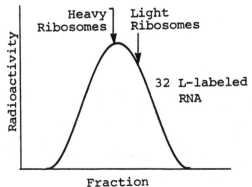

Bacteria were grown in a medium containing heavy isotropes (^{15}P and ^{13}S), infected with phage, and then immediately transferred to a medium containing light isotopes (^{14}P and ^{12}S). Constituents synthesized before and after infection could be separated by density - gradient centrifugation because their densities differed. The new RNA was labeled by the radioisotope ^{32}L or ^{14}N and new protein by ^{35}R. Light ribosomes were absent.

190. These experiments showed that

 (A) ribosomes were synthesized after infection

 (B) ribosomes were not synthesized after infection

 (C) light ribosomes were synthesized

 (D) RNA was not synthesized after infection

 (E) most of the radioactively labeled RNA were "light" ribosome

191. The radioisotope ^{35}R appeared transiently in the "heavy" ribosome peak, this meant that

 (A) ribosomes are specialized structures which synthesize proteins

 (B) tRNA did not transcribe for protein synthesis

 (C) no protein syntheiss was recorded

 (D) new proteins were syntheiszed in preexisting ribosomes

 (E) None of the above is correct

Questions 192 - 195

In Drosophila, the loci for the alleles cut wings (ct), yellow body (y) and vermillion eyes (v) are linked. Each of these alleles are recessive to the corresponding wild type alleles (+) for normal wings, normal body color and normal eye color. The results below were produced for a mating between a heterozygous fly and one that was homozygous recessive.

F₁ Phenotypes

$$\text{F}_1 \quad \textbf{Phenotypes}$$

ct	+	+	−10
+	y	+	− 2
ct	y	v	− 4
ct	y	+	−35
+	+	+	− 4
ct	+	v	− 1

+	y	v	− 7
+	+	v	−37

192. The alleles of the parental chromosome are in the

(A) cis position

(D) trans retinol position

(B) cis retinol position

(E) None of the above positions is correct

(C) trans position

193. The alleles when mapped should align in this way

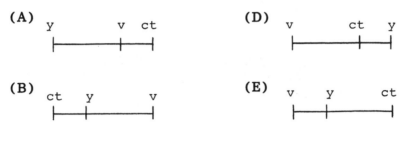

(A)   y   v ct

(D)   v   ct   y

(B)   ct   y   v

(E)   v   y   ct

(C)   y   ct   v

194. The coefficient of coincidence would be

(A) 1.36

(D) 0.136

(B) 2.36

(E) 0.0036

(C) 0.36

195. The interference would be

(A) −0.36

(D) 1.51

(B) 1.00

(E) −0.2

(C) .15

390

A series of experiments were performed as follows:

Experiment	Procedure	Observation
I	A frog limb was exposed to radiation of 6000R of x-rays and then amputated	No nerve regeneration occurred
II	A frog limb was denervated and then amputated before the nerves regenerated	No limb regeneration occurred
III	A frog limb was exposed to radiation and denervated and the limb was amputated before the nerves regenerated	No nerve regeneration occurred
IV	A frog limb was exposed to radiation and denervated. The nerves are then allowed to grow back from the brachial plexus into the limb. The limb is then amputated	Nerve regeneration occurs after amputation

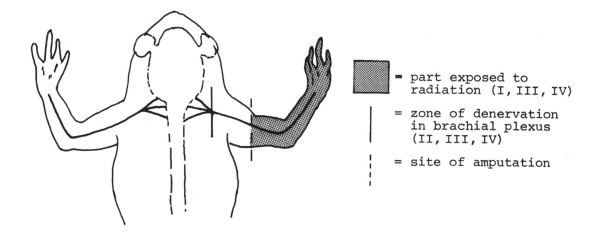

▓ = part exposed to radiation (I, III, IV)

| = zone of denervation in brachial plexus (II, III, IV)

┊ = site of amputation

196. With reference to experiment I which of the following hypotheses could most likely be correct?

(A) Frog regeneration does not require the interaction of nerves.

(B) Limb regeneration is stimulated by x-rays.

(C) Frog tissue when x-rayed does not regenerate.

(D) Unless the nerves are permanently removed x-rayed frog tissue will continue to regenerate.

(E) Limb regeneration is obviously stimulated by nerves.

197. Experiments I, II and III were done because

(A) they aided in the determination of whether denervation or exposure to radiation was a more efficient way to prevent regeneration

(B) they proved that the results of experiment IV were most definitely false

(C) they were controls for experiment IV

(D) they demonstrated that the effect of radiation on limb tissues is opposite to the effect of x-rays on nerves

(E) they revealed the effectiveness of denervation

198. It was possible to do experiment IV and obtain favorable results because

(A) the cell bodies of the limb were irradiated

(B) irradiated nerves were not likely to regenerate

(C) the myelinated nerves are unaffected

(D) the cells bodies of the limb nerves were not irradiated when the limb was exposed to x-rays

(E) irradiated and unirradiated nerves regenerate at the same rate

199. In another experiment, experiment V, a segment of pigmented myelinated nerve which had not been irradiated was implanted

into a limb which had been saturated with radiation. This segment was introduced into an albino host. After allowing some time for growth, the limb was severed at the site of the implant. The resulting regenerate was found to be pigmented. Such results imply that

(A) pigment will not be produced by axons of implanted segments of nerves

(B) the myelinated nerve cells were instrumental in this regeneration

(C) the irradiated areas did not support growth of the implanted nerve segment

(D) cells of the Schwann sheath failed to reproduce

(E) the implant had nothing to do with the limb's regeneration

Questions 200 - 201

In raccoon the following genotypes of two independently assorting autosomal genes determine coat color:

$$A\text{ -}B\text{- } = \text{(grey)}$$
$$A\text{ -}bb = \text{(yellow)}$$
$$aaB\text{- } = \text{(black)}$$
$$aabb = \text{(cream)}$$

A complementary gene pair on a separate autosome determines whether any color will be produced. CC and Cc allow color expression according to the characteristics of the A and B alleles. The cc genotype results in albino raccoons regardless of the presence of the A and B alleles.

200. If a homozygous dominant raccoon for coat color but heterozygous for its complementary C gene was crossed with a raccoon heterozygous for both coat color and its complementary C gene. The most likely phenotypic ratio would be:

(A) $\frac{4}{16}$ grey : $\frac{8}{16}$ yellow : $\frac{4}{16}$ albino

(B) $\frac{9}{16}$ grey : $\frac{3}{16}$ yellow : $\frac{4}{16}$ cream

393

(C) $\frac{1}{2}$ yellow : $\frac{1}{2}$ albino

(D) $\frac{12}{16}$ grey : $\frac{4}{16}$ albino

(E) $\frac{3}{8}$ black : $\frac{4}{8}$ yellow : $\frac{1}{8}$ cream

201. Given the inheritance pattern of coat color in raccoons, according to Question 179 predict the genotype of the parents who produced the F_2 offspring $\frac{3}{4}$ grey : $\frac{1}{4}$ albino.

(A) AAbbCc × AabbCc

(B) aaBBCC × AaBbCc

(C) AABBCc × AABBCc

(D) AaBBcc × AaBbcc

(E) All produced the F_2 phenotypic ratio.

Questions 202 – 203

The gonadal cortical tissue and gonadal medullary tissue were removed from the gonads of a group of male rats. They were removed before full sexual development was completed. These extracted gonadal tissues were then randomly replaced in the gonad of the rats and sexual development proceeded.

202. The result expected from such an experiment would be that

(A) all the embryos will be abnormal

(B) the rudimentary region of the gonads will be intact

(C) sperm cells will not be produced in the rats

(D) some of the rats will receive an original of the part of the gonad that was the same sex as their own, and some will receive one that was of a different sex

(E) the rats sexes will all be changed

203. Such an experiment was done to determine

(A) if gonadal cortical tissue affected reproduction and growth

394

(B)　whether a new species of rat could be formed

(C)　if cortical dominance produces ovarian tissue while medullary dominance results in the development of testes

(D)　whether ovarian tissue was produced by medullary dominance while cortical dominance causes the production of testes

(E)　if the Müllerian duct will not be present in 50% of the embryos

Questions 204 - 206

The following experiments were performed in order to test the role of cytokinins and auxins in stimulating growth in the callus of tobacco pith. the callus consists of parenchyma cells.

Series	Experiments	Results
I	Cytokinins, no auxin	No growth
II	Cytokinins coupled with other plant hormones.	Plant growth and cell division.
III	High cytokinins to auxin ratio.	Development of leaves, no roots.
IV	High auxin to cytokinin ratio.	Development of roots, no leaves.
V	Intermediate cytokinin to low auxin.	Continued growth as callus.

204.　The results obtained from experiments I, II and V indicate that all of the statements are correct except that

(A)　cytokinins inhibit plant growth

(B)　cytokinins stimulate cell division

(C)　for cytokinins to be active it must be coupled with other plant hormones

(D)　cytokinin is an important plant growth requirement

(E)　All of the statements are correct about cytokinin

205.　The experiments indicated that

(A)   cytokinins cause the early aging of plants

(B)   cytokinins have no particular role in cell division or plant growth

(C)   the parenchyma cells contain all the genetic information necessary to develop into a number of kinds of other plant cells

(D)   the parenchyma cells have no bearing on the production of different parts of the plant

(E)   pith tissue requires a high concentration of auxin in order to be differentiated

206.   From the results of the experiment low auxin to cytokinin ratios

(A)   caused the cells to be differentiated into root tissues

(B)   caused the pith to be changed into apical meristem

(C)   caused the ordinary cells of the pith to develop into a variety of tissues and organs which facilitated growth

(D)   All of the above were caused by this ratio.

(E)   None of the above is related to this ratio.

Questions 207 - 210

Irradiation and incubation of an auxotrophic leu$^-$ strain of bacteria until they reached the stationary phase were used to determine the spontaneous and x-ray induced mutation rate of leu$^-$ to leu$^+$. Also studied was the reaction of a control culture which was not irradiated. The cultures were then serially diluted and 0.1 ml of various dilutions were plated on minimal medium plus leucine and on minimal medium.

Culture	Medium	Dilution	Number of Colonies
Irradiated	(I) Minimal medium plus leucine	10^{-9}	24
	(II) Minimal medium	10^{-2}	12
Control	(III) Minimal medium plus leu	10^{-9}	12
	(IV) Minimal medium	10^{-1}	3

207. Each value obtained from medium (I) – (IV) respectively represents the

(A) total number of bacteria in the irradiated culture

(B) total number of bacteria with $leu^{(-)}$ to $leu^{(+)}$ mutations present in the irradiated culture

(C) total number of bacteria present in the control culture

(D) total number of bacteria with $leu^{(-)}$ to $leu^{(+)}$ mutations present in the control culture

(E) All of the above

208. Which values should be approximately equal?

(A) I and II

(B) II and III

(C) II and IV

(D) I and III

(E) III and IV

209. What was the induced rate leading to prototrophic growth?

(A) $\dfrac{12 \times 10^2}{12 \times 10^9}$

(B) $\dfrac{12 \times 10^2}{24 \times 10^9}$

(C) $\dfrac{3 \times 10^1}{12 \times 10^9}$

(D) $\dfrac{24 \times 10^9}{12 \times 10^9}$

(E) None of the above correctly represents the induced rate.

210. The spontaneous mutation rate leading to prototrophic growth (leu^- to leu^+) is most likely represented as

(A) $\dfrac{12 \times 10^9}{24 \times 10^9}$

(B) $\dfrac{3 \times 10^1}{12 \times 10^9}$

(C) $\dfrac{24 \times 10^9}{12 \times 10^9}$

(D) $\dfrac{12 \times 10^9}{3 \times 10^1}$

(E) $\dfrac{12 \times 10^2}{3 \times 10^1}$

THE GRADUATE RECORD EXAMINATION BIOLOGY TEST

MODEL TEST IV

ANSWERS

1.	D	20.	D	39.	A
2.	C	21.	B	40.	C
3.	E	22.	D	41.	B
4.	B	23.	C	42.	A
5.	E	24.	B	43.	C
6.	C	25.	B	44.	B
7.	A	26.	D	45.	B
8.	E	27.	B	46.	C
9.	A	28.	B	47.	E
10.	C	29.	E	48.	B
11.	B	30.	D	49.	B
12.	A	31.	E	50.	D
13.	D	32.	E	51.	C
14.	E	33.	D	52.	A
15.	E	34.	C	53.	D
16.	B	35.	A	54.	C
17.	B	36.	B	55.	B
18.	C	37.	D	56.	E
19.	C	38.	D	57.	A

58.	B		89.	D		120.	E
59.	A		90.	E		121.	A
60.	D		91.	A		122.	E
61.	D		92.	C		123.	D
62.	A		93.	C		124.	B
63.	C		94.	A		125.	E
64.	C		95.	B		126.	C
65.	D		96.	C		127.	A
66.	A		97.	A		128.	B
67.	B		98.	A		129.	E
68.	D		99.	C		130.	A
69.	A		100.	C		131.	C
70.	C		101.	B		132.	D
71.	A		102.	C		133.	D
72.	B		103.	E		134.	B
73.	D		104.	A		135.	C
74.	C		105.	E		136.	A
75.	E		106.	C		137.	E
76.	D		107.	C		138.	D
77.	A		108.	A		139.	A
78.	C		109.	C		140.	C
79.	C		110.	D		141.	B
80.	A		111.	B		142.	D
81.	B		112.	B		143.	C
82.	A		113.	C		144.	B
83.	B		114.	A		145.	A
84.	C		115.	A		146.	E
85.	D		116.	C		147.	C
86.	D		117.	D		148.	A
87.	E		118.	B		149.	B
88.	B		119.	D		150.	E

151.	D	171.	A	191.	D
152.	D	172.	A	192.	C
153.	A	173.	B	193.	C
154.	B	174.	C	194.	A
155.	E	175.	E	195.	A
156.	C	176.	B	196.	C
157.	E	177.	B	197.	C
158.	D	178.	C	198.	D
159.	B	179.	E	199.	B
160.	C	180.	A	200.	D
161.	A	181.	D	201.	C
162.	E	182.	B	202.	D
163.	B	183.	C	203.	C
164.	C	184.	E	204.	A
165.	A	185.	A	205.	C
166.	D	186.	D	206.	B
167.	C	187.	C	207.	E
168.	B	188.	E	208.	D
169.	D	189.	B	209.	B
170.	E	190.	B	210.	B

THE GRE BIOLOGY TEST

MODEL TEST IV

DETAILED EXPLANATIONS
OF ANSWERS

1. (D)
Phosphates form the polar or charged end of phospholipids. These are major components of cell membranes which consist of two layers of lipids, primarily phospholipid and glycolipid.

2. (C)
Hydrolytic cleavage is the process applied to break down sucrose into glucose and fructose, the two monomers of which it is made. It is just the opposite of a dehydration linkage which links glucose and fructose together to form sucrose.

In hydrolytic cleavage a water molecule is added to the linkage breaking the sucrose into its component parts.

3. (E)
Amylopectin is a large molecule consisting of 1-4 glucose chains cross-linked with 1-6 and 1-3 linkages between chains. The individual straight chains are twenty to thirty glucose units long.

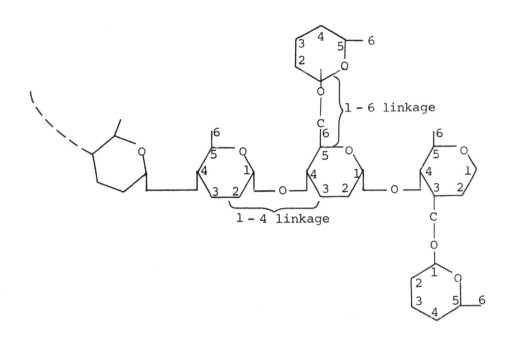

1 – 6 linkage

1 – 4 linkage

4. (B)
The R group is always attached to a carbon atom. Attached to the same carbon atom are three other groups: an amino group which has a positive charge, a carboxyl group which has a negative charge, and a hydrogen.

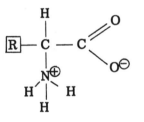

5. (E)
Some of the globular proteins assume a final quaternary structure level. Here two proteins interact to form a dimer. Quaternary forces are the same as those at the tertiary level. The most familiar quaternary is probably the hemoglobin of red blood cells.

6. (C)
As a cell increases in size, the volume grows faster than the surface area (the membrane).

7. (A)
Plant cells, particularily those with large internal vacuoles, can regulate the surface volume ratio by using cytoplasmic or cell streaming in which the cytoplasm whirls around the cell.

8. (E)
In 1940, J.F. Danielli, a British biologist who specialized in lipid biochemistry, observed that fat soluble molecules diffused through the membranes of living cells, while polar molecules generally did not. He found that different kinds of molecules flowed at different rates across the membrane but that the rate of flow could be predicted for most molecules on the basis of their lipid solubility, polarity and size. Some large, polar molecules such as glucose and certain amino acids traveled quite readily across living cell membranes in either direction but always in the direction of greater concentration to lesser concentration. The movement he surmised was due to facilitated diffusion. Danielli postulated that the cell membrane although a phospholipid bilayer in structure, had protein lined pores through which these large, polar molecules entered.

9. (A)
Until recently, it was assumed that O_2 moved from the environment into cells and blood by diffusion only. It has however now been shown that mammals utilized a cell membrane O_2 carrier, a heme protein called cytochrome P_{450} (the 450 refers to the color of the protein – that is the wavelength of light absorbed). The facilitated diffusion of oxygen by cytochrome P_{450} was found to be 80% of the oxygen that enters mammals.

10. (C)
The scanning electron microscope was developed in the 1940's. While its resolving power was not as wide as the standard electron microscope, it has the distinct advantage of producing the illusion of three-dimensional image, with extraordinarily great depth of field.

11. (B)
If plants such as lettuce, celery or carrots lose water by

evaporation their cells shrink. The turgor pressure within the cell diminishes, and the plants wilt. These plants if they have not died can be restored if they are placed in a solution of water. The water on the surface of the lettuce is at a higher concentration than that inside the cytoplasm and vacuole of the plant cell. The water molecules therefore diffuse inwardly, from a higher concentration to a lower concentration until the turgor pressure reaches a point where it balances the diffusion of water. If the produce man then sprayed the lettuce with a solution of salt in which the (number of salt molecules) equaled that in the cytoplasm of the lettuce cells, nothing will happen. The water concentration would be the same inside as outside and the spray would be called isotonic to the lettuce cytoplasm.

12. (A)
There are two types of endoplasmic reticulum; smooth endoplasmic reticulum and rough endoplasmic reticulum. Rough endoplasmic reticulum is seen most commonly in cells that manufacture proteins.

Smooth endoplasmic reticulum is primarily found in cells that synthesize, secrete and/or store carbohydrates, steroid hormones, lipids, or other non-protein products.

13. (D)
Lysosomes are membrane bound sacs that are roughly spherical. They are bags of powerful hydrolytic enzymes which are packaged and synthesized by the golgi apparatus.

If these enzymes were freely floating around in the cytoplasm, the cell would quickly self destruct.

14. (E)
Mitochondria are complex energy-producing organelles found in every eucaryotic cell. Like chloroplasts mitochondria are enclosed in double membranes; have their own circular DNA; have their own ribosomes and other machinery of protein synthesis.

15. (E)
Cilia and flagella are fine, hairlike, movable organelles

found on the surfaces of some cells. They apparently have some sort of cellular sensory function. These organelles have a nine plus two arrangement which can still be seen in the rods and cones of the retina, the olfactory fibers of the nasal epithelium and the sensory hairs of the cochlea and semicircular canals of the internal ear.

16. (B)
Many oxidation-reduction reactions in cells are mediated by specific electron carriers. Of particular significance are two pairs: NAD^+ and NADH (one pair) and $NADP^+$ and NADPH (the other pair). NAD^+ and $NADP^+$ differ only by one phosphate group, but they participate in quite different cellular reactions. NAD^+ is used more in oxidative phosphorylation reactions while $NADP^+$ is used more in photosynthesis.

17. (B)
In the photolysis of water, two hydrogen ions are released into the interior space of the thylakoid, and two electrons enter the electron transport system. Photosystem P680 and P700 are used.

18. (C)
The lamellae extend continuously between the thylakoids, through an amorphous region called the stroma. The thylakoids, with their photosynthetic pigments are involved in the immediate, light related events of photosynthesis - the light reaction. However, carbohydrate formation actually begins in the stroma. This part of photosynthesis does not involve light directly and is termed the dark reaction or the light independent reaction.

19. (C)
The final product of glycolysis is pyruvic acid or pyruvate. Pyruvate can undergo one of three fates. Some organisms in the absence of oxygen carry on fermentation. Pyruvate is reduced by NADH and CO_2 is removed. Two moelcules of ethyl alcohol are formed for each glucose molecule that is fermented.

Glycolysis during muscular activity is also anaerobic. Pyruvate is reduced by NADH to lactate.

Under fully aerobic conditions pyruvate is oxidized by NAD^+ and a CO_2 molecule is removed forming one molecule of acetyl CoA which enters the Krebs cycle.

20. (D)
Chloroplast and mitochondria, even though they are both cell organelles which are independent of the cells they inhabit, are different to each other in their electron transport system.

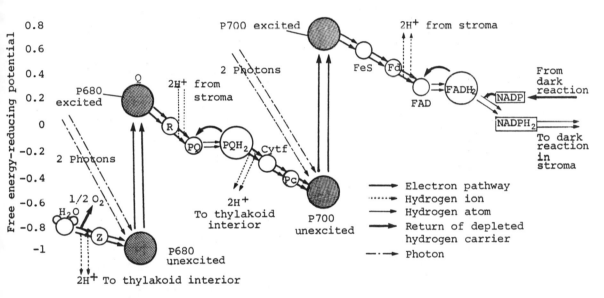

Z-Scheme diagram of the photolysis of water.

Mitochondrion system utilizes stalked F_1 particles which project inwardly from the inner membrane surface. The chloroplast system utilizes CF_1 particles which are on the outer surface of the thylakoid membrane.

The Z scheme of photoshynthesis began with the splitting of water into hydrogen ions, electrons and molecular oxygen while the electron transport of mitochondria ends with the combining of hydrogen ions and molecular oxygen into water.

Another difference is that the photosynthetic electron transport chain ended with the electrons being passed to NADP, while the mitochondrial system begins with electrons being passed from NADH to the first membrane bound carrier.

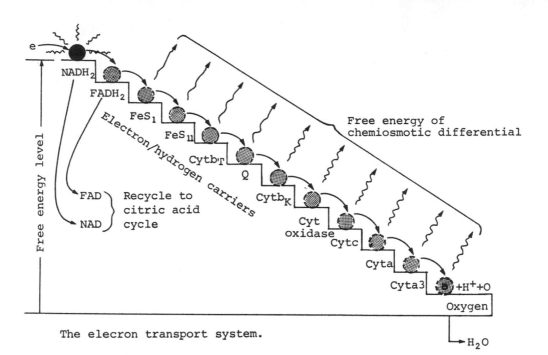

The elecron transport system.

Chloroplasts turn energy into carbon compounds and oxygen while mitochondria turn carbon compounds and oxygen into energy.

21. (B)
In laboratory preparations of tubulin subunits taken from cells arrested in the early prophase of mitosis it was observed that microtubules would not form unless intact centrioles were added. When these intact centrioles were added asters formed around them. It was therefore postulated that centrioles do indeed organize asters.

22. (D)
In comparing procaryotic, higher plant and animal cells it was found that the procaryotic and higher plant cells differed from the animal cell in the formation of their microtubules, centrioles and cytoskeleton.

Animal cells were found to possess protein cytoskeleton while both the higher plant and procaryotic cells possessed polysaccharide cell walls. Animal cells had hollow, membrane bound nine plus two microtubule pattern while the procaryote's was solid and rotating and microtubules were absent from higher plants. Centrioles were found to be present only in animal cells.

23. (C)
A gap junction is an organized system in cells which allows
for the direct exchange of nutrients and intracellular
hormones through channels that pass between cells. They
make the cytoplasm of many multicellular tissues effectively
continuous.

24. (B)
After a period of heavy exertion, the muscle tissue in
humans and other vertebrates will be loaded with lactic acid
and the supply of ATP depleted. To compensate for and
replace the depleted ATP in the muscles, the creatine
phosphate (phosphocreatine) is tapped, thus causing high
energy phosphate to be transferred to ADP causing more
ATP to be produced.

25. (B)
The tRNA molecule is precisely coiled and loops back on
itself to form a distinct pattern. When folded, it has three
loops and a stem. These are the D loop, the T loop, the
anticodon loop and the -CCA stem.

The anticodon loop can recognize specific codons of mRNA.
The -CCA stem - is capable of being covalently bonded to
an amino acid by an enzyme.

26. (D)
Of the 64 codons three are "stop" codons which specify
the end of a protein. These three are UAA, UAG and UGA.
The remaining sixty-one codons specify the twenty amino
acids.

27. (B)
In eucaryotes a special initiator tRNA with a 5'-CAU-3'
anticodon recognizes and pairs with the initiation codon
5'-AUG-3'.

28. (B)
The word somatic refers to body. Somatic cells are body

cells which make up our body tissues. These cells and all their cellular descendants are doomed to perish. They are any cells other than the germicidal cells which are used during sexual reproduction.

29. (E)
At the meiotic stage chromosomes have been at times found to fail to separate correctly. This phenomena is termed nondisjunction.

Nondisjunction can occur with any chromosome. Having only one autosome of a pair instead of the normal two is always fatal. Having three instead of two is almost always fatal. This results in spontaneous abortion or death in infancy. There is however, one exception. A person can survive with three of the chromosome 21. Such an infant may grow to adulthood but may be abnormal. The syndrome is known as trisomy-21 or Down's syndrome.

30. (D)
Translocation involves the movement of a chromosomal segment to a new place. It may occur on a single chromosome or between non-homologous chromosomes. When this happens the mutation is known as translocation.

Deletion is a mutation in which a segment of the chromosome is missing. In duplication, which is a chromosomal aberration, a segment of the chromosome is repeated. Inversion is another chromosomal aberration in which the order of a chromosomal segment has been reversed. The segment is removed and inserted in the same location but in the opposite direction.

31. (E)
A society is made up of a group or groups of animals which belong to the same species and are organized in a cooperative manner. These animals are usually bound together by reciprocal communication which leads to this type of cooperative behavior.

32. (E)
DNA was first studied in 1868 by Friedrick Miescher. It

was he who was able to separate nuclei from the cytoplasm of cells and to then isolate from them nucleic acid.

Avery, MacLeod and McCarty were the ones who by their publication of an article concerning the chemical nature of a "transforming factor" set the stage for the acceptance of DNA as the genetic material.

Fredrick Griffith did the transformation studies using different strains of the bacterium diplococcus pneumoniae. He viewed transformation as a genetic event and demonstrated through his experiments that mice when injected with virulent bacterium died and that those injected with non-virulent bacterium did not die. He however, also discovered that mice injected with heat killed virulent bacteria lived but that those injected with heat killed virulent bacteria as well as living bacteria died. This phenomena he accredited to transformation of genetic material.

The tetranucleotide hypothesis was proposed by Phoebus A. Levene to explain the chemical arrangement of nucleotides in nucleic acids. He assumed a 1:1:1:1 ratio of the four nucleotide unit. This was however disproved by Erwin Chargaff who showed that for most organisms this ratio is not accurate.

33. (D)
Initially the cell was able to maintain a higher concentration of dye in the intracellular environment than extracellular due to active transport. However if the inhibitor interferred with the production of ATP and, as a result active transport, normal diffusion processes would take over resulting in equal intracellular and extracellular concentrations of the dye.

34. (C)
The chordates are divided into three subphyla: Urochordata, Cephalochordata, and the Vertebrata.

35. (A)
Aves is a class. It is the class of the birds.

36. (B)
Ramapithecus is tentatively accepted as being one of the first primates in the phylogenetic tree of humans. This acceptance is based mainly on teeth which constitutes the bulk of the fossils found to date. Fossils of this primate have been found in Africa, India, Pakistan and Greece.

37. (D)
Homo habilis is the name given by Louis Leakey to certain fossils found in Africa that seemed closer to the human line. He placed them in the genus Homo. Homo habilis has a much greater cranial capacity than Australopithecus and appears closer to the human line than Australopithecus. Australopithecus, however, was the one that possessed strong jaws and teeth.

38. (D)
$$6RU - P + 6CO_2 + 12NADPH_2 + 18ATP$$
$$\rightarrow C_6H_{12}O_6 + 18ADP + 18P_i + 6Ru - P + 6H_2O$$

The formula presented is the formula of the Calvin Cycle, also known as the light-independent reactions. The end products are carbohydrates ADP, Ru - P and water.

39. (A)
Endorphins and enkephalins are endogenous opiates. They are the body's natural pain killers and stress antagonists and they are chemical transmitters. Drugs such as alcohol, heroin, cocaine, and opium are external opiates which are introduced into the body by artifical means.

40. (C)
Somatic nervous system innervates skeletal muscle usually and involves some conscious control of the reflex. Blood vessels cannot be stimulated voluntarily to constrict and are under the control of the autonomic nervous system.

41. (B)
When a myelinated neuron is activated the action potential

moves along the axon by jumping from one node to another. This type of transmission is called saltatory transmission and is much faster than point to point propagation.

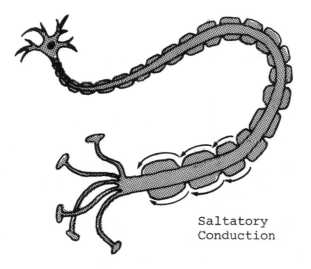

Saltatory Conduction

42 - 45

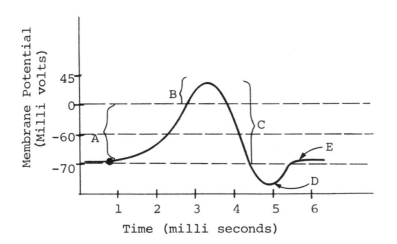

42. (A)
Part A of the curve represents a condition called depolarization.

43. (C)
If a stimulus of -66 millivolts was applied an action potential will not be created since in order for the action potential to be created the stimulus must be greater than or equal to -60 millivolts.

44. (B)
Part C of the curve is due to the outflow of K^+.

45. (B)
Parts A and B of the curve are due to the influx of Na^+.

46. (C)
The most logical evolutionary advantage of multicellularity is that multicellular organisms can cope differently with the environment and its resources.

In multicellular organisms there is a very high degree of specificity. This allows for a high level of efficiency which allows the multicellular organism to cope differently with the environment.

47. (E)
Robert Hooke was the scientist who discovered cells. He focused his microscope on a cut open cork and observed little boxes which he called cells.

Matthias Schleiden and Theodore Schwann said that all living things were composed of cells.

Rudolf Virchow propagated the doctrine that all cells are derived (made) from cells.

48. (B)
Watson and Crick postulated that because of the arrangement and nature of the nitrogenous bases, DNA replication was semi-conservative.

They further went on to predict that each strand served as a template for the synthesis of its complement. They further proposed that if the DNA helix was unwound, each nucleotide from the parental side would have an affinity for its complementary nucleotide.

49. (B)
Most of the electrons from the citric acid cycle are carried as NADH, each electron will pump three hydrogen ions out against the gradient. One of these reactions however, passed a pair of hydrogens directly to a lower-energy FAD coenzyme. This FAD coenzyme now $FADH_2$ passes its hydrogen to the cycle further down the line. These lower-energy hydrogens which do not participate in the first active transport reaction will contribute only two hydrogen ions each to the chemiosmotic differential.

50. (D)
Morphogenesis is the process whereby the embryo develops form or shape.

51. (C)
The nitrogenous bases of opposite chains of nucleotides are electrostatically attracted to each other by the formation of hydrogen bonds. This hydrogen bonding serves to maintain stability in the helical structure.

52. (A)
When a depolarizing wave reaches the synaptic knobs of neurons, the order in which the impulse is transmitted is Ca^{2+} ions being admitted from surrounding fluid, the microtubular system is activated, microtubules attach to vesicles and neurotransmitters are released which then diffuse across the cleft, contacting receptor sites on the postsynaptic membrane of the next effector.

53. (D)
Acetylcholine is the neurotransmitter which is found at neuromuscular junctions in the brain and at junctions in the internal organs.

54. (C)
On being suddenly chased by the large bear the system of the body immediately come into play with the fight-or-

flight mechanism being activated. This causes an increase in heart rate, the pupils dilate, blood is shunted away from the digestive tract and peripheral vessels. These changes are all wrought by the sympathetic system of the autonomic nervous system.

55. (B)
The hypothalamus controls basic drives such as hunger, thirst, sex, and rage. Experimental stimulation of different centers in the hypothalamus by means of electrodes can cause a cat to act hungry, thirsty, cold, angry or hot.

56. (E)
The principal visual pigment in vertebrates is visual purple. This pigment is present in large quantities in stacks in the rods of the vertebrate retina. When visible purple absorbs a photon of light cis-retinal, the not too stable and reactive form of retinal, is transformed to the trans-form and dissociates from its opsin.

57. (A)
Light receptors are sensitive to wavelengths that are at about 430 and 750 nm.

58. (B)
Sensory hair cells push against the tectorial membrane which is relatively rigid.

59. (A)
Cortisol is a glucocorticoid hormone released from the adrenal cortex. It has been found to increase the metabolism of glucose and protein and to reduce inflammation.

60. (D)
The animals apparently experienced "physiological shock". One of the symptoms was high ACTH levels and adrenal

steroid levels. This in turn severely reduced the ability of the deer to reproduce.

61. (D)
The order of the contraction of the heart is

S-A node → atria → A-V node → Bundle of His → ventricles.

The origin of the heart beat is the S-A node (sinoatrial node). The S-A node causes the atrium to contract. The impulse is then transmitted to A-V node (atrioventricular node) and then to the bundle of His which then initiates ventricular contraction.

62. (A)
There are two basic proteins in the clotting process of blood. These are prothrombin and fibrinogen. When the vessel is damaged, the damaged cells release thromboplastins. These are enzymes which break down prothrombin into thrombin. Thrombin in turn breaks apart fibrinogen, thus causing fibrin to be formed. Fibrin fibers along with damaged platelets forms the clot.

63. (C)
When the sperm reaches the egg it releases the enzyme hyaluronidase which breaks down the corona radiata thus allowing the sperm to contact the egg. The attached sperm head is then engulfed in the fertilization cone and soon detaches, thus allowing the nuclear membrane to break down and release the chromosomes into the egg cytoplasm. The sperm pro-nucleus then contacts the ovum pro-nucleus. Fertilization is now complete.

64. (C)
Actomyosin is the protein that serves as an enzyme and breaks down ATP to ADP + P_i.

65. (D)
The growth of roots and shoots is called geotropism. The

417

growth of roots downwards into the soil is called positive geotropism and the upward growth of shoots is called negative geotropism.

66. (A)
Transduction is usually done in the laboratory. In this process genetic material from one bacterium is transferred to another bacterium using a bacteriophage as the carrier.

67. (B)
Evolution in actuality involves a change in allele frequencies. This means that the relative number of one form of the gene increases and the relative number of a different form decreases. This continued change in allele frequencies is a very vital stage in the evolutionary process which occurs between starting point of a new allele by mutation and the final replacement of the first form by newer forms of the gene.

68. (D)
Chemotrophs are organisms which derive their energy from inorganic energy sources such as hydrogen, sulphur, hydrogen sulfide or ammonia, leaving the oxidized remains as wastes.

Chemotrophs use carbon dioxide as a carbon source.

69. (A)
The ecosystem is made up of producers, consumers, and reducers. Consumers are divided into different trophic levels in the food chain. Herbivores are primary consumers, while carnivores which feed on herbivores are secondary consumers.

70. (C)
Releasers are sign stimuli that are usually emitted by a member of the same species. These stimuli are particularly effective in triggering behavioral responce. The responce of

the male European robin to the tuft of red feathers is an instinctive pattern because tufts of red feathers are normally on the breast of competitors. This is an illustration of the action of releasers.

71.   (A)
If the islets of Langerhans of the pancreas are ligated or damaged in any way, the flow of insulin from the pancreas to the circulatory system is affected.

If the pancreatic duct is ligated, the flow of trypsinogen to the small intestine will be hindered.

72.   (B)
In a mating such as this the allelic configuration can be varied.   The mating is between two dogs with black fur color.   In any mating such as this it is impossible to get all grey colored dogs since either S or P must be absent from the mating if grey color is to be expressed in the phenotype.

73.   (D)
An instinctive behavior is one which depends on an inhibitor or a block.   Innate behavior is based on fixed patterns.   Not all innate behavior is instinctive.   The flight of birds to warm places during winter is an innate behavior that is not instinctive.

74.   (C)
The pollen tube produced when the pollen germinates makes its way down through the stigma and into the style and then penetrates the ovule at the micropyle. The sperm's nuclei then move to the embryo sac where it fuses with the egg nuclei and thereby fertilizes the egg cell.   The process of fertilization is now complete.

75.   (E)
Vertebrate muscles are divided into three types according

to their microscopic structure and the nerves that activate them.

Smooth muscle has a smooth, glistening appearance and is also known as involuntary muscle. It is found in many internal organs and it causes the hair to rise under eery conditions or the stomach to make noises. Bands of smooth muscle form the muscle of the hair. They are called arrector pili.

76. (D)

Trypsin and chymotrypsin are the enzymes of the small intestine which decompose peptide chains of proteins into free amino acids.

77. (A)

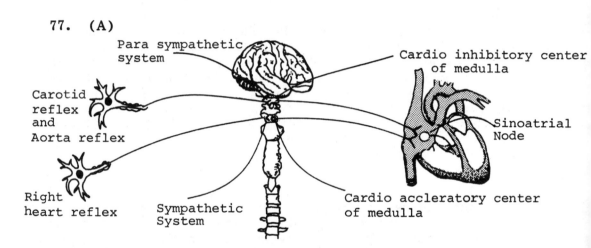

The carotid reflex which is located just above the bifurcation of carotid arteries synapses with the cardio-inhibitory center of the medulla of the parasympathetic system.

78. (C)

All vertebrates have closed circulatory systems. The most efficient closed circulatory systems are found in crocodiles, mammals and birds. Fish also have a closed circulatory system.

Annelids have an extraordinarily well-developed closed system with distinct vessels. The earthworm is one example of an annelid.

79. (C)

The diver at about 100M who breathes compressed air with a total pressure of about 10 atmospheres where the partial pressure is 2.1 atmospheres finds that the partial pressure is too high for normal physiological processes.

The correct percentages of air that the diver should have used should have been 98% inert helium and 2% O_2. At 10 atmospheres total pressure is a partial pressure of 0.20 atmosphere of O_2. This would have been ideal for the continuance of normal physiological processes at this height.

80. (A)

In the capillary network blood entering the alveolar is rich in CO_2 and poor in O_2 while blood leaving the alveoli and entering the circulatory system is poor in CO_2 and rich in O_2.

81. (B)

In the case of gas exchange and respiration, birds, are the most advanced animals not mammals. They have a one way, countercurrent circulation of air in the lung. This does not allow for much mixing of the old and new air.

82. (A)

Blood which leaves the left atrium passes through the valve called the bicuspid valve before it enters the left ventricle.

83. (B)

Vasoconstriction and vasodilation of the arterioles is partly under the control of the sympathetic fibers of the autonomic nervous system. This is augmented by hormonal influence. In addition to hormonal influence, oxygen, carbon dioxide, hydrogen ions and a variety of drugs can cause vasoconstriction and vasodilation.

84. (C)

Overactive and underactive thyroid glands produce

conditions known as hyperthyroidism and hypothyroidism, respectively. The symptoms of hyperthyroidism are weight loss, nervousness, insomnia and hyperactivity.

85. (D)
When the stimulus elicits a response that further increases the stimulus instead of decreasing it, this is a positive feedback.

In the illustration with the mother and baby, the baby instead of ceasing to cry when the mother introduces the feed cries harder. This is an example of negative feedback.

86. (D)
Antidiurectic hormone is the hormone whose stimulations causes the uptake of water in the kidneys by making the walls of the distal convoluted tubule and the collecting ducts permeable to water.

87. (E)
Altruism is an activity which benefits another organism at the individual's own expense.

88. (B)
By far the most useful and precise method for separating amino acids in hydrolysate is chromatography. It is the quantitative separation of the amino acids on a column of an ion-exchange resin, followed by the complete elutriation of each amino acid. When this method is used, the position of each amino acid from the protein hydrolysate is known from its position in the chromatography column or its time of elution.

89. (D)
If a mutation adds a nucleotide to the sequence of the nucleotides a frame shift would occur and the order of the nucleotides would be changed from the point of insertion of the new nucleotides.

e.g. CAT   CA T CAT   CAT   CAT   CAT
              ↑
              A

The sequence would now read:

CAT   CAA   TCA   TCA   TCA   TCA   T

90. (E)
A territory is an area actively defended from other species by an animal or a group of animals. Territories may be recognized as areas where nesting, feeding and mating occur.

Questions 91 - 92

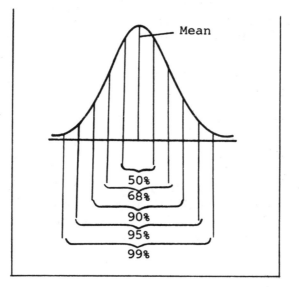

91. (A)
According to the curve most of the population center around the mean.

92. (C)
The curve approximates the idealized population distribution. This is most likely based on a survey taken of the population.

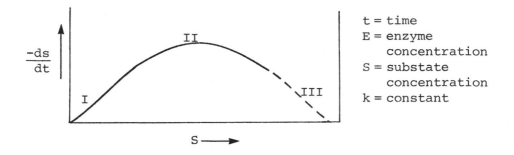

The concentration of the enzyme is held constant and the substrate concentration is varied.

93. (C)
If the reaction velocity $-\dfrac{ds}{dt}$ is plotted against the substrate s, the first portion of the curve which is a straight line may be expressed as

$$-\frac{ds}{dt} = k \ \{ E \cdot S \}$$

94. (A)
The third stage of the reaction (III) occurs when enzymatic activity is inhibited at high concentrations of substrate. The enzyme in combining with substrate molecules has a limited capacity for such binding and becomes saturated. This saturation causes a drop in enzymatic activities.

95. (B)
At the second stage a plateau is reached. This is caused by increasing substrate concentration and suggests that saturation of the enzyme has occurred.

96 - 98 concerns the curve of excretion of sex hormones during pregnancy.

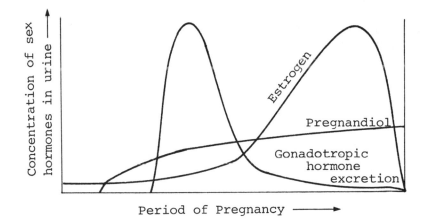

Concentration of sex hormones in urine →

Estrogen

Pregnandiol

Gonadotropic
hormone
excretion

Period of Pregnancy ⟶

96. (C)
The appearance of pregnandiol in the urine indicates the presence of progesterone. Pregnandiol is the metabolic end product of progesterone.

97. (A)
The substance secreted by the placenta which is similar to LH (luteinizing hormone) of the pituitary is gonadotropic hormone. Both are found in the gonads. LH is a pituitary gonadotropic hormone while human chorionic gonadotropin is not found in the pituitary.

98. (A)
The hormone which is produced in large amounts during the first two weeks of pregnancy and which serves as a basis for pregnancy tests is gonadotropic hormone.

99 - 100

Each gas, in a mixture of gases, exerts a partial pressure equal to the pressure it would exert if it existed alone. The total pressure of the mixture is dependent on the partial pressure of each gas.

99. (C)

At sea level (760 mmHg) oxygen makes up 20.93% by volume of the air. Its partial pressure in the inspired air is therefore

$$\frac{20.93}{100} \times 760\,\text{mmHg} = 159.1\,\text{mmHg}$$

100. (C)

If we were however at an altitude of 18,000 ft. where the atmospheric pressure is 380 mmHg, the partial pressure of oxygen would be

$$\frac{20.93}{100} \times \frac{380\ \text{mmHg}}{1} = 79.6\ \text{mmHg}$$

Questions 101 - 103 concern the diagram of the ECG.

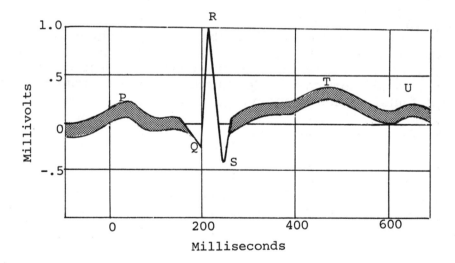

101. (B)

The P wave is the initial depolarization of the SA node. As a result of this depolarization atrial contraction is caused.

102. (C)

The QRS complex of the ECG is depolarization of the myocardial tissue.

103.   (E)
If the R and S peaks were reversed the function affected would be depolarization of the myocardial tissue.

104.   (A)
Epithelial tissue is the tissue which forms the covering or lining of all free body surfaces; external and internal. It makes up the outer portion of the skin, the linings of the digestive tract, the lungs, the blood vessels and the body cavity.

105.   (E)
A prototroph is a bacterium which can grow on the most minimal medium and which synthesizes all the essential organic compounds it needs.

106.   (C)
The enzyme hyaluronidase is released by sperm when it reaches within the vicinity of the egg, in the upper part of the oviduct. This enzyme breaks down the corona radiata, which is the outer protective layer of the ovum thus allowing the sperm to contact the egg. A great number of sperm is required to supply enough hyaluronidase to break down the corona radiata.

107.   (C)
Taxis is a term which refers to the movement of a free-living organism toward the source of an external stimulation. The orientation of a grayling butterfly towards the sun thus causing the pursuing predator to be partly blinded is a typical example of taxis. Tropism is used usually to refer to plant movement toward a stimulus.

108.   (A)
Gibberellins play a very important role in development of the apical meristem. They are formed in young leaves around the growing tip and have been found to be very powerfully involved in stem elongation.

109. (C)
The nucleus is the cell organelle that integrates radioactive deoxyribonucleosides. This ability makes it easily distinguishable and serves as a way of identifying the nucleus from within the cell.

110. (D)
The material responsible for conferring resistance to antibiotics is the R plasmid.

111. (B)
The situation where because of an erected barrier a specific species is split, thus causing the formation of two new sets of species, has distinct similarity to the evolutionary process as influenced by geographic isolation.

112. (B)
The two new sets of species, B and C, most likely appeared because sexual reproduction was impossible due to the lack of interbreeding within the species.

113. (C)
Based on the evolution of the two new species, B and C, it can be expected that species B and C can be incapable of interbreeding. If species B and C were capable of interbreeding there would have been no new species thus signifying that there was no barrier.

114. (A)
From examining the data in the graph, and the graph it is clear that the count of yellow chickens was greatest during the years of longest flooding.

115. (A)
The maximum number of yellow chickens appeared after the maximum days of flooding. From the graph it is seen

that in the year 1967 the maximum number of flooding was reached and it was after that year that the maximum number of yellow chickens appeared.

116. (C)
The low CO_2 content in the atmosphere during summer the entering of CO_2 into the photosynthetic cycle from the atomsphere.

During photosynthesis plants utilize CO_2 and give off O_2.

117. (D)
Observable patterns during winter months show that the CO_2 present in the atmosphere rises. This increase in CO_2 is due in part to the fact that many of the plants are seasonal. Thus in the winter months many have completed their life cycles and die. Decomposers break down the death bodies of plant and CO_2 is released to the atmosphere through the respiration of decomposers.

118. (B)
According to the diagram the seasonally highs and lows reflect variations in carbon fixed during photosynthesis. During the summer months when the plants are photosynthesizing the CO_2 available in the atmosphere is much less than in the winter months when the plants have stopped photosynthesizing.

119. (D)
Inversion is a type of chromosomal aberration in which a segment of a chromosome is turned around 180° and is reinserted into the chromosome. It is produced by two breaks in a chromosome. The presence of an inversion does not imply that any new genes are present.

120. (E)

Isolation most likely contributed the incidence of polydactyl (more than five fingers or toes). This caused the inhabitants off the shore of China to develop characteristics unlike those of the inhabitants. The sea which separates the island from the mainland is the reproductive barrier which prevents interbreeding.

121. (A)

The Hardy-Weinberg principle is one of the fundamental concepts in population genetics. It states that a population remains genetically stable in succeeding generations.

122. (E)

A eutrophic aquatic system is one where the biomass is steadily increasing. Succession occurs here and the system is rich in nutrients and has a high productivity.

Lakes which are choked by much nutrients and overcrowding go to the eutrophic condition, and age rapidly. This aging process is increased by the dumping of the nutrients phosphorus and nitrogen into the system.

123-127
  (D)
  (B)
  (E)
  (C)
  (A)

Each cell cycle is divided into four major parts which are designated M, G_1, S and G_2. The M stage represents mitosis and cell division which includes four stages called prophase, metaphase, anaphase and telophase.

Mitosis is followed by G_1 where the cell busily synthesizes proteins, builds structures, and carries on all types of metabolic activities. The S stage is the time when DNA replication and the formation of the new chromosomes take place. The G_2 stage is the stage when the cell continues to prepare for division.

In prophase of mitosis the chromosomes condense, the nuclear membrane and the nucleoli disappear and spindle fibers form. In metaphase, the chromosomes align on a plane in the middle of the spindle fibers. In anaphase the centromeres divide and separate and the two daughter chromosomes of each pair travel to opposite poles of the spindle. In telophase, new nuclear membranes form around each group of daughter chromosomes, the nucleoli appear, the chromosomes decondense and the constricting ring closes down on the reamins of the spindle fiber to form two daughter cells.

128. (B)
The receptor that detects the relative position of bones is the joint kinesthetic receptor. They are located in capsules around joints.

129. (E)
The Golgi tendon organ detects the degree of stretch and contraction of a muscle. It is located in tendons. Firing of the Golgi tendon body leads to the presynaptical inhibition of acetylcholine release and muscle relaxation is resulted.

130. (A)
The Pia Mater is the meninx closer to the surface of the brain.

131. (C)
Baroreceptors are located at various points in the circulatory system. They convey information about changes in pressure.

132. (D)
Meissner's corpuscles are involved in detecting light or fine touch. They respond to mechanical stimulation.

133. (D)
In the temperate decidious forest rainfall is abundant and the summers are relatively long and warm. Here is found rich vegetation which is covered by a carpet of herbs.

134. (B)
Deserts are places where the annual rainfall is often less than 25 cm. Most of them are nearly barren and there is often many small rapid-growing annual herbs with seeds that will germinate when there is a heavy rainfall.

135. (C)
In the tundra biome which is the most continuous of the earth's biomes, the subsoil is permanently frozen. This frozen soil impedes soil drainage, and the decomposition and activities of soil animals.

136. (A)
In the tropical rain forest where there is abundant rainfall some of the most complex communities are found.

In this biome creepers and epiphytes are dominant and many are found growing on the tall trees of the rain forest.

Questions 137 - 141

- (A) Parasitism
- (B) Commensalism
- (C) Mutualism
- (D) Mimicry
- (E) Allelopathy

137. (E)
Allelopathy is the secretion or excretion of substances which have inhibitory effects on other organisms. One example of this is een in bacteria which fail to reproduce successfully when their colony is seeded with fungi.

432

138. (D)
Species not naturally protected by some unpleasant characteristic of their own or which want to adapt some features that would be advantageous to them often resort to Mimicry.

Mimicry is the adaptation of the appearance, behavior or smell by a species of another species. A typical example is the appearance of the Oncidium orchid like a male bee to encourage visitation by a male bee.

139. (A)
A parasite passes most of its life on or in the body of a living host, from which it derives food while harming the host.

140. (C)
Mutualism is symbiotic relationships that are beneficial to both species. The relationship between a termite or a cow and the cellulose-digesting microorganisms in its digestive tract is a mutualistic relationship.

141. (B)
The situation where a species derive shelter, support, transport or food from its association with its host is called commensalism. Commensalism is a kind of symbiosis in which one species benefits while the other is unaffected nor benefited.

Questions 142 - 146

    (A)   Vitamin B_1 (thiamine)
    (B)   Vitamin B_6 (pyridoxal)
    (C)   Vitamin A (retinol)
    (D)   Vitamin D (calciferol)
    (E)   Folic acid

142. (D)
Vitamin D is involved in calcium absorption and metabolism.

Its deficiency causes skeletal deformity because the bones which lack calcium are soft.

143. (C)
Vitamin A is a fat soluble compound whose deficiency causes excessive keratinization and degeneration of columnar and cuboidal epithelia into stratified squamous epithelia. The most serious result of deficiency of Vitamin A is xerophthalmia which is a keratinization of tissues of the eye that can lead to permanent blindness.

144. (B)
Vitamin B_6 (pyridoxal) is a co-enzyme which is involved in reactions that transfer amino acids from one compound to another.

145. (A)
Thiamine (Vitamin B_1) is a part of the coenzyme that catalyzes pyruvic acid oxidation.

146. (E)
Folic acid which is another B vitamin is important in the synthesis of some of the nucleotides which are needed for the building of nucleic acids. It is therefore important for cell division.

Questions 147 - 151

    (A)   Colchicine
    (B)   Actinomyosin D
    (C)   Cytochalasin B
    (D)   Cycloheximide
    (E)   Puromycin

147. (C)
Cytochalasin B is a reagent which alters the properties

of actin. It is therefore used quite regularly to reveal cellular processes in which actin or actin-like filaments are involved. This reagent can affect cells in this way because microfilaments which contain actin are found in the region of movement that brings about the changes in the cell.

148. (A)
Colchicine is a substance that causes cessation of microtubular activities and prevention of the assembly of new microtubes thus causing the cell to lose its shape.

Colchicine produces functional disruptions in the microtubules. These disruptions prevent the expression of the specific roles of the microtubules.

149. (B)
Actinomysin D is known as an inhibitor of ribonucleic acid synthesis because it binds to DNA and blocks the movement of RNA polymerase.

150. (E)
It has been discovered that neuronal activities increase the amount of RNA in the nerve cells; this increase in RNA may reflect an increase in protein synthesis.

The antibiotic puromycin when injected into mice and fish was found to block long term memory but not short term memory. Puromycin affects protein synthesis by causing the premature release of nascent polypeptide chain.

151. (D)
Cycloheximide in a drug which affects protein synthesis by blocking the peptidyl transferase reaction on ribosomes in eukaryotic cells.

Questions 152 - 156

(A) Endoplasmic reticulum
(B) Golgi-apparatus or dictyosome

(C)   Cell wall
(D)   Cell membrane
(E)   Nuclear membrane

152.   (D)
The cell membrane is the cell structure which is responsible for maintaining stability in the cytoplasm of the cell by selectively allowing certain substances to pass through.   Thereby, it maintains an unequal concentration of ions on either side and let nutrients to get in while waste products to get out of the cell.

153.   (A)
The endoplasmic reticulum is the cell organelle which is continuous with the cell membrane and the outer membrane of the nucleus.

154.   (B)
The organelles which become obvious when treated with silver salts were the Golgi bodies.   They were discovered in a variety of secretory cells.

155.   (E)
The nucleus is enclosed in a double membrane which is made up of lipids, associated proteins and carbohydrates. This membrane is known as the nuclear membrane.

156.   (C)
Whereas cell membranes are considered living structures, there is another structure in which bacteria and the cells of plants are encased.   This is a dead structure and it is known as the cell wall.   Cell walls are stiff semirigid, extracellular inclusions which give cells a definite size and shape.

Questions 157 - 161

    (A)   Apical meristem
    (B)   Cortex
    (C)   Phloem
    (D)   Vascular cambium
    (E)   Parenchyma cells

157. (E)
Parenchyma cells are cells that surround the large leaf veins and which contain few or no chloroplasts.

158. (D)
The vascular cambium is a thin layer of undifferentiated cells in vascular tissue.

159. (B)
The cortex is made up of parenchyma and collenchyma and is found just outside the phloem.

160. (C)
The phloem tissue composes of sieve-tube members and sieve tubes and also companion cells, phloem fibers and phloem parenchyma.

161. (A)
The root tips contain the apical meristem tissue which is located just above the root cap. The apical meristem is an area of small, rapidly dividing cells.

Questions 162 - 166

    (A)   Echinodermata
    (B)   Annelida
    (C)   Mollusca

        (D)   Chordata
        (E)   Coelenterata

162.   (E)
    Phylum Coelenterata includes the hydra, jellyfish, corals
and anemones.   They possess a sac-like gut called the
gastrovascular cavity because it serves both digestive and
circulatory functions.   A simple opening serves as both
the mouth and anus.

163.   (B)
    Phylum Annelida includes earthworm and leeches.   They
are distinguished by their segmented bodies.

164.   (C)
    Phylum   Mollusca   is   the   second   largest   in   the   animal
kingdom.   Included in this phylum are snails, slugs, clams
and oysters.   The body of the mollusca consists of three
principal parts.   There are a large ventral muscular foot
which contains sensory and motor systems, a visceral mass
above the foot, a heavy fold of tissue called a mantle.

165.   (A)
    Phylum Echinodermata is characterized by the fact that
the   echinoderms   in   this   phylum   are   exclusively   marine.
Almost all the members possess an internal skeleton which
has a spiny crusted covering.

166.   (D)
    The   phylum   Chordata   is   usually   divided   into   three
subphyla - Brochordata, Cephalochordata and Vertebrata.
They all share three important characteristics.   These are
they all have a structure called a notochord which serves
as a skeletal support, pharyngeal gill slits and a dorsal
hollow nerve cord.

    (A)   Bryophyta
    (B)   Spermophyta
    (C)   Angiosperms
    (D)   Pterophyta
    (E)   Chlorophyta

167. (C)
The angiosperms are flowering plants which can be divided taxonomically into two major groups. These are monocots and dicots.

168. (B)
The subdivision spermophyta are advanced vascular plants which produce seeds and pollen.

169. (D)
Subdivision pterophyta contain the ferns which all have well developed vascular tissue. The ferns alternate generations and are spore formers.

170. (E)
Chlorophyta or the green algae is a group which includes both unicellular and multicellular species. The algae contains both chlorophylls a and b. In this group there is little differentiation.

171. (A)
The division Bryophyta which contains mosses, liverworts and hornworts are multicellular plants with well-differentiated tissues and considerable cell specialization. They contain chlorophyll a and b.

Questions 172 - 176

    (A)   Edward Jenner

(B)   Louis Pasteur
(C)   Sir Alexander Fleming
(D)   J. Watson
(E)   John Tyndall

172. (A)
Edward Jenner was the first to successfully vaccinate against smallpox. Using virus isolated from cowpox lesions. Jenner innoculated smallpox victims.

173. (B)
The theory of spontaneous generation was finally disproved by Louis Pasteur. He used a flask with a long, narrow, gooseneck opening. When he heated a nutrient solution it was found that the germs and dust settled in the gooseneck. No microbes appeared in the solution. This confirmed that living things did not suddenly evolve but came from other living things.

174. (C)
Sir Alexander Fleming discovered and initiated the isolation of penicillin from the mold penicillin sp. This discovery opened the era of antibiotics.

175. (E)
John Tyndall was the scientist who proved that dust carries germs. He conducted experiments in a specially designed box and discovered that if no dust is present, sterile broth remains free of microbial for indefinite periods of time.

176. (B)
Louis Pasteur isolated the germ responsible for chicken cholera by growing the bacteria in pure culture.

(A) Density-independent
(B) Abiotic control
(C) Age profile
(D) Carrying capacity
(E) Programmed death

177. (B)
Physical factors which depress population numbers are known as abiotic controls. Examples of abiotic controls are drought, cold weather and extreme heat.

178. (C)
Age profiles show the ages of people in a particular group, country or people. They break the population down according to sex and age groups. India, for example, is a typical country whose birth rate is high and whose death rate is declining.

179. (E)
Programmed death is the time it takes for a living thing to be born, mature, reproduce and die. Many plants and animals are annuals. Their life cycles are dependent on the seasons. Brazilian fish live in temporary ponds that exist only during rainy seasons. Shortly before the pond dries up the fish spawn and lay cyst-like drought resistant eggs and then die.

180. (A)
Density-independent influences such as severe weather, environmental conditions and diseases, are population depressing. Mortality brought about by these means is independent of the density of the population.

181. (D)

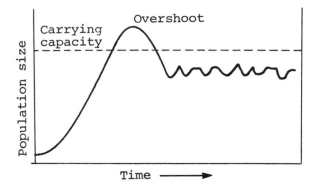

The full biotic potential of any species is unlikely to be reached due to the limitations imposed by the environment. This phenomena which is known as environmental resistance is encountered when an essential property in the environment comes into short supply and adversely affects the population's rate of increase. This limit of the biotic potential set by environmental resistance is known as the carrying capacity of the environment.

Questions 182 – 186

    (A)   erythrocytes
    (B)   neutrophils
    (C)   eosinophils
    (D)   lymphocytes
    (E)   basophils

182. (B)
Neutrophils are phagocytic cells which are concerned primarily with local infections. They indicate localized infections such as appendicitis or abcesses in other parts of the body.

183. (C)
Eosinophils are blood cells which generate protein substances which inactivate foreign protein. High eosinophil counts may indicate allegeric conditions or invasion of parasitic roundworms such as trichinella spiralis.

442

184. (E)
Basophils are leukocytes which probably produce heparin, histamine and serotonin.

185. (A)
Erythrocytes are cells that transport O_2 and CO_2. This ability is related to their ability to carry hemoglobin.

186. (D)
Lymphocytes are phagocytic cells which are concerned primarily with generalized infections. High lymphocyte counts are present in whooping cough and some viral infections.

187. (C)
The curling back of the lips of a dog to expose his teeth is a sign of communicaton.

The ultimate function of communication is to increase reproductive success in those animals able to send and receive signals.

188. (E)
There are two types of mechanisms for storing information. This is evidenced in the differences between long-term memory and short-term memory. The short-term memory can be overloaded while the long-term memory cannot. The consolidation hypothesis states that memory is stored in two ways. The transformation from short-term to long-term memory is termed consolidation.

189. (B)
An oligotrophic system is one with meager nutrients and low productivity. It is characterized by sparse littoral vegetation and is the opposite to an entrophic system.

Explanations for questions 190 and 191 concern the diagram.

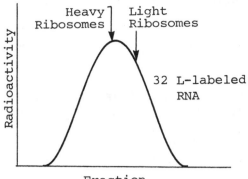

190. (B)
According to the experiments ribosomes were not synthesized after infection. This was evidenced by the absence of "light" ribosomes.

191. (D)
The radioisotope ^{35}R appeared transiently in the heavy ribosome peak; this showed that new proteins were synthesized in pre-existing ribosomes.

Questions 192 - 195 concerns the data below of F_1 phenotypes.

```
+   +   v  -  37 ⎫ parental chromosomes
ct  y   +  -  35 ⎭
ct  +   +  -  10 ⎫ single recombinant
+   y   v  -   7 ⎭ Class I

ct  y   v  =   4 ⎫ Single recombinant
+   +   v  =   4 ⎭ Class II
+   y   +  =   2 ⎫ Double recombinant
ct  +   v  =   1 ⎭ or double crossover
```

192. (C)
    Allelic configuration: $\left.\begin{array}{ccc} + & + & + \\ ct & y & v \end{array}\right\}$ cis position

Alleles in the cis-position are homozygous

$$\left.\begin{array}{ccc} + & + & v \\ ct & y & + \end{array}\right\} \text{trans-position}$$

Alleles in the trans-position are heterozygous.

193. (C)
The difference between $\frac{++v}{+y+}$ and $\frac{+++}{+yv}$ is the change in position of y in single recombinant class I. Therefore, allele y is on one end of the chromosome. Difference between $\frac{++v}{ct\,y\,+}$ and $\frac{ctyv}{+++}$ is the change in position of v. Therefore, allele v is on the other end of the chromosome. Hence, the arrangement of these three alleles should be y ct v. The frequency of recombination for the y ct interval is (10 + 7 + 2 + 1) ÷ 100 = 0.2 = 20 map units. That for the ct v interval is (4 + 4 + 2 + 1) ÷ 100 = 0.11 = 11 map units. The linkage map should be

$$\begin{array}{ccc} y & ct & v \\ \vdash & + & \dashv \\ \leftarrow 20 \rightarrow & \leftarrow 11 \rightarrow \end{array}$$

194. (A)
The coefficient of coincidence

$$= \frac{\text{\% actual double crossover}}{\text{\% expected double crossover}}$$

$$= \frac{0.03}{(0.17 + 0.03)(0.08 + 0.03)} = 1.36$$

195. (A)
Interference = 1 - coefficient of coincidence

$$= 1 - 1.36 = -0.36$$

196. (C)
According to experiment I where the frog limb was exposed to radiation of 6000R of x-rays and then amputated,

regeneration did not occur. This result for experiment I shows that radiated frog tissue does not regenerate.

197. (C)
Experiments I, II and III were done because they were controls for experiment IV.

In experiment I the limb was exposed to radiation only and the limb amputated.   In experiment II the limb was denervated and amputated before nerve regeneration and was not exposed to radiation.   In experiment III the limb was exposed to radiation and denervated and the limb was amputated before nerve regeneration occurred.

In the fourth experiment the nerves were allowed to grow back from the brachial plexus.

198. (D)
In experiment IV the frog limb was not exposed up to the brachial plexus to radiation.   The nerve cell bodies which were not radiated grows back into the limb.   After the limb is amputated, they still continue to grow.

199. (B)
The resulting pigmented regenerate implies that the myelinated nerve cells were instrumental in the regeneration. The myelinated nerve cells which were pigmented were introduced into the albino host. If the resulting regenerate was pigmented it is obvious that the pigmented myelinated nerve cells had a great influence on the results.

Questions 200 - 201

The following genotypic and phenotypic configurations show the independent assorting autosomal genes which determine coat color.

       A - B - = (grey)
       A - bb   = (yellow)
       aa B-   = (black)
       aa bb   = (cream)

446

CC and Cc allow color expression according to the characteristic of the A and B alleles. The cc genotype results in albino raccoons regardless of the presence of the A and B alleles.

200. (D)

Mating between AABBCc and AaBbCc

$$AA \times Aa \qquad BB \times Bb \qquad Cc \times Cc$$

$$= \tfrac{1}{2}AA + \tfrac{1}{2}Aa \quad = \tfrac{1}{2}BB + \tfrac{1}{2}Bb \quad = \tfrac{1}{4}CC + \tfrac{2}{4}Cc + \tfrac{1}{4}cc$$

When the three matings were performed 12/16 were dominated by the grey gene with either the CC or Cc as complementary gene and 4/16 were dominated by the cc gene which produces albino raccoons.

201. (C)

F_1 offsprings are all grey.

$$AABBCc \times AABBCc$$

Step (1) $AA \times AA = 1\ AA$

$BB \times BB = 1\ BB$

$Cc \times Cc = \tfrac{1}{4}CC + \tfrac{2}{4}Cc + \tfrac{1}{4}cc$

Step (2) $AA \times BB = 1\ AB$

Step (3) $1\ AABB \times \left(\tfrac{1}{4}CC + \tfrac{2}{4}Cc + \tfrac{1}{4}cc\right)$

$$= \tfrac{1}{4}AABBCC + \tfrac{2}{4}AABBCc + \tfrac{1}{4}AABBcc$$

202. (D)

The result expected from the experiment would be that some of the rats would receive an original of the part of the gonad that was the same sex as their own and some would receive one that was of a different sex. This would be because in the random replacement of the gonadal tissue, some of the gonadal cortical tissue and gonadal medullary tissue would be aligned in their correct positions.

203. (C)
This experiment was done to determine if cortical dominance produces ovarian tissue while medullary dominance results in the development of testes.

There are two types of tissue which arise in the rudimentary region of the gonads; the gonadal cortical tissue and the gonadal meduallry tissue. These two tissues are hypothesized to be antagonistic to each other. Cortical dominance produces ovarian tissue while medullary dominance produces testes.

204. (A)
Cytokinins are plant cell hormones which stimulate cell division in plants. They however cannot alone stimulate cell division but have to be coupled with other plant hormones. They are an important plant growth requirement.

205. (C)
The experiments indicated that the parenchyma cells contain all the genetic information necessary to develop into a number of kinds of other plant cells. The callus which consists of tobacco pith (parenchyma) and a combination of auxin and cytokinin produced either a stem shoot or roots according to the ratio of this combination.

206. (B)
From the results of the experiment low auxin to cytokinin ratios caused the pith to be changed into apical meristem.

207. (E)
Each value obtained from medium I-IV respectively represents the total number of bacteria in the irradiated culture, the total number of bacteria with $leu^{(-)}$ to $leu^{(+)}$ mutations present in the irradiated culture, total number of bacteria present in the control culture and the total number of bacteria with $leu^{(-)}$ to $leu^{(+)}$ mutations present in the control culture.

208. (D)

Values obtained from medium I and III should be approximately equal because both the medium and the dilution are the same; the only difference is the culture where some were irradiated and some were not.

209. (B)

The induced rate is the fraction value of the colonies formed on the minimal medium irradiated culture only divided by the colonies formed on the minimal medium plus leucine irradiated culture. The positive reciprocal of the dilution factor is used. The induced rate is therefore

$$\frac{12 \times 10^2}{24 \times 10^9} = 5 \times 10^{-8}$$

210. (B)

The spontaneous mutation rate which led to prototrophic growth ($leu^{(-)}$ to $leu^{(+)}$) is represented as

$$\frac{3 \times 10^1}{12 \times 10^9} = 2.5 \times 10^{-9}$$

The spontaneous mutation rate is the value obtained from the minimal medium of the control culture divided by the value obtained from the minimal medium plus leucine control culture.

GRE

BIOLOGY TEST

MODEL TEST V

THE GRADUATE RECORD EXAMINATION

BIOLOGY TEST

ANSWER SHEET

1. Ⓐ Ⓑ Ⓒ Ⓓ Ⓔ	25. Ⓐ Ⓑ Ⓒ Ⓓ Ⓔ	49. Ⓐ Ⓑ Ⓒ Ⓓ Ⓔ
2. Ⓐ Ⓑ Ⓒ Ⓓ Ⓔ	26. Ⓐ Ⓑ Ⓒ Ⓓ Ⓔ	50. Ⓐ Ⓑ Ⓒ Ⓓ Ⓔ
3. Ⓐ Ⓑ Ⓒ Ⓓ Ⓔ	27. Ⓐ Ⓑ Ⓒ Ⓓ Ⓔ	51. Ⓐ Ⓑ Ⓒ Ⓓ Ⓔ
4. Ⓐ Ⓑ Ⓒ Ⓓ Ⓔ	28. Ⓐ Ⓑ Ⓒ Ⓓ Ⓔ	52. Ⓐ Ⓑ Ⓒ Ⓓ Ⓔ
5. Ⓐ Ⓑ Ⓒ Ⓓ Ⓔ	29. Ⓐ Ⓑ Ⓒ Ⓓ Ⓔ	53. Ⓐ Ⓑ Ⓒ Ⓓ Ⓔ
6. Ⓐ Ⓑ Ⓒ Ⓓ Ⓔ	30. Ⓐ Ⓑ Ⓒ Ⓓ Ⓔ	54. Ⓐ Ⓑ Ⓒ Ⓓ Ⓔ
7. Ⓐ Ⓑ Ⓒ Ⓓ Ⓔ	31. Ⓐ Ⓑ Ⓒ Ⓓ Ⓔ	55. Ⓐ Ⓑ Ⓒ Ⓓ Ⓔ
8. Ⓐ Ⓑ Ⓒ Ⓓ Ⓔ	32. Ⓐ Ⓑ Ⓒ Ⓓ Ⓔ	56. Ⓐ Ⓑ Ⓒ Ⓓ Ⓔ
9. Ⓐ Ⓑ Ⓒ Ⓓ Ⓔ	33. Ⓐ Ⓑ Ⓒ Ⓓ Ⓔ	57. Ⓐ Ⓑ Ⓒ Ⓓ Ⓔ
10. Ⓐ Ⓑ Ⓒ Ⓓ Ⓔ	34. Ⓐ Ⓑ Ⓒ Ⓓ Ⓔ	58. Ⓐ Ⓑ Ⓒ Ⓓ Ⓔ
11. Ⓐ Ⓑ Ⓒ Ⓓ Ⓔ	35. Ⓐ Ⓑ Ⓒ Ⓓ Ⓔ	59. Ⓐ Ⓑ Ⓒ Ⓓ Ⓔ
12. Ⓐ Ⓑ Ⓒ Ⓓ Ⓔ	36. Ⓐ Ⓑ Ⓒ Ⓓ Ⓔ	60. Ⓐ Ⓑ Ⓒ Ⓓ Ⓔ
13. Ⓐ Ⓑ Ⓒ Ⓓ Ⓔ	37. Ⓐ Ⓑ Ⓒ Ⓓ Ⓔ	61. Ⓐ Ⓑ Ⓒ Ⓓ Ⓔ
14. Ⓐ Ⓑ Ⓒ Ⓓ Ⓔ	38. Ⓐ Ⓑ Ⓒ Ⓓ Ⓔ	62. Ⓐ Ⓑ Ⓒ Ⓓ Ⓔ
15. Ⓐ Ⓑ Ⓒ Ⓓ Ⓔ	39. Ⓐ Ⓑ Ⓒ Ⓓ Ⓔ	63. Ⓐ Ⓑ Ⓒ Ⓓ Ⓔ
16. Ⓐ Ⓑ Ⓒ Ⓓ Ⓔ	40. Ⓐ Ⓑ Ⓒ Ⓓ Ⓔ	64. Ⓐ Ⓑ Ⓒ Ⓓ Ⓔ
17. Ⓐ Ⓑ Ⓒ Ⓓ Ⓔ	41. Ⓐ Ⓑ Ⓒ Ⓓ Ⓔ	65. Ⓐ Ⓑ Ⓒ Ⓓ Ⓔ
18. Ⓐ Ⓑ Ⓒ Ⓓ Ⓔ	42. Ⓐ Ⓑ Ⓒ Ⓓ Ⓔ	66. Ⓐ Ⓑ Ⓒ Ⓓ Ⓔ
19. Ⓐ Ⓑ Ⓒ Ⓓ Ⓔ	43. Ⓐ Ⓑ Ⓒ Ⓓ Ⓔ	67. Ⓐ Ⓑ Ⓒ Ⓓ Ⓔ
20. Ⓐ Ⓑ Ⓒ Ⓓ Ⓔ	44. Ⓐ Ⓑ Ⓒ Ⓓ Ⓔ	68. Ⓐ Ⓑ Ⓒ Ⓓ Ⓔ
21. Ⓐ Ⓑ Ⓒ Ⓓ Ⓔ	45. Ⓐ Ⓑ Ⓒ Ⓓ Ⓔ	69. Ⓐ Ⓑ Ⓒ Ⓓ Ⓔ
22. Ⓐ Ⓑ Ⓒ Ⓓ Ⓔ	46. Ⓐ Ⓑ Ⓒ Ⓓ Ⓔ	70. Ⓐ Ⓑ Ⓒ Ⓓ Ⓔ
23. Ⓐ Ⓑ Ⓒ Ⓓ Ⓔ	47. Ⓐ Ⓑ Ⓒ Ⓓ Ⓔ	71. Ⓐ Ⓑ Ⓒ Ⓓ Ⓔ
24. Ⓐ Ⓑ Ⓒ Ⓓ Ⓔ	48. Ⓐ Ⓑ Ⓒ Ⓓ Ⓔ	72. Ⓐ Ⓑ Ⓒ Ⓓ Ⓔ

73. Ⓐ Ⓑ Ⓒ Ⓓ Ⓔ
74. Ⓐ Ⓑ Ⓒ Ⓓ Ⓔ
75. Ⓐ Ⓑ Ⓒ Ⓓ Ⓔ
76. Ⓐ Ⓑ Ⓒ Ⓓ Ⓔ
77. Ⓐ Ⓑ Ⓒ Ⓓ Ⓔ
78. Ⓐ Ⓑ Ⓒ Ⓓ Ⓔ
79. Ⓐ Ⓑ Ⓒ Ⓓ Ⓔ
80. Ⓐ Ⓑ Ⓒ Ⓓ Ⓔ
81. Ⓐ Ⓑ Ⓒ Ⓓ Ⓔ
82. Ⓐ Ⓑ Ⓒ Ⓓ Ⓔ
83. Ⓐ Ⓑ Ⓒ Ⓓ Ⓔ
84. Ⓐ Ⓑ Ⓒ Ⓓ Ⓔ
85. Ⓐ Ⓑ Ⓒ Ⓓ Ⓔ
86. Ⓐ Ⓑ Ⓒ Ⓓ Ⓔ
87. Ⓐ Ⓑ Ⓒ Ⓓ Ⓔ
88. Ⓐ Ⓑ Ⓒ Ⓓ Ⓔ
89. Ⓐ Ⓑ Ⓒ Ⓓ Ⓔ
90. Ⓐ Ⓑ Ⓒ Ⓓ Ⓔ
91. Ⓐ Ⓑ Ⓒ Ⓓ Ⓔ
92. Ⓐ Ⓑ Ⓒ Ⓓ Ⓔ
93. Ⓐ Ⓑ Ⓒ Ⓓ Ⓔ
94. Ⓐ Ⓑ Ⓒ Ⓓ Ⓔ
95. Ⓐ Ⓑ Ⓒ Ⓓ Ⓔ
96. Ⓐ Ⓑ Ⓒ Ⓓ Ⓔ

97. Ⓐ Ⓑ Ⓒ Ⓓ Ⓔ
98. Ⓐ Ⓑ Ⓒ Ⓓ Ⓔ
99. Ⓐ Ⓑ Ⓒ Ⓓ Ⓔ
100. Ⓐ Ⓑ Ⓒ Ⓓ Ⓔ
101. Ⓐ Ⓑ Ⓒ Ⓓ Ⓔ
102. Ⓐ Ⓑ Ⓒ Ⓓ Ⓔ
103. Ⓐ Ⓑ Ⓒ Ⓓ Ⓔ
104. Ⓐ Ⓑ Ⓒ Ⓓ Ⓔ
105. Ⓐ Ⓑ Ⓒ Ⓓ Ⓔ
106. Ⓐ Ⓑ Ⓒ Ⓓ Ⓔ
107. Ⓐ Ⓑ Ⓒ Ⓓ Ⓔ
108. Ⓐ Ⓑ Ⓒ Ⓓ Ⓔ
109. Ⓐ Ⓑ Ⓒ Ⓓ Ⓔ
110. Ⓐ Ⓑ Ⓒ Ⓓ Ⓔ
111. Ⓐ Ⓑ Ⓒ Ⓓ Ⓔ
112. Ⓐ Ⓑ Ⓒ Ⓓ Ⓔ
113. Ⓐ Ⓑ Ⓒ Ⓓ Ⓔ
114. Ⓐ Ⓑ Ⓒ Ⓓ Ⓔ
115. Ⓐ Ⓑ Ⓒ Ⓓ Ⓔ
116. Ⓐ Ⓑ Ⓒ Ⓓ Ⓔ
117. Ⓐ Ⓑ Ⓒ Ⓓ Ⓔ
118. Ⓐ Ⓑ Ⓒ Ⓓ Ⓔ
119. Ⓐ Ⓑ Ⓒ Ⓓ Ⓔ
120. Ⓐ Ⓑ Ⓒ Ⓓ Ⓔ

121. Ⓐ Ⓑ Ⓒ Ⓓ Ⓔ
122. Ⓐ Ⓑ Ⓒ Ⓓ Ⓔ
123. Ⓐ Ⓑ Ⓒ Ⓓ Ⓔ
124. Ⓐ Ⓑ Ⓒ Ⓓ Ⓔ
125. Ⓐ Ⓑ Ⓒ Ⓓ Ⓔ
126. Ⓐ Ⓑ Ⓒ Ⓓ Ⓔ
127. Ⓐ Ⓑ Ⓒ Ⓓ Ⓔ
128. Ⓐ Ⓑ Ⓒ Ⓓ Ⓔ
129. Ⓐ Ⓑ Ⓒ Ⓓ Ⓔ
130. Ⓐ Ⓑ Ⓒ Ⓓ Ⓔ
131. Ⓐ Ⓑ Ⓒ Ⓓ Ⓔ
132. Ⓐ Ⓑ Ⓒ Ⓓ Ⓔ
133. Ⓐ Ⓑ Ⓒ Ⓓ Ⓔ
134. Ⓐ Ⓑ Ⓒ Ⓓ Ⓔ
135. Ⓐ Ⓑ Ⓒ Ⓓ Ⓔ
136. Ⓐ Ⓑ Ⓒ Ⓓ Ⓔ
137. Ⓐ Ⓑ Ⓒ Ⓓ Ⓔ
138. Ⓐ Ⓑ Ⓒ Ⓓ Ⓔ
139. Ⓐ Ⓑ Ⓒ Ⓓ Ⓔ
140. Ⓐ Ⓑ Ⓒ Ⓓ Ⓔ
141. Ⓐ Ⓑ Ⓒ Ⓓ Ⓔ
142. Ⓐ Ⓑ Ⓒ Ⓓ Ⓔ
143. Ⓐ Ⓑ Ⓒ Ⓓ Ⓔ
144. Ⓐ Ⓑ Ⓒ Ⓓ Ⓔ

145. Ⓐ Ⓑ Ⓒ Ⓓ Ⓔ	167. Ⓐ Ⓑ Ⓒ Ⓓ Ⓔ	189. Ⓐ Ⓑ Ⓒ Ⓓ Ⓔ
146. Ⓐ Ⓑ Ⓒ Ⓓ Ⓔ	168. Ⓐ Ⓑ Ⓒ Ⓓ Ⓔ	190. Ⓐ Ⓑ Ⓒ Ⓓ Ⓔ
147. Ⓐ Ⓑ Ⓒ Ⓓ Ⓔ	169. Ⓐ Ⓑ Ⓒ Ⓓ Ⓔ	191. Ⓐ Ⓑ Ⓒ Ⓓ Ⓔ
148. Ⓐ Ⓑ Ⓒ Ⓓ Ⓔ	170. Ⓐ Ⓑ Ⓒ Ⓓ Ⓔ	192. Ⓐ Ⓑ Ⓒ Ⓓ Ⓔ
149. Ⓐ Ⓑ Ⓒ Ⓓ Ⓔ	171. Ⓐ Ⓑ Ⓒ Ⓓ Ⓔ	193. Ⓐ Ⓑ Ⓒ Ⓓ Ⓔ
150. Ⓐ Ⓑ Ⓒ Ⓓ Ⓔ	172. Ⓐ Ⓑ Ⓒ Ⓓ Ⓔ	194. Ⓐ Ⓑ Ⓒ Ⓓ Ⓔ
151. Ⓐ Ⓑ Ⓒ Ⓓ Ⓔ	173. Ⓐ Ⓑ Ⓒ Ⓓ Ⓔ	195. Ⓐ Ⓑ Ⓒ Ⓓ Ⓔ
152. Ⓐ Ⓑ Ⓒ Ⓓ Ⓔ	174. Ⓐ Ⓑ Ⓒ Ⓓ Ⓔ	196. Ⓐ Ⓑ Ⓒ Ⓓ Ⓔ
153. Ⓐ Ⓑ Ⓒ Ⓓ Ⓔ	175. Ⓐ Ⓑ Ⓒ Ⓓ Ⓔ	197. Ⓐ Ⓑ Ⓒ Ⓓ Ⓔ
154. Ⓐ Ⓑ Ⓒ Ⓓ Ⓔ	176. Ⓐ Ⓑ Ⓒ Ⓓ Ⓔ	198. Ⓐ Ⓑ Ⓒ Ⓓ Ⓔ
155. Ⓐ Ⓑ Ⓒ Ⓓ Ⓔ	177. Ⓐ Ⓑ Ⓒ Ⓓ Ⓔ	199. Ⓐ Ⓑ Ⓒ Ⓓ Ⓔ
156. Ⓐ Ⓑ Ⓒ Ⓓ Ⓔ	178. Ⓐ Ⓑ Ⓒ Ⓓ Ⓔ	200. Ⓐ Ⓑ Ⓒ Ⓓ Ⓔ
157. Ⓐ Ⓑ Ⓒ Ⓓ Ⓔ	179. Ⓐ Ⓑ Ⓒ Ⓓ Ⓔ	201. Ⓐ Ⓑ Ⓒ Ⓓ Ⓔ
158. Ⓐ Ⓑ Ⓒ Ⓓ Ⓔ	180. Ⓐ Ⓑ Ⓒ Ⓓ Ⓔ	202. Ⓐ Ⓑ Ⓒ Ⓓ Ⓔ
159. Ⓐ Ⓑ Ⓒ Ⓓ Ⓔ	181. Ⓐ Ⓑ Ⓒ Ⓓ Ⓔ	203. Ⓐ Ⓑ Ⓒ Ⓓ Ⓔ
160. Ⓐ Ⓑ Ⓒ Ⓓ Ⓔ	182. Ⓐ Ⓑ Ⓒ Ⓓ Ⓔ	204. Ⓐ Ⓑ Ⓒ Ⓓ Ⓔ
161. Ⓐ Ⓑ Ⓒ Ⓓ Ⓔ	183. Ⓐ Ⓑ Ⓒ Ⓓ Ⓔ	205. Ⓐ Ⓑ Ⓒ Ⓓ Ⓔ
162. Ⓐ Ⓑ Ⓒ Ⓓ Ⓔ	184. Ⓐ Ⓑ Ⓒ Ⓓ Ⓔ	206. Ⓐ Ⓑ Ⓒ Ⓓ Ⓔ
163. Ⓐ Ⓑ Ⓒ Ⓓ Ⓔ	185. Ⓐ Ⓑ Ⓒ Ⓓ Ⓔ	207. Ⓐ Ⓑ Ⓒ Ⓓ Ⓔ
164. Ⓐ Ⓑ Ⓒ Ⓓ Ⓔ	186. Ⓐ Ⓑ Ⓒ Ⓓ Ⓔ	208. Ⓐ Ⓑ Ⓒ Ⓓ Ⓔ
165. Ⓐ Ⓑ Ⓒ Ⓓ Ⓔ	187. Ⓐ Ⓑ Ⓒ Ⓓ Ⓔ	209. Ⓐ Ⓑ Ⓒ Ⓓ Ⓔ
166. Ⓐ Ⓑ Ⓒ Ⓓ Ⓔ	188. Ⓐ Ⓑ Ⓒ Ⓓ Ⓔ	210. Ⓐ Ⓑ Ⓒ Ⓓ Ⓔ

THE GRE BIOLOGY TEST

MODEL TEST V

Time: 170 Minutes
        210 Questions

DIRECTIONS: *Choose the best answer for each question and mark the letter of your selection on the corresponding answer sheet.*

1. Which of the following statements defines meiosis?

    (A) The number of chromosomes in the diploid nucleus is reduced by half.

    (B) The fusion of two haploid nuclei to form a diploid nucleus.

    (C) The process by which four haploid cells or nuclei are transformed into a diploid cell.

    (D) The separation of chromosomes.

    (E) The fusion of chromosomes.

2. The hydrostatic pressure of the blood within the glomerular capillaries is

    (A) less than the mean blood pressure in the large arteries of the body.

    (B) about half the mean arterial pressure.

    (C) usually about 50 mmHg.

    (D) All of the above

457

(E) None of the above

3. The apical meristem is responsible for growth and differentiation in the

(A) stem, root tips and branches.

(B) vascular cambium.

(C) stem, sepal and root tips.

(D) blastula.

(E) branches.

4. In 3 point gene mapping, the occurrence of the first crossover reduces the chance of a second crossover nearby. This phenomenon is called

(A) encroachment.

(B) enhancement.

(C) linkage reduction.

(D) interference.

(E) advancement.

5. In eukaryotes, the DNA duplex is wrapped around protein particles. These structures are referred to as

(A) solinoids.

(B) nucleosomes.

(C) peroxisomes.

(D) ribosomes.

(E) nucleolus.

6. The site of protein construction (synthesis) is in the

(A) chloroplast.

(B) lysosome.

(C) ribosome.

(D) nucleus.

(E) cell wall.

7. Which of the following is/are capable of cutting the DNA duplex to relieve twisting forces (torques), and reuniting the DNA strands during replication?

(A) DNA ligase

(B) Initiator protein

(C) Gyrase (swivelase)

(D) Helix-destabilizing protein

(E) All of the above

8. The genetic disorder that is best expressed as (47,21+) is known as

(A) Turner's Syndrome.

(B) Down's Syndrome.

(C) Trisomy.

(D) Klinefelters' Syndrome.

(E) Patau's Syndrome.

9. Structure(s) which enable(s) eukaryotic cells to be motile is/are

(A) cilia.

(B) flagella.

(C) pseudopods.

(D) All of the above

(E) None of the above

10. Which of the following neuroglia is responsible for forming the myelin sheath around the processes of neurons of the peripheral nervous system?

(A) Schwann cells

(B) Digodendrites.

(C) Microglia

(D) Astrocytes

(E) Oligodendrites

11. During fertilization which of the following enzymes is released by the sperm cells?

(A) calcium carbonase

(B) hyaluronidase

(C) acid phosphatase

(D) chrondroitin sulfatase

(E) chorionase

12. During development, the term used to describe an offspring

that has developed to a point to which its species can be identified is

(A) blastocyst.

(D) morphogenesis.

(B) fetus.

(E) embryo.

(C) zygote.

13. The process whereby cells and tissues become specialized during development is called

(A) neutralization.

(D) actualization.

(B) differentiation.

(E) predestination.

(C) specialization.

14. Which of the following chemical transmitters may be chemically similar to heroin, and may serve as the body's own pain killer?

(A) the endorphins

(D) serotonin (5-HT)

(B) norepinephrine

(E) dopamine

(C) acetyl choline

15. The process whereby the embryo develops form is called

(A) morphogenesis.

(D) ontogeny.

(B) catharsis.

(E) epiboly.

(C) parthenogenesis.

16. Disappearance of nuclear membrane and nucleoli, and migration of centrioles to opposite poles to form the mitotic spindle apparatus occur during

(A) anaphase.

(D) telophase.

(B) prophase.

(E) G_1 stage.

(C) metaphase.

17. In which part of the brain are the cardioacceleratory and cardioinhibitory centers located?

(A) medulla

(D) cerebral cortex

(B) cerebellum

(E) cerebrum

(C) thalamus

18. The parasympathetic nerve that innervates the heart releases

(A) epinephrine.

(D) acetyl choline.

(B) norepinephrine.

(E) serotonin.

(C) GABA.

19. Which of the following membranes does a sound wave strike first?

(A) Tympanic membrane

(D) Tectorial membrane

(B) Membrane of the round window

(E) None of the above

(C) Membrane of the oval window

20. Individuals with (47,xxy) are afflicted by

(A) Cri-du Chat Syndrome.

(D) Turner's Syndrome.

(B) Edward's Syndrome.

(E) Down's Syndrome.

(C) Klinefelter's Syndrome.

21. If you are sitting with your feet under a bed and you are asked to point to it (your feet), despite you cannot see it, you are able to point to its location. This is due to

(A) chemoreception.

(D) thermoreception.

(B) protoreception.

(E) proprioception.

(C) application.

22. Ovum and follicle development are stimulated by

    (A) increased FSH production.

    (B) low estrogen levels.

    (C) high estrogen levels.

    (D) high progesterone levels.

    (E) low progesterone levels.

23. Progesterone which is secreted by the corpus luteum has the effect of

    (A) suppressing LH production.

    (B) increasing LH production.

    (C) stimulating FSH production.

    (D) suppressing estrogen production.

    (E) decreasing the thickness of the uterine lining.

24. A biology instructor wanted a microscope that would provide him with a three-dimensional image which could be viewed on a viewing screen. Which of the following would be applicable?

    (A) Brightfield microscope

    (B) Darkfield microscope

    (C) Phase contrast microscope

    (D) Fluorescence microscope

    (E) Scanning electron microscope

25. The word SPONCH represents 99% of the living matter. Which of the following combinations best defines it.

    (A) Sulphur, phosphorus, oxygen, nitrogen, carbon, hydrogen

    (B) Sodium, potassium, oxygen, neon, calcium, helium

    (C) Sulphur, phosphorus, oxygen, nitrogen, calcium, hydrogen

    (D) All of the above

    (E) None of the above

26. All of the following are found in prokaryotic cells except

(A) cytoplasm.

(D) chromosome.

(B) nuclear membrane.

(E) enzymes.

(C) peptidoglycan in cell wall.

27. The relationship between fungi and the algae in lichens is known as

(A) mutualism.

(D) saprophytism.

(B) parasitism.

(E) All of the above

(C) commensalism.

28. Associate a medicinal property for each of the following families of compounds: (a) phenothiazines, (b) barbiturates, and (c) amphetamines with any of the statements below.

  I. used to treat psychoses
  II. used for sedative and hypnotic purpose
  III. used to bring about an increase in alertness, elevation of blood pressure and increase in heart action
  IV. act as depressants of the central nervous system with little action in brain function

(A) I

(D) III, IV and II

(B) I and IV

(E) I, II and III

(C) II and I

29. Which of the following pairs represents ecological equivalents?

(A) Squirrel and rattlesnake

(D) Wild horse and zebra

(B) House cat and lion

(E) All of the above

(C) Seagull and codfish

30. Choose the statement that best describes the climax stage of an ecological succession.

(A) It is usually populated only by plants.

(B) It remains until there are severe changes in the environment.

(C) It represents the initial phases of evolution.

(D) It changes rapidly from season to season.

(E) It is usually populated only by animals.

31. Organisms which break down the compounds of dead organisms are called

(A) phagotrophs.   (D) producers.

(B) parasites.   (E) autotrophs.

(C) saprophytes.

32. A sequence of species related to one another as predator and prey comprises a(an)

(A) trophic level.   (D) food chain.

(B) ecosystem.   (E) None of the above

(C) climax.

33. The ecological unit composed of organisms and their physical environment is known as a(an)

(A) niche.   (D) community.

(B) ecosystem.   (E) nation.

(C) population.

34. Which must be present in an ecosystem if it is to be maintained?

(A) Producers and carnivores   (D) Herbivores and carnivores

(B) Carnivores and decomposers

(C) Producers and decomposers   (E) Herbivores and decomposers

35. A Doberman pinscher and a cocker spaniel are structurally similar because they are members of the same

(A) genus.

(D) class.

(B) species.

(E) kingdom.

(C) family.

36. Plants are unable to use the pure nitrogen from the atmosphere. It has to be first converted into a form which can be used by them. This process of conversion is enacted by

(A) legumes.

(D) denitrifying bacteria.

(B) bacteria of decay.

(E) algae.

(C) nitrogen fixing bacteria and cyanobacteria.

37. Neisseria gonorrhoeae occurs as pairs of spherical organisms. Its morphological characteristic would be that it is

(A) cocci.

(D) spiral.

(B) filamentous.

(E) All of the above

(C) bacilli.

38. The phases of bacterial growth can be characterized by which of the following sequences?

(A) Lag, exponential, stationary, death

(B) Lag, exponential, death, stationary

(C) Exponential, lag, death, stationary

(D) Stationary, exponential, lag, death

(E) Exponential, stationary, death, lag

39. In poodles, assume that the heterozygous genotype produces a beige fur color. Perform a mating between two heterozygous poodles for fur color. What is the genotypic ratio?

Let: B = brown fur
b = white fur

(A) $\frac{1}{4}$BB:$\frac{1}{2}$Bb:$\frac{1}{4}$bb

(B) $\frac{1}{4}$Bb:$\frac{1}{4}$bb:$\frac{1}{2}$BB

(C) $\frac{1}{4}$bB:$\frac{1}{2}$bb:$\frac{1}{4}$BB

(D) All of the above ratios are correct

(E) None of the above ratios are correct

40. Using the data from the previous problem, what would the phenotypic ratio be?

(A) $\frac{1}{4}$ brown fur:$\frac{1}{2}$beige fur:$\frac{1}{4}$ white fur

(B) $\frac{1}{4}$ beige fur:$\frac{1}{4}$ white fur:$\frac{1}{2}$ brown fur

(C) $\frac{1}{4}$ beige fur:$\frac{1}{2}$ white fur:$\frac{1}{4}$ brown fur

(D) All of the above ratios are correct

(E) None of the above ratios are correct.

41. An enzyme was added to a chemical reaction which after being monitored for some time showed a significant change. The action of the enzyme was one of

(A) a substrate.

(B) an amino acid.

(C) a catalyst.

(D) a hormone.

(E) a nucleic acid.

42. The below reaction is representative of which of the following process(es)?

$$C_6H_{12}O_6 + C_6H_{12}O_6 \longrightarrow C_{12}H_{22}O_{11} + H_2O$$

(A) hydrolysis

(B) decomposition

(C) dehydration synthesis

(D) osmosis

(E) All of the above

43. A heart transplant operation was performed on a patient whose recovery was favorable. After a few days the patient became severely ill. The body's immunity system "attacked" the cells of the new heart. This peculiar reaction was due to the action of

(A) B (bursa of fabricus) lymphocytes

(B) T (thymus dependent) lymphocytes

(C) monocyte

(D) granulocyte

(E) fibrinocyte

44. The blood plasma contains a number of inorganic cations and anions. If NaCl increases in the concentration in the plasma, thus increasing the number of ions, what will be the expected result?

(A) Blood pressure will decrease

(B) Blood pressure will increase

(C) Blood pressure will remain unchanged

(D) All of the above

(E) None of the above

45. Which of the following proteins is/are critical for binding miscellaneous impurities and toxins in the blood?

(A) Albumins

(B) Globulins

(C) Figrinogen

(D) All of the above

(E) None of the above

46. Which of the following proteins is required for the fibrinogen to fibrin conversion?

(A) Thromboplastin

(B) Platelets

(C) Thrombin

(D) Fibrin

(E) Prothrombin

47. The transport of lipids and fat soluble vitamins across cell membranes is enabled by antibodies and plasma proteins. These plasma protein are called

(A) fibrinogen.

(D) All of the above

(B) globulins.

(E) None of the above

(C) albumins.

48. Which of the following represents the normal sequence of stimulation in the normal heart beat?

(A) AV node → SA node → Purkinje fibers → bundle of His → cardiac muscle

(B) SA node → AV node → bundle of His → Purkinje fibers → cardiac muscle

(C) AV node → SA node → bundle of His → Purkinje fibers → cardiac muscle

(D) SA node → AV node → Purkinje fibers → bundle of His → cardiac muscle

(E) AV node → SA node → Purkinje fibers → bundle of His → cardiac muscle

49. The diffusion of CO_2 from the tissue spaces to the blood is called

(A) external respiration.

(D) inspiratory capacity.

(B) ventilation.

(E) respiratory capacity.

(C) internal respiration.

50. Which of the following blood cells produces antibodies?

(A) Monocytes

(D) Basophil

(B) B-lymphocytes

(E) Thrombocytes

(C) Eosinophil

51. Organic molecules which contain nitrogen atoms as part of a single ring, 6 membered structure are best described as

(A) steroids.

(B) pyrimidines.

(C) triglycerides.

(D) purines.

(E) None of the above

52. Which of the following is involved with initiating blood clotting?

(A) Lymphocytes

(B) Monocytes

(C) Thrombocytes

(D) Eosinophil

(E) Basophil

53. The electrical impulse that causes a muscle fiber to contract is quickly carried to the interior of the myofiber by the

(A) regulator proteins.

(B) T-tubules.

(C) sarcoplasmic reticulum.

(D) myofilaments.

(E) B-tubules.

54. During muscle contraction, the protein that also serves as an enzyme that breaks down ATP $\rightarrow$ ADP + Pi is

(A) tropomyosin.

(B) fibrinogen.

(C) actomyosin.

(D) troponin.

(E) None of the above

55. During muscle contraction, the Ca^{++} that is released combines with

(A) troponin.

(B) actomyosin.

(C) tropomyosin.

(D) fibrinogen.

(E) None of the above

56. Which of the following ions or molecules is bound to myosin when the muscle fiber is not contracting (at rest)?

(A) ATP (D) ADP

(B) Ca^{++} (E) ADP + Pi

(C) Na^{+}

57. Which of the following gives the correct composition of the thin myofilaments that are attached to the Z lines of a sarcomere?

(A) Actin, myosin, troponin

(B) Myocin, tropomyosin, troponin

(C) Actin, troponin, tropomyosin

(D) Myocin, actin, tropomyosin

(E) Tropomyosin, myocin, troponin

58. Which of the following molecules is thought to block the myosin (cross bridge) binding site on actin when a muscle fiber is not contracting?

(A) Calcium (D) ATP

(B) Troponin (E) ATP + Pi

(C) Tropomyosin

59. In myelinated axons and dendrites, the reversed potential leaps from one node of Ranvier to the next by a process called

(A) hydrophobic induction. (D) electrophilic conduction.

(B) saltatory conduction.

(C) electron transport. (E) None of the above

60. After a neuron fires an impulse, the original balance of Na^{+} (outside) and K^{+} (inside) is restored by the process called

(A) diffusion. (D) active transport.

(B) filtration. (E) None of the above

(C) dialysis.

61. The inner ear contains small crystals of calcium carbonate that are involved with the vestibular process of maintaining static equilibrium. These crystals are called

(A) otoliths.

(D) vestibular clefts.

(B) protostones.

(E) earoliths.

(C) vestibular crystaline apparatus.

62. Which of the following is more involved with the process of static equilibrium?

(A) Utricle.

(D) Cupula.

(B) Cristae.

(E) static canals.

(C) Semi-circular canals.

63. The organ of corti has a gel-like membrane that extends over the auditory hair cells and takes part in the deformation of the hairs to allow sound detection. This membrane is called

(A) eustacian.

(D) tympanic.

(B) tectorial.

(E) None of the above

(C) oval.

64. Stimulation of baroreceptors in the aortic arch will cause

(A) an ultimate inhibition of the parasympathetic pathway to the heart.

(B) a positive chromotropic effect.

(C) an ultimate increase in blood pressure.

(D) stimulation of the cardioinhibitory center.

(E) All of the above

Assume that the composition of air is as follows:

N_2 = 78% CO_2 = .04%

O$_2$ = 21%   Other = .96%

For questions 65-66 refer to the information above.

65. If a scuba diver encounters a pressure of 762 mmHg, what would be the partial pressure of N$_2$ at this depth?

(A) 59.436 mmHg   (D) 549.63 mmHg

(B) 594.36 mmHg   (E) 600.20 mmHg

(C) 59436.0 mmHg

66. Referring to the previous problem what would be the partial pressure of O$_2$ at this depth?

(A) 160.02 mmHg   (D) 1600.2 mmHg

(B) 60.00 mmHg   (E) 160 mmHg

(C) 16002.0 mmHg

67. Which of the following prevents blood from flowing back into the right ventricle?

(A) Aortic semilunar valve   (D) Bicuspid (mitral) valve

(B) Pulmonary semilunar valve   (E) None of the above

(C) Tricuspid valve

68. The heart's contraction is initiated by a tissue which serves as a pacemaker. The tissue is called the

(A) purkinje fibers.   (D) atrio-ventricular node.

(B) sino-atrial node.   (E) initiation tissue.

(C) bundle of His.

69. A persistent change in behavior resulting from experience is a phenomenon called

(A) learning.

(B) imprinting.

(C) motivated behavior.

(D) All of the above

(E) None of the above

70. Which of the following sequences describes the blood clotting process correctly?

(A) Damaged platelets + Ca^{++} $\xrightarrow{\text{enzyme}}$ prothrombin $\xrightarrow{\text{enzyme}}$ fibrinogen →fibrin.

(B) Damaged platelets + Ca^{++} → prothrombin $\xrightarrow{\text{enzyme}}$ fibrinogen → fibrin

(C) Damaged platelets + Ca^{++} $\xrightarrow{\text{enzyme}}$ prothrombin → enzyme$\xrightarrow{}$ fibrin → fibrinogen

(D) Damaged platelets + Na^{+} → thromboplastin → thrombin → fibrinogen → fibrin

(E) None of the above sequences are correct.

71. Roots have evolved many structural adaptations which have increased their absorptive capacity. The order of these structural components in a dicot plant from the outer surface to the interior is

(A) epidermis – endodermis – cortex – pericycle – phloem – xylem

(B) endodermis – cortex – epidermis – pericycle – xylem – phloem

(C) epidermis – cortex – endodermis – pericycle – phloem – xylem

(D) cortex – epidermis – endodermis – xylem – phloem – pericycle

(E) cortex – pericycle – endodermis – epidermis – xylem – phloem

72. The two strands of a DNA helix unwind and separate when treated by heating or by adding acids or alkalis to a solution of DNA. This unwinding and separation of the double helix is

called melting. Single-stranded DNA has a higher absorbency at 260nm than double-stranded DNA. Below is a graph showing the DNA melting curves. Which of the following is or are correct conclusions?

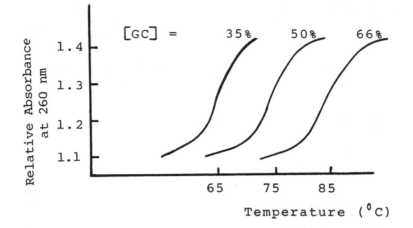

I. DNA double helix is a highly cooperative structure.

II. DNA double helix with a high GC content melts at a higher temperature.

III. The A-T rich regions of DNA melts earlier on heating.

IV. A-T base pairs are more stable than the G-C base pairs.

(A) I, II

(B) II, III

(C) III, IV

(D) I, II, III

(E) I, II, III, IV

Question 73 is based on the diagram below:

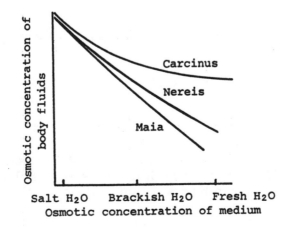

474

73. From examination of the above diagram, it would be correct to say that

   (A) the spider crab (Maia) fares equally well in salt H_2O and brackish H_2O.

   (B) Nereis (the clam worm), evolved a considerable degree of osmoregulation which enabled it to live in both sea water and brackish water.

   (C) Carcinus (the shore crab), can maintain relatively concentrated body fluids even in very dilute external medium.

   (D) Both A and B are correct.

   (E) Both B and C are correct.

74. During electron transport in the mitochondria, the energy which is released is directed by components of the respiratory chain to move protons (H^+) from the matrix space to the intermediate space. The consequence(s) of this movement of protons is(are)

   I. generation of a pH gradient across the inner mitochondrial membrane.

   II. a significant rise in the H^+ concentration in the matrix than in the rest of the cell.

   III. generation of a voltage gradient across the inner mitochondrial membrane.

   IV. the inside of the inner mitochondrial membrane potential becomes more positive and the outside more negative.

   V. the pH in the rest of the cell is generally close to 7.

   (A) I, III and V   (D) II and IV

   (B) II and III   (E) I and IV

   (C) I only

75. X-ray diffraction photographs of a hydrated DNA fiber were taken by Franklin and Wilkins and analyzed by Watson and Crick who proposed a model of DNA structure. Among the following statements, which correctly describes the features of the DNA model?

I. Two helical polynucleotide strands running in opposite direction, and coiled with respect to a common axis.

II. The purine and pyrimidine bases are located on the inside of the double helix.

III. The two polynucleotide strands are held together by hydrogen bonding between adenine and thymine, guanine and cytosine.

IV. The base sequence of a polynucleotide strand is restricted by its helical structure.

(A) I, II   (D) I, II, III

(B) II, III   (E) I, II, III, IV

(C) III, IV

76. The genetic code is described as "degenerate" because many amino acids are designated by more than one codon. Those codons that specify the same amino acid differ mainly in the third base of the triplet code. Which of the following is or are the biological significance of degeneracy of the genetic code?

I. The probability that a codon will mutate to a stop signal is minimized.

II. Some mutations at the third base of the triplet codon can be tolerated.

III. The same genetic information is allowed to vary in the DNA composition.

(A) I   (D) I, II

(B) II   (E) I, II, III

(C) III

77. The newly formed polypeptides are usually not the final protein products utilized by the cell but rather the accomplishment of some post-translational modification of polypeptides. From the statements below select the processes which apply to the post-transitional modification of polypeptides.

I. Removal of the formyl group at the amino terminus of bacterial polypeptides.

II. Formation of disulfide bonds.

III. Specific cleavage of polypeptide chains.

IV. Hydroxylation of some amino acids in a polypeptide chain.

V. Attachment of sugars to some amino acids.

(A) I, II

(B) I, II, III

(C) II, III, IV

(D) I, II, III, IV

(E) I, II, III, IV, V

78. Lambda phages are being used as a vector to deliver recombinant DNA into bacteria for the purpose of DNA cloning. Which of the following is or are the advantage(s) of choosing λ phage as a vector?

I. There is no limitation on the length of foreign DNA being inserted into λ genome.

II. λ phage infects bacteria with a high frequency.

III. Large segments of the λ genome are not essential for its lytic and lysogenic life cycles.

(A) I

(B) II

(C) III

(D) I, II

(E) II, III

79. In an enzyme purification experiment, you are provided with a whole cell extract, equipment, and chemicals required for the below three kinds of chromatography. Arrange these chromatography in the appropriate order that the experiment should be conducted.

I. Affinity chromatography

II. Ion-exchange chromatography

III. Gel-filtration chromatography

(A) I, II, III

(B) II, III, I

(C) II, I, III

(D) III, I, II

(E) III, II, I

80. Which of the following statements concerning the X chromosomes in female mammalian cells are true?

I. Only one of the two X chromosomes expresses.

II. The inactive X chromosome is faithfully inherited.

III. The inactive X chromosome exists in the form of euchromatin.

(A) I, II

(B) II, III

(C) I, III

(D) I, II, III

(E) None of the above

81. Which one of the following is used for the study of membrane potential?

(A) Geiger counter

(B) Electrophoresis

(C) Cathode ray oscilloscope

(D) Immunoradioassay

(E) None of the above

82. The first law of thermodynamics states that the total energy of a system and its surroundings is a constant. This law is expressed by the formula

$$\Delta E = E_B - E_A = Q - W.$$

Analyzing this equation and recognizing that E_A is the energy of the system at the start and E_B is the energy of the system at the end of the reaction, Q is the heat absorbed by the system and W is the work done by the system, one can conclude that

(A) the change in the energy of a system depends on the path of the transformation.

(B) the change in the energy of a system depends on the initial state.

(C) the change in the energy of a system depends on the final state.

(D) All are correct.

(E) Only B and C are correct.

83. Covalent modification and allosteric interactions are known to control regulatory enzymes and their activities. Which of the following is true about the action of covalent modification?

   (A) The catalytic activity of glycogen phosphorylase is diminished while that of glycogen synthetase is enhanced.

   (B) The catalytic activities of glycogen phosphorylase and glycogen synthetase are enhanced.

   (C) The anabolic activity of glycogen phosphorylase is increased whereas that of glycogen synthetase is decreased.

   (D) The catalytic activity of glycogen phosphorylase is enhanced while that of glycogen synthetase is diminished.

   (E) None of the above are correct.

84. Which of the following statements is(are) true of desert biomes

   (A) They receive very little annual rainfall (10 inches or less).

   (B) They experience dramatic fluctuations in temperature from night to day.

   (C) Some deserts are produced and maintained by high mountain ranges that block coastal precipitation.

   (D) They exist on all continents except Europe and Antarctica and tend to be created along the 30° north and south latitude lines.

   (E) All of the above

85. According to the Hardy-Weinberg Law, under certain conditions of stability both allelic frequencies and genotypic ratios remain constant from generation to generation in sexually reproducing populations.

   Which of the following are among the "certain conditions" which must be satisfied if the gene pool of a population is to be in genetic equilibrium?

   (A) The population must be large enough to insure that chance alone could not significantly alter allelic frequencies.

   (B) Mutations must not occur, or else there must be mutational equilibrium.

(C) There must be no immigration or emigration.

(D) Reproduction must be totally random.

(E) All of the above

86. The flow of glucose 6-phosphate in the different modes of pentose phosphate pathways is dependent on the need for

(A) NADPH.

(D) None of the above

(B) Ribose 5-phosphate.

(E) All of the above

(C) ATP

87. Considering the situation involving the hypothalamus where an animal detects the lengthening of days in spring and its gonads secrete more sex hormones, we can conclude that its perception of the stimulus-increasing day length affects its endocrine glands through

(A) nervous connections to the anterior pituitary which relay the information to the gonads.

(B) nervous connections directly to endocrine glands such as the gonads.

(C) nervous tissue which stimulates the endocrine glands by chemical messengers.

(D) All of the above are correct.

(E) None of the above are correct.

88. The oleander (Nerium) which lives in a very dry habitat has stomata which are located in

(A) shallow hair-lined depressions in the upper epidermis.

(B) deep hair-lined depressions in the lower epidermis.

(C) very thin upper epidermis.

(D) deep hair-lined depressions in the upper epidermis.

(E) None of the above are correct.

89. The example of inhibition of malonic acid with succinic acid, an intermediary in the Krebs cycle for the enzyme succinic acid dehydrogenase is termed

(A) non-competitive inhibition.

(B) allosteric inhibition.

(C) inactivation of reactive groups.

(D) competitive inhibition.

(E) Only B and D are correct.

90. In one of his early experiments, Mendel cross-pollinated pea plants that were true breeding for round seeds with those that were true breeding for wrinkled seeds. (Round seeds were the dominant traits). These represented the P_1 generation. What was the composition of the first filial (F_1) generation?

(A) $\frac{1}{2}$ round:$\frac{1}{2}$ wrinkled   (D) All wrinkled

(B) $\frac{3}{4}$ round:$\frac{1}{4}$ wrinkled   (E) $\frac{3}{4}$ wrinkled:$\frac{1}{4}$ round

(C) All round

91. Taking two seeds from the F_1 generation Mendel performed a second cross. The progeny of the F_2 generation was found to be

(A) $\frac{3}{4}$ round:$\frac{1}{4}$ wrinkled   (D) All round

(B) $\frac{1}{2}$ round:$\frac{1}{2}$ wrinkled   (E) All wrinkled

(C) $\frac{1}{4}$ round:$\frac{3}{4}$ wrinkled

92. A living body can change in many ways to meet the challenge of unsuitable environmental temperatures. This may be accomplished by

(A) piloerection.   (D) B and C only

(B) vasoconstriction.   (E) All of the above

(C) vasodilation.

93. More gene transcription occurs in the absence of the regulatory protein than in its presence. This kind of genetic control system is called

(A) positive regulation.

(B) negative regulation.

(C) equilibrium regulation.

(D) thermodynamic regulation.

(E) allosteric regulation.

94. When inserting a restriction fragment of foreign DNA into a plasmid vector, what appropriate step must be taken to prevent the recyclization of linearized plasmid DNA?

(A) Introduction of a linker sequence.

(B) Treatment with alkaline phosphatase.

(C) Methylation of restriction sites.

(D) λ exonuclease digestion

(E) S_1 nuclease digestion

95. The character of an ecosystem is determined by the environmental factor that is in shortest supply. This is

(A) the law of minimum.

(B) Borty's law.

(C) the law of diminishing returns.

(D) Dollo's law.

(E) the limiting reagent.

96. Choose the correct statement:

(A) The earth is a closed system for materials.

(B) In an ecosystem energy is usually not in balance.

(C) All energy in an ecosystem is not ultimately lost as heat.

(D) Food relations are in chains, not in webs.

(E) None of the above are correct.

97. A mRNA molecule is translated by several ribosomes at the

same time. In this way, more protein is produced before the mRNA is degraded. The structure which has a group of ribosomes bound to a mRNA molecule is called a

(A) granular body.
(D) peroxisome.

(B) polymer.
(E) polyoma.

(C) polysome.

98. Which of the following is the correct order for the classification of living things, from the broadest category to the most specific?

(A) Kingdom, order, class, phylum, family, genus, species

(B) Kingdom, class, family, order, phylum, genus, species

(C) Phylum, kingdom, class, family, order, genus, species

(D) Kingdom, class, order, genus, family, phylum, species

(E) Kingdom, phylum, class, order family, genus, species

99. Among the choices provided below, which is or are involved in the gene expression in eucaryotes?

(A) The synthesis of primary RNA transcripts

(B) The processing of primary RNA transcripts

(C) The translation of messenger RNA

(D) The degradation of messenger RNA molecules in the cytoplasm

(E) All of the above

100. Select the correct sentence which shows the relationship between size and metabolic rate in poikilothermic animals and plants.

(A) Larger size and smaller surface-to-volume ratio increases heat loss and slows down metabolic rate.

(B) Normal metabolic rate is directly related to body size.

(C) The smaller the organism, the lower the relative metabolic rate.

(D) Metabolic heat is promptly lost to the environment and is normally replaced by increased metabolism.

(E) Smaller size and larger surface-volume ratio increases heat loss and the relative metabolic rate.

101. The geologic time scale can be divided into four main eras. Which of the following is the correct order of eras, from most ancient to most recent?

(A) Paleozoic, Precambrian, Mesozoic, Cenozoic

(B) Precambian, Paleozoic, Mesozoic, Cenozoic

(C) Paleozoic, Mesozoic, Cenozoic, Precambrian

(D) Precambrian, Cenozoic, Mesozoic, Paleozoic

(E) Cenozoic, Mesozoic, Paleozoic, Precambrian

102. The three periods within the mesozoic era are the

(A) Permian, Carboniferous, and Devonian.

(B) Quaternary, Tertiary, and Cretaceous.

(C) Triassic, Jurassic, and. Cretaceous.

(D) Silurian, Ordovician, Cambrian.

(E) Tertiary, Jurassic, Devonian.

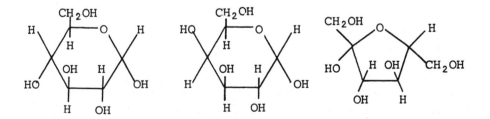

103. The three molecules shown above are best described as

(A) isomers.

(B) isotopes.

(C) ions.

(D) polymers.

(E) fatty acids.

104. In bacteria, which of the following facilitates the transcription of genes?

(A) Regulatory proteins

(B) DNA helices

(C) Z-form DNA

(D) Promoter

(E) DNA supercoiling

105. The orderly change from one ecological community to another in an area is called

(A) convergence.

(B) climax.

(C) succession.

(D) dispersal.

(E) progression.

106. All ecosystems have three basic living components. Which one of the following is not necessarily found in all ecosystems?

(A) Producer plants

(B) Consumers (animals)

(C) Decomposers

(D) Parasites and commensals.

(E) None of the above

107. Which of the following molecules has an N group, O group, H, and a variable group, all attached to the same carbon?

$$\begin{array}{cc} \text{N group,} & \text{O} \\ | & || \\ \text{H}-\text{N} & \text{C}-\text{OH} \end{array}$$

(A) Nucleotides.

(B) Steroid

(C) Amino acid

(D) Fatty acid

(E) Glycerol

108. Which of the following choices is the most common population distribution pattern for plants and animals in nature?

(A) Uniform

(B) Random

(C) Clumped

(D) All of the above

(E) None of the above

109. The first event in visual excitation is the

(A) conversion of trans-retinal to cis-retinal.

(B) absorption of a photon of light by the rods and cones.

(C) isomerization of 11-cis-retinal.

(D) entering of the reactive trans-retinal in the bloodstream.

(E) None of the above

110. Regardless of the mechanism, enzymes are thought to increase the rate of the reaction by

(A) rearranging the tertiary structure of the reactants.

(B) lowering the required energy of activation.

(C) increasing the need for thermal energy.

(D) decreasing the concentration of the reactants.

(E) All of the above

111. The biological membrane consists of a continuous double layer of lipid molecules with various membrane proteins embedded in it. Which one of the following lipid molecules is found only on the outer monolayer?

(A) Glycolipid

(B) Cholesterol

(C) Phosphatidyl ethanolamine

(D) Phosphatidyl serine

(E) Sphingomyelin

112. Certain derivatives of minerals represent small inorganic mole-

cules that are bound to enzymes during chemical reactions to
make the enzymes functional. These are called

(A) hydrophobic determinants.   (D) quarternary groups.

(B) cofactors.   (E) All of the above

(C) coenzymes.

113. Which one of the following factors contributes to the tendency
of compartment formation for lipid bilayers?

(A) Kinetic   (D) Static

(B) Energetic   (E) None of the above

(C) Equilibrium

114. Which one of the following does not involve bilayer adherence
and bilayer joining of membrane fusion?

(A) Endocytosis   (D) Cell division

(B) Exocytosis   (E) Plasmolysis

(C) Cell fusion

115. Which one of the following is used to identify membrane proteins exposed at the cell surface?

(A) Vectorial labeling   (D) Scanning electron
                                      microscopy
(B) Freeze fracture electron
    microscopy   (E) None of the above

(C) 2-d polyacrylamide gel
    electrophoresis

116. According to the fluid-mosaic model of cell membrane structure,
which components of an animal cell are located at the outer
membrane surface, exposed directly to the extracellular fluid?

(A) Lipids only   (D) Lipids and proteins

(B) Proteins only   (E) Lipids, proteins, and
                                      glycocalyx
(C) The glycocalyx only

117. In actively secreting cells, vacuoles may form or be carried to the plasma membrane, fuse with it, then release their contents to the extracellular environment. This process is called

(A) pinocytosis.   (D) phagocytosis.

(B) cyclosis.   (E) endocytosis.

(C) exocytosis.

118. Sugar passes through the plasma membrane by combining with protein carrier molecule (a permease). This combination is then able to pass through the membrane from a region of high to one of low concentration of sugar, without the use of cellular energy. This process is called

(A) filtration.   (D) dialysis.

(B) active transport.   (E) simple diffusion.

(C) facilitated diffusion.

119. Organelles found in animal cells that probably give rise to spindle fibers and assist in cell division are called

(A) flagella.   (D) cilia.

(B) centrioles.   (E) ribosomes.

(C) peroxisomes.

120. Which one of the following is not a density-dependent limitation on population growth?

(A) Intraspecific competition   (D) Predator-prey system

(B) Interspecific competition   (E) Environmental changes

(C) Physiological mechanism

121. Which of the following acts as the final hydrogen acceptor in the mitochondrial electron transport system?

(A) NAD

(B) carbon

(C) acetyl co-enzyme A

(D) oxygen

(E) None of the above

122. Which of the following is a true statement with regard to the differences between monocotyledon and dicotyledon plants?

(A) The pattern of veins in the leaf is netlike in the dicot and parallel in the monocot.

(B) In the seeds, one cotyledon is present in the dicot and two cotyledons are present in the monocot.

(C) The flowers of dicots have three floral parts, or multiples of three; the flowers of monocots have four or five floral parts, or multiples of four or five.

(D) The vascular system of dicots occurs in scattered bundles, while the vascular system of monocots is generally a neat ring of vascular bundles arranged in a circle around the stem.

(E) In dicots, there is no vascular cambium; in monocots vascular cambium is present.

123. In protein-driven chemiosmotic phosphorylation, ATP energy is generated as protons pass from a region of high concentration to one of low concentration, through complex structures of the thylakoid membranes called the

(A) FAD enzyme complex.   (D) cytochromes.

(B) CF-1 particles.   (E) F-1 particles.

(C) Fe S_1 proteins.

124. Most of the carbons are removed as CO_2, and most of the NAD are reduced during which of the following stages of respiration?

(A) Kreb's cycle   (D) Calvin cycle

(B) Glycolysis   (E) Fermentation

(C) Electron transport system

125. Select from the following those factors that contribute to the divergence in geographically separated population systems.

   I. It is possible for a population to be geographically separated in a way that the population systems carry different initial gene frequencies.

   II. Separated population systems may be subject to different environmental selection pressures.

   III. Separated population systems may experience different mutations.

   IV. Non-random reproduction is practiced.

   (A) I, II

   (B) II, III

   (C) I, II, III

   (D) II, III, IV

   (E) I, II, III, IV

126. Which of the following events in the functioning of the P680 photosystem is not correctly stated?

   (A) Light energy excites electrons in antenna pigments in photosynthetic unit.

   (B) The excitation is passed along to a moelcule of P680 chlorophyll in the photosynthetic unit.

   (C) The P680 chlorophyll donates a pair of electrons directly to a molecule of NADP, which then picks up $2H^+$ to become $NADP \cdot H_2$.

   (D) The electron vacancies in the P680 chlorophyll are filled by electrons derived from water.

   (E) All of the above are true.

127. Succession refers to the progressive replacement of plant and animal life in a given area. It is a result of

   (A) modification of physical environment.

   (B) climatic changes.

   (C) environmental toxification.

   (D) All of the above

   (E) None of the above

128. Electrons lost from the P700 reaction center of photosynthesis are replaced by electrons that escape from

(A) NADP.

(D) P680 chlorophyll.

(B) water.

(E) P640 chlorophyll.

(C) carbon dioxide.

129. In which of the following structures or areas would the light reactions of photosynthesis occur?

(A) Stroma of chloroplast

(D) Matrix of mitochondria

(B) Thylakoid disc

(E) None of the above

(C) Cristae

130. When CO_2 combines with a component of the erythrocyte instead of water, it forms a molecule called

(A) carbonic acid.

(D) carbaminohemoglobin.

(B) carboxypeptidase.

(E) None of the above

(C) carbonic anhydride.

131. The genetic experiment which was instrumental in establishing that DNA molecules are reproduced by semiconservative replication was carried out by which of the following scientists?

(A) Louis Pasteur

(D) Oswald Avery, Colin MacLeod, and Maclyn McCarty

(B) Gregor Mendel

(C) Messelson and Stahl

(E) Hugo de Unes

132. Which of the following organisms utilizes a countercurrent device to make gas exchange more efficient?

(A) Fish

(D) Insects

(B) Flatworms

(E) Echinoderms

(C) Lungless salamanders

133. A small group of birds left the main flock and established a new feeding and breeding community in an unexploited area. The individuals that found the new area were a random group which may not have represented the same genetic makeup as the original population. Thus one population may establish itself with a new genetic composition by chance alone. This is an example of

(A) genetic drift.

(D) A and B only

(B) the founder effect.

(E) None of the above

(C) natural selection.

134. The concept of "evolution by the inheritance of acquired characteristics" is most often associated with which of the following biologists?

(A) Darwin

(D) Wallace

(B) Pasteur

(E) Pavlov

(C) Lamarck

135. Species not naturally protected by some unpleasant characteristic of their own, may closely resemble (mimic) in appearance and behavior, some warningly colored unpalatable species. This phenomenon is called

(A) Batesian mimicry.

(D) parasitism.

(B) Müllerian mimicry.

(E) None of the above

(C) mutualism.

136. Fermentation reactions may produce any of the following products except

(A) CO_2.

(D) lactic acid.

(B) O_2.

(E) CO.

(C) ethyl alcohol.

137. Vast areas of the temperate and subarctic regions are vegetated by a gymnosperm that contributes to the production of

paper, plastic and wood. It has a characteristic of shedding the older leaves as they grow, thus allowing the leaves to remain green or evergreen. These seedless plants are examples of

(A) Pteredophytes.   (D) monocots.

(B) Ginkgos.   (E) dicots.

(C) conifers.

138.   The end products of the light reactions of photosynthesis are

(A) H_2O, ADP, NADP.

(B) PGAL, ADP, ribulose.

(C) O_2, ATP, $NADPH_2$

(D) CO_2, PGAL, $2H^+$.

(E) O_2, ATP, ribulose.

139.   Which one of the following is employed to cut out the genome in a specific way?

(A) Ribonuclease   (D) Topoisomerase

(B) Exonuclease   (E) DNA polymerase

(C) Restriction endonuclease

140.   The initiation phase of protein synthesis in bacteria is determined by two kinds of interactions. One is the pairing of the start codon with the charged initiator tRNA. The other is

(A) the base pairing of an upstream region of mRNA with the 3' end of 16s rRNA.

(B) the base pairing of an upstream region of mRNA with the 5' end of 16s rRNA.

(C) the base pairing of an upstream region of mRNA with the 3' end of 23s rRNA.

(D) the base pairing of an upstream region of mRNA with the 5' end of 23s rRNA.

(E) the base pairing of an upstream region of mRNA with the 3' end of 5s rRNA.

141. Which of the following makes it possible for eucaryotic genes to be expressed by bacteria?

(A) A universal genetic code

(B) The cDNA

(C) The same composition of 20 amino acids.

(D) Plasmid vectors

(E) None of the above

142. Protein synthesis in bacteria starts with

(A) alanine.

(B) leucine.

(C) methionine.

(D) formylmethionine.

(E) cysteine.

143. Bromthymol blue is a blue indicator that turns yellow in the presence of CO_2. A solution of yellow bromthymol blue and an Elodea plant were placed in two test tubes. Test tube I was kept in the dark and test tube II was kept in the light. What was probably observed after 24 hours?

(A) The solution in test tube II turned blue.

(B) The solution in test tube I turned blue.

(C) The solution in both test tubes turned blue.

(D) No change in color occurred in either test tubes.

(E) The solution in test tube II turned yellow while the solution in test tube I remained unchanged.

144. Hybrid cells were formed by fusing human cells and mouse cells together. Two sets of antibodies were prepared. The one that had an affinity for human membrane proteins was coupled to rhodamine, a red fluorescent dye. The other which bound to mouse membrane protein was coupled to fluorescein, a green fluorescent dye. At first, the hybrid cells had one-half of their surface bright red and the other half green. About one hour later, it was found that the red and green fluorescent dyes had mixed over the entire surface of most cells. Which of the following statements is incorrectly related to this experimental result.

(A) Some plasma membrane proteins are mobile in the plane of the membrane.

(B) The mobility of membrane proteins is made possible by the fluidity of the lipid bilayer.

(C) Membrane molecules are evenly divided between daughter cells at cell division.

(D) Membrane proteins are capable of lateral diffusion, rotational diffusion and flip-flop.

(E) Membrane molecules can be distributed from sites of insertion to sites of utilization.

145. Nitrogen metabolism of completely terrestrial animals is identified with

(A) the correlation of uric acid excretion with viviparity.

(B) the correlation of urea excretion with egg laying.

(C) the correlation of uric acid excretion with egg laying and viviparity.

(D) the correlation of urea excretion with egg laying and viviparity.

(E) the correlation of uric acid excretion with egg laying.

146. The scientist(s) who discovered the hormone secretin, which is released by the mucosal cells of the duodenum when they are stimulated by the acidity of food was(were)

(A) W.M. Bayliss and E.H. Starling.

(B) Ivan P. Pavlov.

(C) J.S. Edkins.

(D) Von Mering and Minkowski.

(E) F.G. Banting and C.H. Besf.

Questions 147 and 148 are concerned with the below dissociation curves.

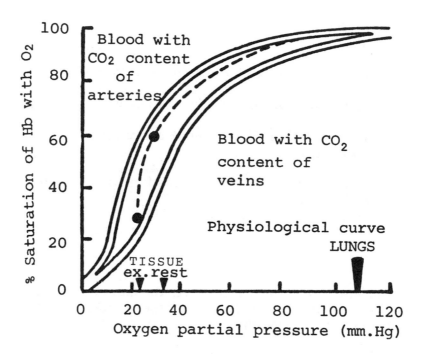

147. Upon examination of the graph one notices that the hemoglobin is about 98 percent saturated with the oxygen at the partial pressure of oxygen typical of the lungs (108 mm), while it is only about 58 percent saturated at the partial pressure of oxygen typical of the tissues at rest (32 mm). What therefore does the difference (40 percent) represent?

(A) The percentage of carbon dioxide released by the tissues.

(B) The log phase between oxygen uptake and carbon dixoide release.

(C) The approximate percent of carbon carried by hemoglobin in the form of carboxyhemoglobin.

(D) The approximate percentage of oxygen carried by hemoglobin that is actually released to the tissues.

(E) The relative amounts of oxygen and carbon dioxide in venous blood.

148. In the two different dissociation curves given for human hemoglobin, it is seen that an increase in the concentration of CO_2 shifts the curve to the right. This means that

(A) a higher partial pressure of oxygen is needed to load the hemoglobin.

(B) the blood concentration becomes more alkaline thus reducing the affinity of hemoglobin for oxygen.

(C) oxyhemoglobin releases oxygen more readily.

(D) A and C are correct.

(E) All are correct.

For questions 149-153, refer to the graph below.

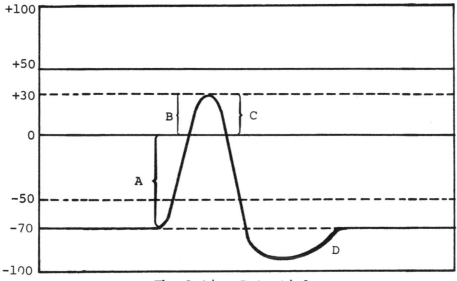

The Action Potential.

149. Part A of the curve represents a condition called

(A) destabilization.   (D) hyperpolarization.

(B) reversed potential.   (E) depolarization.

(C) hypopolarization.

150. What would happen if a stimulus of -66 millivolts was applied?

   (A) An action potential would be generated.

   (B) There would be no action potential generation.

   (C) There would be a slight generation of an action potential.

   (D) An action potential would be generated but there will be no depolarization.

   (E) None of the above

151. Part C of the curve is due to the

   (A) outflow of K^+.

   (B) influx of K^+.

   (C) influx of Na^+.

   (D) outflow of Na^+.

   (E) inflow of Na^+ and K^+.

152. Part D of the curve represents a condition called

   (A) destabilization.

   (B) reversed potential.

   (C) hypopolarization.

   (D) hyperpolarization.

   (E) depolarization.

153. Parts A and B of the curve are due to the

   (A) outflow of Na^+.

   (B) influx of Na^+.

   (C) outflow of K^+.

   (D) influx of K^+.

   (E) influx of Na^+ and K^+.

Questions 154-158. Select the answer which matches the sentence, statements or groups of words below.

   (A) Nitrogen

   (B) Phosphorus

   (C) Potassium

   (D) Magnesium

   (E) Calcium

154. Structural component of chlorophyll which is also a co-factor for many enzymes involved in carbohydrate metabolism.

155. Plays a role in the ionic balance of cells; co-factor for enzymes involved in protein synthesis and carbohydrate metabolism.

156. Structural component of amino acids, hormones and co-enzymes.

157. Influences permeability of membranes, component of pectic salts in middle lamellae and necessary for wall formation.

158. Structural component of nucleic acids, phospholipids, co-enzymes.

159-163.

Match each of the following biomes with the appropriate description.

(A) Taiga	(D) Deciduous forest
(B) Tundra	(E) Grassland
(C) Tropical rain forest	

159. These are vast inland plains where rainfall is seasonal and limited. They support huge populations of animals, including large, grazing herbivores.

160. Of all the biomes, this one has the heaviest rainfall and the greatest number of animal species. It is also characterized by its very warm climate.

161. This biome is notable for its several species of tree, an average annual rainfall of forty inches, pronounced seasonal changes, and abundant, diverse plant and animal life. Plants adapt to winter by losing their leaves.

162. This biome exists only in the northern regions of the Northern Hemisphere. The dominant plants are conifers. Large herbivores and carnivores still survive in these regions, and insect populations are large, but seasonal.

163. This is the northernmost biome. Plant growth is limited because of the scarcity of water (which is frozen during most of the year) and light (which is seasonal). Ground cover includes mosses, lichens, grasses, and dwarfed trees. Animal life consists of herbivores and carnivores, large and small.

(A) Blastula   (D) Corona radiata

(B) Zygote   (E) Transpiration

(C) Amnion

For questions 164-168 choose a letter from the above list that best matches the statements.

164. The mechanism used by plants to rapidly evaporate the water it absorbs.

165. During this stage embryo resembles a hollow fluid-filled ball that is only 1 cell layer thick.

166. An extraembryonic membrane that covers the embryo in its fluid-filled sac.

167. A one-cell fertilized ovum.

168. Dense covering of follicle cells.

169-171.

Match each of the following purification techniques with the appropriate description of its method.

    I. Selective precipitation

II.   Ion-exchange chromatography

III. Gel filtration

IV.   Affinity chromatography

169. This technique utilizes the solubility differences among proteins
      to "salt-out" a particular protein from solution. The precipita-
      tion of proteins at high ionic strength is termed "salting out,"
      and different proteins salt-out at different ionic strengths.

      (A) I   (D) IV

      (B) II   (E) None of the above

      (C) III

170. This technique depends on the ionic association of proteins with
      charged groups bound to an inert supporting material, often
      cellulose.

      (A) I   (D) IV

      (B) II   (E) None of the above

      (C) III

171. This technique of protein purification depends most heavily on
      the physical properties of proteins, such as molecular weight
      and shape. The means for separation is a cylindrical column
      of porous beads through which the solution of proteins slowly
      passes. While passing through the column, proteins of various
      molecular weight are sieved in such a way that they are eluted
      from the column after different volumes of buffer have been
      added.

      (A) I   (D) IV

      (B) II   (E) None of the above

      (C) III

Questions 172-176.

      (A) S phase

501

(B) Telophase I

(C) Prophase I

(D) phloem

(E) interferon

For questions 172-176 choose a letter from the above list that best matches the statements.

172. Both synapsis and crossing-over occur during this stage.

173. The synthesis of DNA occurs during this stage.

174. Cytokinesis produces daughter cells which have the haploid number of double stranded chromosomes during this stage.

175. Consists of sieve tubes and sieve tube members. Its primary function is to conduct products of photosynthesis to other parts of the body.

176. A group of rapidly producing proteins manufactured by infected viral cells.

Questions 177-180.

(A) Pia mater

(B) Baroreceptor

(C) Meissner's corpuscle

(D) Golgi tendon organ

(E) Dura mater

For questions 177-180 choose a letter from the above list that best matches the statement.

177. This menix is closest to the surface of the brain.

178. This receptor detects changes in the blood pressure.

179. This receptor detects light or fine touch.

180. This receptor detects the degree of stretch (and contraction) of a muscle.

For questions 181-183 choose a letter from the list below that best matches the statement.

(A) Respiration

(B) Fermentation

(C) Photosynthesis

(D) Light-independent reaction

(E) Light-dependent reaction

181. For each NADH + H$^+$ molecule entering this chain, 3 ATP molecules are produced.

182. Decomposition of carbohydrates by bacteria, and the production of ethanol in the abscence of oxygen.

183. Chlorophyll a has to be activated or elevated to a higher energy level to offset the occurrence of this reaction.

For questions 184-188 choose a word from the list below that best describes the statement.

(A) Microtubules

(B) Nucleus

(C) Endoplasmic reticulum

(D) Lysosomes

(E) Chloroplasts

184. A spherical or oblong body which controls the activities of the cell.

185. Membrane bound organelle which converts solar energy to chemical energy.

186. Digestive enzymes are found within these spherical or irregularly shaped bodies within the cytoplasm.

187. The basic substance in the cilia and flagella of motile cells.

188. This lipid and protein contained membrane provides transport channels within the cell.

For questions 189–190 choose the letter that describes the statement from the list below.

(A) Active site

(B) Apoenzyme

(C) Coenzyme

(D) Activation

(E) Regulatory enzymes

189. During the enzymatic activity, the substrate binds to a portion of the enzyme.

190. It has been determined that these enzymes have two binding sites.

DIRECTIONS: *The following groups of questions are based on laboratory or experimental situations. Choose the best answer for each question and mark the letter of your selection on the corresponding answer sheet.*

Questions 191–193 relate to the experiment and diagram below.

A certain experiment was conducted by some scientists on a group of people to test variations in the respiratory quotient under conditions of starvation. The results were then graphed. The graph below shows these results.

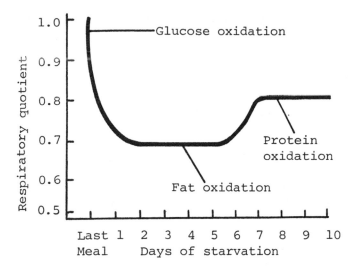

191. After a person has eaten a meal rich in carbohydrates his respiratory quotient (RQ) rises almost to 1.0. This indicates that

   (A) the food is being digested.

   (B) his blood level count increases.

   (C) carbohydrates are being used up.

   (D) glycogen which is a respiratory outlet is being generated.

   (E) All of the above

192. If the same person eats no more in about 12 hours, what will happen?

   (A) His RQ will drop to about 0.7.

   (B) His body will revert to using his fat reserves.

   (C) Protein oxidation will take place, as well as carbohydrate oxidation.

   (D) All three will occur.

   (E) Only A and B will occur.

193. It was noticed that after not eating for about one week a peculiar thing happened. The RQ dropped to a certain level and then rose to about 0.80. This showed that

   (A) the fat reserve was exhausted.

505

(B) the protein of the muscles and other organs were used.

(C) nitrogenous wastes were given off.

(D) A and B only

(E) All of the above are correct.

A couple desiring genetic counseling undergoes genetic typing: The couple's human leukocyte antigen (HLA) karyotypes are:

A1 B7 C4 D8
A3 B7 C3 D4

and A3 B5 C2 D9
A3 B1 C7 D6

The A3 allele has been associated with multiple sclerosis (MS). Assume independent assortment of the HLA alleles.

194. What percent of this couple's children can be expected to develop MS?

(A) 0   (D) 75

(B) 25   (E) 100

(C) 50

195. In later years, it is found that none of the offspring develop MS despite their homozygosity of the recessive allele. This is an illustration of:

(A) incomplete penetrance   (D) chance

(B) mutation   (E) incomplete dominance

(C) selection

196. The following F_1 recombinations occur at the given percentages:

A: A1 B1 C7 D8   27%
B: A3 B1 C3 D4   17%
C: A3 B7 C2 D9   14%
D: A3 B5 C4 D6   42%

Which alleles have a recombination frequency of 41%?

(A) A and B

(D) A and C

(B) B and C

(E) B and D

(C) C and D

Five sections of fresh, small intestinal tissue of a rat were inverted, tied at both ends, filled with concentrated glucose solution, aerated, and dialyzed in flasks of the following media for twenty minutes at 37°C.

#1 Na^+ only

#2 glucose

#3 glucose + Na^+

#4 maltose

#5 maltose + Na^+

197. Which intestinal sac will show the largest increase in concentrations of glucose?

(A) 1

(D) 4

(B) 2

(E) 5

(C) 3

198. The function of the sodium ions is:

(A) as the glucose carrier molecules

(B) as the maltose carrier molecules

(C) the maintainance of osmolarity

(D) the maintainance of tonicity

(E) the maintainance of the electrochemical gradient

199. The process illustrated in the experiment is that of:

(A) active transport

(B) passive transport

(C)   simple diffusion

(D)   facilitated diffusion

(E)   osmosis

E. coli were pipetted into culture tubes of lactose, glucose and lactose + glucose broth using standard techniques. 10 ml of .05% phosphate buffer and 20 µg/ml chlorophenical were added, the tubes centrifuged, the supernatants discarded. An additional 5 ml of buffer were added to resuspend the cells. 5 drops of toluene were added to 1 ml of the suspension which was then incubated at 37°C for twenty minutes. 2 ml of 1% orthonitrophenyl β-D-galactoside was added to each tube. After an additional fifteen minutes of incubation (at 37°C), 3 ml of 2% K_2CO_3 were added to indicate the pH. The following table shows the degree of change in each tube.

Tube	color change
glucose	–
lactose	+++
glucose+lactose	+

200.   β-D-galactoside is required for the synthesis of β-D-galactosidase. In the tube(s) containing which of the following sugars were the E. coli actively synthesizing this enzyme in appreciable amounts?

(A)   glucose

(B)   lactose

(C)   glucose + lactose

(D)   both glucose and lactose, but not glucose + lactose

(E)   both glucose and glucose + lactose, but not lactose

201.   This experiment demonstrates the process of:

(A)   enzyme promotion   (D)   enzyme substitution

(B)   enzyme repression   (E)   enzyme induction

(C)   enzyme operation

202. No change occurred in the tube that contained only glucose due to:

(A) catabolite induction

(D) catabolite substitution

(B) catabolite repression

(E) catabolite operation

(C) catabolite promotion

Dinitrosalicylic acid is used as a test reagent for the determination of glucose in urine. If glucose is present, the solution turns dark brown (in the presence of NaOH). The density of color produced is proportional to the concentration of glucose present in the sample. The change in color density may be measured using a spectrophotometer set at 540 μm.

203. In the reaction, the glucose undergoes:

(A) reduction

(D) hydrolysis

(B) oxidation

(E) dehydration

(C) phosphorylation

204. Why is the wavelength set at 540 μm?

(A) This is the wavelength at which maximum absorption occurs.

(B) This is the wavelength at which minimum absorption occurs.

(C) This is the wavelength at which minimum transmission occurs.

(D) Both (A) and (B)

(E) Both (A) and (C)

205. Glucose's property as a(n) _____ is responsible for its activity in this reaction.

(A) aldose

(D) alcohol

(B) ketose

(E) weak electrolyte

(C) hexose

Four strains of bacteria, Salmonella, were tested for their ability to grow in the absence of certain amino acids. Petri dishes with thin layers of agar containing LB medium (a rich nutritious medium for bacterial growth), minimal medium (containing carbon, nitrogen, salts and other basic essential elements), minimal + Leu medium, minimal + Pro, minimal + Tyr were prepared. Each of these four strains of Salmonella were streaked on each of the four quadrants of a petri dish. This step was repeated for each of the different media tested. The seeded media were incubated at 37°C overnight. Results are shown in the following table.

Media

Bacterial strains	LB	Minimal	Minimal + Leu	Minimal + Pro	Minimal + Tyr
K	+	+	+	+	+
L	+	-	+	-	-
M	+	-	-	+	-
N	+	-	-	-	+

+: growth
-: no growth

206. What can one conclude from the above experiment?

(A) Salmonella strain K is a wild type.

(B) Strains L, M and N are mutants which depend on exogenous amino acids for growth.

(C) Leucine is an essential amino acid for mutant Salmonella L.

(D) Tyrosine is a non-essential amino acid for mutant Salmonella N.

(E) Only A, B and C are correct.

207. The purpose of including Salmonella strain K in this experiment is:

(A) to show that strain K is a wild type.

(B) to serve as a control.

(C) to serve as a reference for the rate of growth of the other strains.

(D) to check for possible reversion.

(E) None of the above

208. What can one infer from the growth phenomenon of mutant Salmonella L, M and N?

(A) Mutations occur on the enzymes involved in the biosynthesis of these exogenously required amino acids.

(B) Mutations give rise to mutant Salmonella strains' inability to utilize endogenous amino acids.

(C) There exists an abnormal degradation system that breaks down those particular endogenous amino acids.

(D) The cell membrane of these mutant Salmonella bacteria is impermeable to any amino acids except the one they depend on.

(E) None of the above

209. From the above table, which strain(s) of bacteria can be considered fastidious?

I. L   II. M   III. N

(A) I only   (D) II and III

(B) I and III   (E) I, II and III

(C) II and I

210. It was noticed that after incubation of the bacteria at 37° overnight, all of the strains registered growth on the medium LB. This was most likely because LB was a/an

(A) Amies medium.   (D) transport medium.

(B) selective and differential   (E) None of the above
    medium.

(C) selective medium.

THE GRADUATE RECORD EXAMINATION BIOLOGY TEST

MODEL TEST V

ANSWERS

1.	A	20.	C	39.	A
2.	D	21.	E	40.	A
3.	A	22.	A	41.	C
4.	D	23.	A	42.	C
5.	B	24.	E	43.	B
6.	C	25.	A	44.	B
7.	C	26.	B	45.	A
8.	B	27.	A	46.	C
9.	D	28.	E	47.	B
10.	A	29.	D	48.	B
11.	B	30.	B	49.	C
12.	B	31.	C	50.	B
13.	B	32.	D	51.	B
14.	A	33.	B	52.	C
15.	A	34.	C	53.	B
16.	B	35.	B	54.	C
17.	A	36.	C	55.	A
18.	D	37.	A	56.	A
19.	A	38.	A	57.	C

58.	C	89.	D	120.	E
59.	B	90.	C	121.	D
60.	D	91.	A	122.	A
61.	A	92.	E	123.	B
62.	A	93.	B	124.	A
63.	B	94.	B	125.	C
64.	D	95.	A	126.	C
65.	B	96.	A	127.	A
66.	A	97.	C	128.	D
67.	C	98.	E	129.	B
68.	B	99.	E	130.	D
69.	A	100.	E	131.	C
70.	A	101.	B	132.	A
71.	C	102.	C	133.	D
72.	D	103.	A	134.	C
73.	C	104.	E	135.	A
74.	A	105.	C	136.	B
75.	D	106.	D	137.	C
76.	E	107.	C	138.	C
77.	E	108.	C	139.	C
78.	E	109.	C	140.	A
79.	B	110.	B	141.	B
80.	A	111.	A	142.	D
81.	C	112.	B	143.	A
82.	E	113.	B	144.	D
83.	D	114.	E	145.	E
84.	E	115.	A	146.	A
85.	E	116.	E	147.	D
86.	E	117.	C	148.	D
87.	C	118.	C	149.	E
88.	B	119.	B	150.	B

151.	A	171.	C	191.	C
152.	D	172.	C	192.	E
153.	B	173.	A	193.	E
154.	D	174.	B	194.	C
155.	C	175.	D	195.	A
156.	A	176.	E	196.	D
157.	E	177.	A	197.	C
158.	B	178.	B	198.	E
159.	E	179.	C	199.	A
160.	C	180.	D	200.	B
161.	D	181.	A	201.	E
162.	A	182.	B	202.	B
163.	B	183.	C	203.	B
164.	E	184.	B	204.	E
165.	A	185.	E	205.	A
166.	C	186.	D	206.	E
167.	B	187.	A	207.	B
168.	D	188.	C	208.	A
169.	A	189.	A	209.	E
170.	B	190.	E	210.	E

THE GRE BIOLOGY TEST

MODEL TEST V

DETAILED EXPLANATIONS
OF ANSWERS

1. (A)
Meiosis is the process by which the number of chromosomes in the diploid cell or nucleus is reduced by half.

2. (D)
The hydrostatic pressure of the blood within the glomerular capillaries is less than the mean blood pressure in the large arteries of the body. It is usually about 50 mmHg, which is about half the mean arterial pressure.

3. (A)
The apical meristem near the tips of root, stem and branches is important in growth differentiation in these areas of a plant. The meristems in the buds of a stem are responsible for outgrowths of the stem, and the roots give rise to cells that eventually differentiate into all of the cell types present in the cell.

4. (D)
In 3 point gene mapping the occurrence of the first crossover reduces the chances of a second crossover nearby via interference.

5. (B)
Nucleosomes or chromatin subunits are composed of DNA and histones. These histone octamers in eukaryotic cells are wrapped by DNA.

6. (C)
Ribosomes are small particles composed of RNA and proteins. They are found throughout the cell cytoplasm and are sometimes attached to the endoplasmic reticulum. They provide sites for protein synthesis.

7. (C)
During DNA replication the DNA molecule is tightly coiled or supercoiled and is thus under some degree of tension. This tension is relieved by a topoisomerase which uncoils the duplex. The enzyme that enables uncoiling or negative supercoiling is called gyrase or swivelase.

8. (B)
Down's Syndrome is a genetic disorder which is characterized by individuals having small round heads, protruding, furrowed tongues, and an IQ seldom above 70. This disorder is genetically expressed as (47,21+).

9. (D)
Motility in eukaryotic cells is enabled by two microtubules, cilia, and flagella. Pseudopods also enable motility.

10. (A)
The myelin sheath around the processes of the neurons

of the peripheral nervous system is formed by the Schwann cells. These Schwann cells are wrapped around each neuron so that each is several layers thick.

11. (B)
During fertilization, when the sperm reaches the vicinity of the egg, the sperm releases the enzyme hyaluronidase to penetrate the corona radiata to enable contact with the egg.

12. (B)
During development the term used to describe an offspring that has developed to the point at which its species can be identified is fetus.

13. (B)
During development, differentiation is the stage at which the cells and tissues become specialized.

14. (A)
The human body has its own natural pain killers which also act as chemical transmitters. These pain killers, endorphins can be imitated or replaced (if used constantly) by heroin.

15. (A)
Morphogenesis is the process whereby the embryo develops form.

16. (B)
Prophase is the first of the four substages of mitosis. It is characterized by the condensation of chromosomes, disappearance of nuclear membrane and nucleoli, and the migration of centrioles to opposite poles to form the mitotic spindle apparatus.

17. (A)
The cardioacceleratory and cardioinhibitory centers, are centers that control the rate of the heart beat. These centers are located in the medulla of the brain which is specialized as a control center for breathing and heartbeat.

18. (D)
The parasympathetic nervous system mediates responses of "rest and recuperation". It also has ganglion (nerve cell bodies) that secretes acetyl choline.

19. (A)
Sound is collected by the pinna of the outer ear, which channels it into the external auditory canal. This causes the tympanic membrane to vibrate.

20. (C)
Normal human males' chromosomal order is expressed as (46,XY), but if there is a sexual abberrance, that is, underdeveloped testis that fail to produce sperms, and feminine sexual development. The chromosomal order reads (47, XXY). This is an example of Klinefelters' Syndrome.

21. (E)
Proprioception (the kinesthetic sense) is the ability to convey information about the relative position of body parts, location of joints, tendons and muscles.

22. (A)
Increased production of follicle stimulating hormone (FSH) stimulates the development of ovum and follicle.

23. (A)
Corpus lutem which is formed from ruptured follicle cells secretes progesterone and estrogen. Progesterone has the effect of suppressing luteinizing hormone (LH) production.

24. (E)
Scanning electron microscope produces 3-dimensional images. Due to the fact that it is an electron microscope, the image can be viewed on a viewing screen.

25. (A)
Of the 92 naturally occurring elements, six: sulphur, phosphorus, oxygen, nitrogen, carbon and hydrogen (SPONCH), make up 99% of all living matter.

26. (B)
Prokaryotic cells contain a few membrane bound organelles. They do not possess nucleus, so the need for a nuclear membrane does not exist. The prokaryotic cell replaces the nucleus by a nucleoid.

27. (A)
In a mutualistic relationship both species benefit. The algae interspersed among fungi to obtain protection and moisture while providing photosynthetic nutrients to the heterotrophic fungus.

28. (E)
Phenothiazines are one major type of tranquilizers used to treat psychoses. In general, tranquilizers are compounds which act as depressants of the central nervous system with highly selective action on brain function.

There exist drugs that induce relaxation and sleep, which are sometimes used to treat patients with organic and emotional disorders. The barbiturate family is the most common and widely used for this sedative and hypnotic purpose.

Amphetamines belong to the class of anti-depressant drugs. These drugs, as their name implies, are used to treat depression disorders, one of the conditions associated with mental illness. They stimulate the central nervous system, and are used to bring about an increase in alertness, elevation of blood pressure and increase in heart action.

29. (D)
Two species of organism that occupy the same or similar ecologic niche in different geographical locations are termed ecological equivalents.

30. (B)
Ecological succession is the gradual change of a biological community from one stage to another. It slows when it reaches the climax stage, which remains stable for long periods or until there are severe changes in the environment.

31. (C)
Organisms which break down the compounds of dead organisms are called saprophytes (saprobes).

32. (D)
A food chain is most commonly a sequence of organisms that are related to each other as prey and predators. One species is eaten by another, which is eaten in turn by a third and it progresses in such fashion.

33. (B)
Ecology is the study of the total pattern of complex interactions among populations of organisms and between them and their environments. The unit for the study of ecology is ecosystem.

34. (C)
An ecosystem's maintenance or nutrient cycle is manned by producers and decomposers.

The producers, mainly photosynthetic autotrophs, provide nutrients for herbivores who are primary consumers. The herbivores are then eaten by the secondary consumers who are usually carnivores. The producers and consumers then die and become nutrients for the decomposers (bacteria and fungi).

35. (B)
Species is defined as a group of organisms that are closely related structurally and functionally, which interbreeds and produces fertile offspring in a natural environment.

36. (C)
Plants are unable to use pure nitrogen from the atmosphere, so they are aided by two classes of microbes. Bacteria and some cyanobacteria "fix" the nitrogen, that is converts nitrogen to nitrates. The plants can then absorb the nitrates from the soil.

37. (A)
Neisseria gonorrhoea is the etiological agent for gonorrhoea; a sexually transmitted disease. It is a gram negative diplococcus which occurs as pairs of spherical organisms and has the morphological characteristic of cocci bacteria.

38. (A)
Bacterial growth can be measured graphically, with results being represented on a curve (growth curve). There are four phases represented by the growth curve in this sequence; lag phase during which there is no growth. This is the adaptation period for the organisms. After becoming accustomed to their environment the organisms begin to grow (by division). This growth continues until there is a reduction in the supply of essential nutrients or an accumulation of toxic products from the microbial metabolism, which reaches inhibitory levels – exponential phase. The cells' division is then reduced and the cells transcend into a stationary phase, where there is no cell growth.

39. (A)
Genotypic ratio is the genetic makeup of an individual expressed mathematically. This is repsented by symbols.

mating: Bb × Bb

Using the probability method

$$\frac{1}{2}B + \frac{1}{2}b$$
$$\times \underline{\frac{1}{2}B + \frac{1}{2}b}$$
$$\frac{1}{4}BB + \frac{1}{4}Bb$$
$$\frac{1}{4}Bb + \frac{1}{4}bb$$

Genotypic ratio: $\frac{1}{4}BB:\frac{1}{2}Bb:\frac{1}{4}bb$

40. (A)
Phenotypic ratio is the physical appearance of an individual expresed mathematically.

Using the genotypic ratio from the previous problem

Genotype: $\frac{1}{4}BB:\frac{1}{2}Bb:\frac{1}{4}bb$

Phenotype: $\frac{1}{4}$ Brown fur:$\frac{1}{2}$ Beige fur:$\frac{1}{4}$ white fur.

41. (C)
A catalyst is a substance that can speed up or slow down a reaction without itself becoming one of the products.

Enzymes are biological catalysts in that they are composed of protein that are synthesized by cells. They can either speed up or slow down a biological process once they are attached to their appropriate substrate.

42. (C)
Dehydration synthesis is a biochemical process in which two smaller molecules are united to form a larger one with a concurrent loss of one water molecule. The equation shows the unison of two molecules of simple sugar into one molecule of double sugar with the loss of one molecule of water.

43. (B)
T (thymus—dependent) lymphocytes are responsible for cell mediated immunity. They are developed in the bone marrow stem cells, and the thymus gland. That is the stem cells migrate to the thymus to differentiate into T cells.

The T lymphocytes have antibodies that are firmly bound to the cell membrane. These antibodies react with antigens to form the antibody-antigen complex. The T lymphocytes antigens are able to lyse foreign cells and phagocytize them. They are able to recognize incipient cancer cells and destroy them.

44. (B)
The blood plasma contains a number of inorganic cations and anions: sodium (Na^+), calcium (Ca^{++}), potassium (K^+), magnesium (Mg^{++}) [cations], chlorine (Cl^-), phosphate $(PO_4)^{-2}$, sulfate (SO_4^{2-}) and bicarbonate (HCO_3^-) [anions].

The concentration of sodium and sodium bicarbonate determines the osmotic pressure of the plasma. If the concentration of sodium chloride is increased, water would diffuse into the blood by osmosis, increasing its volume, which in turn increases the blood pressure.

45. (A)
Albumins are proteins found in the blood plasma. It gives the blood osmotic pressure and viscosity (thickness and resistance to flow). One of its major functions is to bind toxins and impurities in the blood.

46. (C)
During the clotting process, which is initiated by damaged platelets, several proteins are utilized. The process is concluded with fibrinogen being converted to fibrin (a clot is formed). The protein which enables this conversion is called thrombin.

47. (B)
Globulins are proteins found in the blood plasma. Like the albumins they contribute to the osmotic pressure by regulating the general water balance in the body. These proteins' major functions are to fight infections as antibodies and transport lipids and fat soluble vitamins.

48. (B)
The initiation of the heartbeat originates from a small strip of specialized muscle in the wall of the right atrium called the sino-atrial (SA) node, more commonly called the pacemaker.   This region is influenced by the autonomic nerve fibers, which enable the generation of a rhythmic self-excitatory impulse, thus causing a wave of contraction across the wall of the atria.   This impulse is then transmitted across the atrial walls to a second mass of nodal muscles called the atrioventricular node (AV).   A simultaneous reaction then occurs, the sinoatrial node causes atrial contraction while the atriovertricular node initiates the contraction of ventricles.   Impulses from the atrioventricular node are rapidly transmitted through a strand of specialized muscle in the ventricular septum known as the bundle of His (AV)bundle.   The bundle branches into right and left halves of the heart where an impulse is transmitted via the Purkinje fibers, thus causing a contraction of the cardiac muscles.

49. (C)
Internal respiration takes place throughout the body. There is an exchange of gases between blood and other tissues of the body with oxygen passing from the blood to the tissue cells and carbon dioxide diffusion from the tissue spaces to the blood.

50. (B)
B (Bursa of fabricus) lymphocytes produce antibodies (protein produced by cells) that incapacitates pathogens and are responsible for humoral (blood born) immunity.

51. (B)
Pyrimidines are organic molecules consisting of a single six-cornered structure which contains two nitrogenous bases called thymine and cytosine.

52. (C)
The initiation of blood clotting is done by the damaged thrombocytes or platelets which release the protein thromboplastin.   Thrombplastin in the presence of calcium converts prothrombin to thrombin.   Thrombin in turn acts on the fibrinogen molecules to produce the fibrin clot.

53.   (B)
The electrical impulse that causes a muscle fiber to contract is quickly carried to the interior of the myofiber by transverse tubules (T-tubules); vesicles which run perpendicular to the length of the muscle cell.

54.   (C)
During muscular contraction, the myosin cross-bridges bind to actin to form a functional protein called actinomyosin, which serves as an enzyme in the breakdown of ATP to ADP + Pi.

55.   (A)
During the initiation of muscular contraction the action potentials traveling down the T-tubules and stimulate the sarcoplasmic reticulum to release its stored calcium, which binds to the troponin binding sites.

56.   (A)
When the myofiber is not contracting, ATP molecules attach to the myosin cross-bridges heads.

57.   (C)
The thin myofilaments attached to the Z lines of a sarcomere contain three proteins: a thin protein called actin, a globular protein called troponin which attaches to actin and contains binding sites for Ca^{++} and tropomyosin, a linear protein which is also attached to actin and which covers the myosin binding sites when the muscle is at rest but uncovers them during contraction.

58.   (C)
Tropomyosin, a linear protein attches to actin, and covers the myosin binding sites when the muscle is not contracting.

59.   (B)
In myelinated neurons (the sheath of fatty substance

called myelin is present), axons and dendrites are insulated thus inhibiting the flow of ions between the cell interior and the exterior of the membrane. The only site of action potential occurrence is at the Nodes of Ranvier. Impulses generated at the nodes leap from one node to the next. This conduction is known as saltatory transmission.

60. (D)
After a neuron fires an impulse, the original balance of Na^+ (outside) and K^+ (inside) is restored by the active transport process which enables a cell to maintain a lower concentration of Na^+ inside the cell.

61. (A)
In the inner ear, there are two small, hollow chambers known as the utricle and saccule. Each chamber is linked with sensitive hair cells, upon which rests small crystals of calcium carbonate called otoliths that are involved with the vestibular process of maintaining static equilbriium.

62. (A)
The utricles and saccules are part of the vestibular system. They convey information about static equilibrium; they sense head position at rest.

63. (B)
The organ of Corti has a gel-like membrane that extends over the auditory hair cells and takes part in the deformation of the hairs to allow sound detection. This membrane is called the tectorial membrane.

64. (D)
The cardioinhibitory center is located in the center of the medulla and is stimulated by the aortic baroreceptor which is located in the wall of the aortic arch.

Assume that the composition of air is as follows:

$$N_2 = 78\% \qquad CO_2 = .04\%$$

$$O_2 = 21\% \qquad Other = .96\%$$

65. (B)
If a scuba diver encounters a pressure of 762 mmHg, what would be the pN_2 at this depth?

p means partial pressure

pN_2 = 78% × 762 mmHg

= .78 × 762 mmHg

= 594.36 mmHg

66. (A)
Referring to the previous problem, what would be pO_2 at this depth?

pO_2 = 21% × 762 mmHg

= .21 × 762

= 160.02 mmHg

67. (C)
Blood enters the heart through the superior vena cava, inferior vena cava and the coronary sinus. The blood then enters the right atria where it is squeezed past the tricuspid valve and enters the right ventricle. The backflow of blood is prevented by the tricuspid valve, which opens outward to allow the blood to flow through. Any backward flow of blood causes the closure of the valve.

68. (B)
The heart's contraction is initiated by a tissue which serves as a pacemaker. This strip of specialized muscle in the wall of the right atrium is called the sino-atrial node.

69. (A)

Learning is a persistent change in behavior resulting from experience. Its major factor is memory: long term and short term memory.

70. (A)

Blood clotting is initiated by damaged platelets + calcium → thromboplastin which acts on prothrombin → thrombin → which acts on fibrinogen molecules to produce → fibrin clot, an insoluable mass.

71. (C)

The order of the structural components of roots in a dicot plant from the outer surface to the interior is:

epidermis – cortex – endodermis – pericycle – phloem – xylem

72. (D)

During heating or ionization by acids or alkalis, the hydrogen bonding of base pairs is disrupted and the double helix is melted into single strands. Referring to the graph, a relative absorbance of below 1.1 corresponds to a double helical structure while a value of above 1.4 corresponds to a single stranded DNA. The suddenness of the transition from a double helical structure to a single stranded one as evidenced from the sigmoid curve suggests that the formation of double helix is a cooperative process.

The graph clearly demonstrates that DNA with a higher GC content melts at a higher temperature. This is because each GC base pair is held by three hydrogen bonds which are more stable than the two hydrogen bonds in each AT base pair.

73. (C)

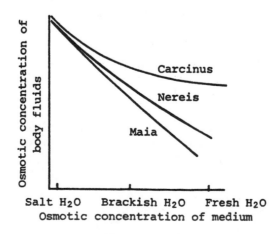

Osmotic concentration of body fluids

Carcinus

Nereis

Maia

Salt H₂O Brackish H₂O Fresh H₂O
Osmotic concentration of medium

From looking at the diagram it is clear that the spider crab (Maia) has no osmoregulatory capacity and that its body fluids fall proportionately according to the fall in concentration of the water. The clam worm (Nereis) has very little osmoregulatory capacity, as obvious by the fact that the concentration of its body fluids does not form a straight line against the osmotic concentration of the medium. The shore crab (Carcinus) however, has considerable osmoregulatory capacity and can maintain concentrated body fluids even in very dilute medium.

74. (A)
During electron transport in the mitochondria, the released energy is directed by components of the respiratory chain to move protons (H^+) from the matrix space to the intermediate space. The consequences of this movement of protons are:

I.   generation of a pH gradient across the inner mitochondrial membrane.

II. generation of a voltage gradient across the inner mitochondria membrane.

III. the pH in the rest of the cell becomes generally closer to 7.

75. (D)
The base sequence along a polynucleotide strand is not restricted. This property allows DNA to be a good candidate to carry genetic information.

76. (E)
Among the 64 codons, only 3, UAA, UAG and UGA, code for chain termination. The probability of having the other 61 codons mutate to these 3 stop codons is much lower than having 20 condons (one for each amino acid for a non-degenerate genetic code) mutate to 44 stop codons. Mutation at the third base of a triplet codon which changes one codon synonym to another does not change the amino acid sequence of a protein. For the same reason, one protein can be coded by many different DNA sequences.

77. (E)
Disulfide bonds, formed through the oxidation of two

cysteine residues, contribute to the 3-dimensional structure of proteins. Specific cleavage of polypeptide chains can be found in the conversion of procollagen into collagen and of proinsulin into insulin. One example of hydroxylation of amino acids is the hydroxyproline and hydroxylysine found in collagen. Insufficient hydroxylation of collagen causes the skin lesions and blood vessel fragility that are typical of scurvy. Finally, attachment of sugars to amino acids such as asparagine and serine causes the formation of glycoproteins.

78. (E)

Recombinant DNA has to be on the range of length from 75% to 105% of a unit λ genome in order to be encapsidated. Therefore the length of the foreign DNA insert is not unlimited. λ phage infects bacteria at a higher frequency than plasmid transformation into bacteria. Large segments of the λ genome can be replaced with foreign DNA leaving its lytic and lysogenic life cycles unaffected.

79. (B)

Chromatography is a commonly used technique for protein fractionation. A mixture of proteins in solution is applied to a chromatography column. Different proteins in the mixture interact with the column matrix at different degrees, and hence, their movement through the column is slowed down to different extents. Proteins can therefore be separately collected as they flow out of the column.

Ion-exchange chromatography utilizes the ionic association between proteins and the charged insoluble matrix. The commonly used matrices are diethylaminoethyl-cellulose (DEAE-cellulose), which is positively charged; and carboxymethyl-cellulose (CM-cellulose), which is negatively charged. Proteins flow out separately as the column is washed with a solution of increasing salt concentration. The tightest binding protein requires the highest concentration of salt in order to dissociate from the charged matrix. Fractions of protein collected is tested for enzyme activity.

Those fractions containing the desirable proteins are pooled and applied to a gel-filtration chromatography column. The matrix in the column is inert but porous so that proteins of different sizes can be separated. Small molecules enter through the spaces in the beads while large molecules stay between the beads. Large molecules are therefore found to flow out more rapidly than small molecules. Again, fractions are collected, tested for activity, pooled and applied to the next chromatography column.

Affinity chromatography makes use of the biological interaction between enzyme and its substrate. An enzyme substrate is immobilized onto an inert matrix such as a polysaccharide bead. Only enzymes with configurations complementary to their substrate remain in the column.

80. (A)
One of the two X chromosomes in each mammalian female cell is permanently inactivated. In mice, it has been found that one or the other of the two X chromosomes is condensed into heterochromatin during the period between the third and the sixth day of development. Heterochromatin is a highly condensed chromatin and is inactive in transcription. Euchromatin is the opposite of heterochromatin. It consists of all of the genome in the interphase nucleus excluding heterochromatin.

The selection of either the paternally inherited X chromosome or the maternally inherited X chromsome to be inactivated is a random process. Once a decision has been made, a clonal group of cells inherits the same inactive X chromosome. This explains why every female is a mosaic with some groups of cells having the paternally inherited X chromosome expressed and other groups of cells having the maternally inherited X chromosome expressed.

81. (C)
Cathode ray oscilloscope is an instrument that can follow instantaneous potential changes. Since signals given off by living cells are small, they are amplified before being fed into the cathode ray oscilloscope.

Electrophoresis is a method used to separate proteins based on the migration of charged molecules to the poles of an electric field. Proteins with different isoelectric points migrate differently in an electric field.

Immunoradioassay is a sensitive method to detect minute amount of chemicals such as nanograms of hormone in the blood. This technique utilizes the competition between the radioactively labelled chemical and the unlabelled one for their antibody binding sites.

Geiger counter is an instrument used to detect the source of and measure the amount of nuclear emission.

82. (E)

The first law of thermodynamics states that the total energy of a system and its surroudnings is a constant and is expressed by the formula

$$\Delta E = E_B - E_B = Q - W.$$

This means that energy is conserved.

Given E_A as the initial energy of the system at the start of the process and E_B as the final energy of the system at the end of the process, we can conclude that the change in the energy of the system (ΔE) is dependent on the initial and final state and not on the path of transformation.

83. (D)

Covalent modification is a control mechanism which is augmented by the insertion of a small group on an enzyme.

Allosteric interactions enable enzymes to detect certain structural changes and respond to them. They cause non-competitive inhibition and control reversible reactions in a pathway. Both covalent modification and allosteric interactions control some regulatory enzymes. The enhancing of the catalytic activity of glycogen phosphorylase and the diminishing of the catalytic activity of glycogen synthetase are controlled by covalent modification and allosteric interactions.

84. (E)

[Each answer provides a correct description of the desert biome.] The desert consists of all the characteristics given in answers A through D. Therefore, answer E (all of the above) is correct.

85. (E)

All of the conditions given must apply in order for the gene pool to remain in equilibrium. As long as these conditions are met, the Hardy-Weinberg law will apply.

86. (E)

The need for NADPH, ribose 5-phosphate and ATP

initiates the flow of glucose 6-phosphate in the four different modes of pentose phosphate pathway.

87. (C)

Perception of any stimulus involves the nervous system. There are however no nervous connections to the endocrine glands such as the gonads. But nervous tissue is capable of hormonal secretion and the endocrine glands are stimulated by chemical messengers directly from the nervous system.

In the situation where the animal detected the change in day length, the hypothalamus is the part of the brain most involved. The hypothalamus is located above the pituitary and they are connected by their blood supplies. Thus the animal detected the light with its sense organs; nervous impulses were transmitted from the sense organs to the hypothalamus, the hypothalamus secreted gonado-tropic releasing hormones (GnRH) into the bloodsecretion of gonadotropic hormones which in turn stimulate the gonads to secrete more sex hormone. The sex hormone helps prepare the animal to commence activities appropriate for the breeding season.

88. (B)

In most plants the stomata are located mainly in the lower eipdermis. The lower epidermis is the side of the leaf usually turned away from the sun's rays where the drying tendency is less harsh. It is also covered with short hairs which reduce direct air currents across the stomatal openings.

The oleander (Nerium) which lives in a very dry habitat has stomata which are located in deep hair-lined depressions in the lower epidermis.

89. (D)

Enzymes control the myriad chemical reactions within living organisms and are themselves controlled by a variety of mechanisms. Competitive inhibition is a form of enzyme control which involves an enzyme substrate that is structurally similar to the normal substrate of the enzyme, which binds reversibly to the enzyme's active site, but which differs from the substrate by not being chemically changed in the process. By binding with the site the inhibitor (I) masks the site and prevents the normal substrate molecules from gaining access to it. This can be demonstrated by the reactions below

E + I = EI competes with the reaction

E + S → ES → E + P.   Both involves the same enzyme which is present in small amounts.

Malonic acid and succinic acid both compete for binding with the enzyme succinic acid dehydrogenase.   This is termed competitive inhibition.

90. (C)
Assuming that the parental types followed the rule of Dominance/Recessiveness, and labeling the true breeding round seeds, WW; and true breeding wrinkled seeds ww we see that

$$WW \times ww = P_1 \text{ generation}$$

Using the Punnet square to find the F_1 generation we derive

	W	W
w	Ww	Ww
w	Ww	Ww

All the F_1 generation was round.

91. (A)
If two seeds of the F_1 generation are crossed (Ww × Ww), the results would be:

	W	w
W	WW	Ww
w	Ww	ww

The phenotype of the F_2 generation would therefore be 3/4 round and 1/4 wrinkled.

92. (E)

Piloerection, vasoconstriction and vasodilation are all ways in which a living body can respond to unsuitable environmental temperatures. Piloerection may aid as an insulator in birds and mammals. The hair or feathers are raised above the skin by contraction of erectile muscles, thus producing air spaces which serve as insulation.

Vasoconstriction and vasodilation are under the control of the sympathetic fibers, of the autonomic nervous system. The degree of dilation or constriction determines the extent of the blood supply to tissues. A major form of heat loss from the body is by radiation from the blood in the superficial capillaries of the skin; altering the amount of blood flow by constriction and dilation of the capillaries help to regulate heat loss.

93. (B)

One example of negative regulation in gene expression is the lac operon in E. coli. Lactose repressor is a regulatory protein that turns off gene expression of the lac operon. It carries out its function by binding to the operator which overlaps the promoter. Thereby, the binding of lactose repressor blocks the access of promoter to RNA polymerase and prevents transcription. The lactose repressor is itself regulated by a signaling ligand called allolactose. Allolactose, when it reaches a high enough concentration in the cell, binds to the lactose repressor and causes a conformational change and a loosing off of the lactose repressor from the operator.

Positive regulation is the opposite of negative regulation in that higher rate of gene transcription takes place in the presence of the regulatory protein than in its absence. Both positive and negative regulation can be grouped under the category of allosteric regulation because both involves regulatory proteins that are themselves regulated by signaling ligands that change the conformation and hence, the affinity of regulatory proteins to specific DNA sequences.

94. (B)

In the linearization of a circular plasmid, a phosphodiester bond is broken leaving a phosphate at the 5' end and a hydroxyl group at the 3' end of each DNA strand. These 5'-phosphate and 3'-hydroxyl are capable of reforming a phosphodiester bond in the presence of ligase unless the 5'-phosphate is removed by alkaline phosphotase.

Linker is an oligonucleotide fragment of duplex DNA containing a recognition site for some restriction endonuclease. It is used commonly in the reconstruction of recombinant DNA.

95. (A)

Any single factor such as the amount of sunlight, temperature, humidity, or the concentration of trace elements is capable of determining the presence or absence of species and thus the characteristics of the entire biome. The Law of Minimum is a generalization that is sometimes used to explain this phenomenon. It states that the factor that is most deficient is the one that determines the presence and absence of species. It does not matter, for example, how favorable temperature and sunlight are in a given locality, or how rich the nutrients and trace elements of the soil are if the precipitation is very low, because the result will still be a desert.

96. (A)

The earth is a closed system for materials because it does not receive matter from space, nor does matter escape the earth's atmosphere. The earth is an open system for energy – life on this planet depends upon the input of light and heat from the sun.

97. (C)

Polysome, which is the same as polyribosome, is a structure with a group of ribosomes bound to an mRNA. Peroxisome is an organelle present in eucaryotic cells which contains several oxidative enzymes such as D-amino acid oxidase, urate oxidase, and catalase. 'Polyoma' is taken from the name of polyoma virus. This virus contains a small circular DNA duplex within an icosahedral coat.

98. (E)

There is no trick here but to memorize the proper hierarchy.

99. (E)

Any step involved in the synthesis of protein is subjected

to control. DNA carries the genetic information of how a protein is made. This information however needs to be transcribed and translated. The first level of control rests on how or when a gene is transcribed. Newly formed transcripts should be processed, with the addition of a poly A tail to the 3' end and a methylated G nucleotide (called a cap) to the 5' end of mRNA molecules in eucaryotes. Following transcription and processing, a transcript is selected to be transported across the nuclear membrane to the cytoplasm. Once in the cytoplasm, mRNA is either translated by ribosomes or degraded. Various levels and accumulation of different sets of proteins in different cell types although the same genome is present in every cell.

100. (E)

In poikilothermic animals and plants the normal metabolic rate is inversely proportional to the size of the body. The smaller the animal the higher the metabolic rate. This is because the smaller the animal, the larger the surface–volume ratio; this causes an increase in heat loss and consequently an increase in the relative metabolic rate.

101 & 102. (B), (C)

The following chart shows the official categorization of Era, Period and Epoch.

Era	Period	Epoch
Cenozoic	Quaternary	Recent Pleistocene
	Tertiary	Pliocene Miocene Oligocene Eocene Paleocene
Mesozoic	Cretaceous Jurassic Triassic	
Paleozoic	Permian Carboniferous Devonian Silurian Ordovician Cambrian	
Precambrian		

103. (A)
Isomers are compounds that have the same molecular formula but different structural formulas.

104. (E)
Bacterial DNA, being circular, has no free end for rotation as DNA helix unwinds during transcription. Positive supercoils accumulate and hinder further unwinding of DNA helix and therefore transcription. This problem is solved by DNA gyrase which introduces negative supercoils into the DNA and causes the circular bacterial DNA to be under tension. Local unwinding of a region of DNA helix by RNA polymerase introduces positive supercoils which in effect cancels out negative supercoils formed by DNA gyrase. Relaxation of tension is an energetically favorable event which enhances the transcription of DNA.

Regulatory proteins can be positive and negative. Negative regulatory protein turns genes off and prevents transcription. It is therefore a wrong choice.

Z-form DNA is a duplex DNA that has a left-hand double helical structure. The effect of Z-DNA on local transcription remains a mystery.

105. (C)
Succession is a fairly orderly process of changes of communities in a region. The change in species composition is continuous, but is usually more rapid in the earlier stages than in the later ones. The number of species, the total biomass in the ecosystem, and the amount of non-living organic matter all increase during the succession until a more stable stage is reached. Toward the final stages, the food webs become more complex, and the relationships between species in them become better defined.

106. (D)
All ecosystems have producers (plants), consumers (animals) and decomposers (bacteria). These elements keep the cyclic balance in check. Parasites and commensals are organisms which are involved in very specific relationships with other organisms. Parasites use another animal as a host, deriving nourishment and sometimes shelter from it. All parasites are harmful to their hosts. Commensals derive benefit from association with another species without harming

that other species.   Neither commensals nor parasites must be present in any given ecosystem.

107.   (C)
An amino acid molecule consists of an amine group, carboxyl group and a variable group all attached to the same carbon.

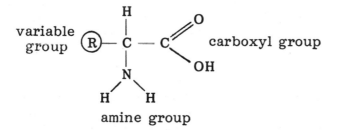

108.   (C)
Uniform and random distributions of population patterns are rare.   Uniform distributions occur in places where the environmental conditions are fairly uniform and competition among individuals is severe.   Random distributions also occur in places with uniform environmental conditions but there is no severe competition among individuals and no tendency for individuals to group together.   Clumped distributions are the most common since environmental conditions are usually non-uniform.   Organisms select the most favorable habitat to live in.   Some animals group together as colonies, schools, or herds which have the advantage of being able to withstand predators.

109.   (C)
The first event in visual excitation is the isomerization of the chromophore 11-cis-retinal by light.   This cis-retinal is converted to all trans-retinal when visual purple (the principal visual pigment) absorbs a photon of light.

110.   (B)
Regardless of the mechanism, enzymes are thought to increase the rate of a reaction by lowering the required energy of activation.

111. (A)
Membrane lipid molecules are amphipathic, that is they contain both hydrophilic and hydrophobic ends. The three major classes of membrane lipid molecules are phospholipids, cholesterol and glycolipids. Different lipid compositions on the inner and outer monolayers contribute to the asymmetry of plasma membrane. Particularly, glycolipids present only in the outer monolayer with their sugar groups exposed at the cell surface. The function of these oligosaccharide side chains is unknown. But their structural complexity and positions on cell surface may suggest a role in intercellular communication. Phosphatidyl ethanolamine, phosphatidyl serine, and sphingomyelin are under the category of phospholipids.

112. (B)
Cofactors are the non-protein part of enzymes. They are derivatives of minerals represented by small inorganic molecules that are bound with enzymes during chemical reactions to make the enzymes functional.

113. (B)
Lipid bilayers tend to close on themselves so that no hydrocarbon chains are exposed to the aqueous medium. As the lipid bilayer closes, water molecules are released from the hydrocarbon chains and a favorable increase in entropy is resulted in the system. Van der Waals forces between hydrophobic tails and electrostatic interactions between the polar head groups lower the energy content of the system which is also favorable.

114. (E)
Bilayer adherence and bilayer joining are membrane phenomena seen in electron micrographs during cell division, cell fusion, endocytosis and exocytosis. Cell fusion occurs during the fertilization of sperm and egg. Endocytosis is a kind of cell ingestion by which extracellular fluid with dissolved macromolecules is taken up. In exocytosis, the content of the intracellular vesicle is secreted to the extracellular environment. Plasmolysis is the shrinkage of plant protoplast due to the osmotic loss of water.

115. (A)
Membrane proteins exposed at the cell surface can be

identified by vectorial labeling technique. The commonly used reagent is lactoperoxidase which oxidizes radioactive iodide to radioactive iodine in the presence of hydrogen peroxide. Radioactive iodine binds to the large enzyme molecule and prevents penetration of the cell membrane, to label proteins exposed at the cytoplasmic side of plasma membrane. The limitation of this reagent is that only membrane proteins with tyrosine or histidine residues on the exposed parts of protein become iodinated.

Freeze fracture electron microscopy is employed in the study of the hydrophobic interior of membranes and intramembrane particles.

2-d polyacrylamide gel electrophoresis is used for isolation and separation of all constituent membrane proteins in the plasma membrane.

Scanning electron microscopy gives a 3-d image reflected from the surface of the specimen. The size range of specimen is limited to single cell and small organisms.

116. (E)
Danielli's basic model of the cell membrane structure postulated that lipids, proteins and glycocalyx (glycoprotein) are the components of an animal cell's outer membrane surface directly exposed to the extracellular fluid.

117. (C)
In actively secreting cells, vacuoles may form, be carried to the plasma membrane, fuse with it, then release their contents to the extracellular environment by a process called exocytosis.

118. (C)
Sugar passes through the plasma membrane by combining with a protein carrier molecule (a permease). This combination is then able to pass through the membrane from a region of higher to one of lower concentration of sugar without the use of cellular energy. This process is called facilitated diffusion.

119. (B)
Centrioles are organelles found in animal cells that give rise to spindle fibers and assist in cell division.

120. (E)
Competition is the chief density-dependent limiting factor, because a limited supply of resources can only support a population of certain size. Organisms of the same species having the same living requirements experience competition. Organisms of different species can also experience competition, but the intensity of competition depends on the similarity of their living conditions. Physiological mechanism is also a density-dependent limiting factor since a density-reduced resistance to pathogens and parasites has been evidenced. Environmental changes such as flooding, fire and abnormal weather changes affect the population size regardless of population density.

121. (D)
The mitochondrial electron transport system's electron flow ends with oxygen combining with hydrogen to form water.

122. (A)
Angiosperms are divided into monocots and dicots according to five major structural differences. Answer (A) was correct: The pattern of veins in the leaf is netlike in the dicot and parallel in the monocot. Answers (B) through (E) had the characteristics of dicots and monocots reversed.

123. (B)
In protein-driven chemiosmotic phosphorylation, ATP energy is generated as protons pass from a region of high concentration to one of low concentration through CF-1 particles, complex structures of the thylakoid membranes.

124. (A)
The citric acid cycle (Krebs cycle) occurs in the mitochrondria. As reactions of the cycle occur carbons

are removed in the form of CO_2. Most of its NAD is reduced.

125. (C)
Non-random reproduction upsets the Hardy-Weinberg equilibrium and causes changes in gene frequencies of a population, it is a universal rule and is not only applied to geographically separated populations.

126. (C)
In the P680 photosystem, light energy excites electrons in antenna pigments in the photosynthetic unit. The excitation is passed along to a molecule of P680 chlorophyll. The P680 chlorophyll then picks up two electrons derived from water.

127. (A)
The most definitive cause of succession is a change in the physical environment, brought about by living organisms, such as a depletion of certain essential nutrients, pH change of the soil, and physiographic changes such as flooding. Climatic changes take a part in selecting what kinds of species will survive and prosper in a certain area. Succession, however, still occurs even when the climate remains the same. Environmental toxification may reach the point where no organisms can grow in the affected area.

128. (D)
Electrons lost from the P700 reactions center of photosynthesis are replaced by electrons that escaped from the P680 chlorophyll.

129. (B)
Light reactions of photosynthesis occurs in the thylakoid disc.

130. (D)
Hemoglobin is a protein that is found in the blood. It is a component of red blood cells (erythrocyte). When combined with CO_2 it forms a molecule called carbaminohemoglobin.

131. (C)
The genetic experiment which was instrumental in establishing that DNA molecules are reproduced by semiconservative replication was carried out by Messelson and Stahl.

The semiconservative mechanism showed that each replicated DNA molecule would consist of an "old" and a "new" strand.

132. (A)
The exchange of gases in fish gills is made more efficient by a countercurrent device. The capillaries of the gills are arranged so that the direction of blood flow is opposite to the direction of water movement. Because the blood that crosses the capillary continually encounters a new supply of water with O_2, the O_2 gradient corresponds to the movement of oxygen into the blood. O_2 can consequently enter the circulatory system until the blood leaves the capillary. Carbon dioxide removal is similar but opposite in concentration gradient between blood and water with respect to the oxygen exchange. The moving current of water thus never becomes saturated, and therefore permits a more efficient exchange.

133. (D)
The random fluctuation of gene frequencies due to chance processes in a finite population is known as genetic drift. Genetic drift may occur by what is commonly referred to as the Founder Principle. This principle operates by chance factors alone. If a group of individuals becomes isolated because of either migration or natural catastrophe, the allelic frequencies of that group may be different from the original population, and eventually a new species may emerge.

This is not natural selection because the survival of the smaller group is due to chance, and not due to being "selected" because of their fitness.

134. (C)
This theory of evolution is most often identified with Jean Baptiste de Lamarck (1744-1829), who was one of its more prominent supporters in the early 1800's.

The Lamarckian hypothesis was that somatic characteristics acquired by an individual during its lifetime could be transmitted to its offspring. This theory has now been disproven, and the theory of natural selection is predominant, held widely as fact, not theory.

135. (A)
This phenomenon is known as Batesian mimicry. The palatable species benefits because it is spared from predation by other organisms who have sampled the unpalatable species.

Müllerian mimicry involves the evolution of similar appearances by two or more different species who have different defense mechanisms. They both benefit because predators need to learn only one avoidance response rather than two (or more). Therefore, the predator learns more quickly, and such species are avoided. Müllerian mimicry is therefore mutually beneficial.

136. (B)
Fermentation is the breaking down of glucose in the absence of oxygen into alcohol and carbon dioxide.

137. (C)
Conifers are a sub group of gymnosperm. They vegetate vast areas of the temperate and subarctic regions. These bare-seeded plants are characterized by their ability to remain green, that is evergreen. They enable the production of paper, plastic and wood.

138. (C)
The light reaction phase of photosynthesis which occurs in the thylakoid concludes with the yielding of the following products: O_2, ATP, NADPH.

139. (C)
Restriction endonuclease is a group of enzymes that recognizes specific short sequences of unmethylated DNA and cuts DNA at either the recognition site or elsewhere. Type I restriction endonuclease recognizes specific sequence but does not cut within sequence while type II restriction endonuclease cuts within recognition sequence. For example, the restriction site for ECoRI is a palindromic sequence of 6 bases, 5'-GAATTC
                      CTTAAG-5'.

Ribonucleases are enzymes that have the ability to cleave phosphodiester bonds in RNA. They participate in a variety of reactions such as the maturation of tRNA, rRNA and the degradation of unwanted RNA.

Topoisomerases are enzymes that have the ability to break the DNA strand, rotate it and seal the break again. The effect of this process is a changing of the linking number in DNA thus forming topological isomers.

140. (A)
The 70S ribosome in procaryotes consists of a 30S and a 50S subunit. The 30S subunit is made up of a 16S rRNA with about 21 proteins while the 50S subunit consists of a 5S rRNA, a 23S rRNA and about 34 proteins. The 16S rRNA has near its 3' end a short base sequence complementary to the purine-rich region in the initiation site upstream of the start codon of mRNA. The binding of initiator tRNA to the mRNA-30S ribosomal subunit forms an initiation complex for the subsequentphases of protein synthesis.

141. (B)
The cDNA is a single stranded DNA synthesized by reverse transcriptase in vitro using mRNA as template. Eucaryotic genes containing non-coding intervening sequences are unsuitable for putting into bacteria for expression since bacteria do not have the machinery to remove intervening sequences. Though the other choices contribute to some extent to the success of having eucaryotic genes expressed in bacteria, they are not the most important ones.

142. (D)
All known bacterial mRNA has start and stop signals to define the length of polypeptide to be synthesized. The

start signal is read by the initiator tRNA which brings formylmethionine to the first amino acid position of a polypeptide chain. Initiator tRNA is different from the tRNA that puts methionine to internal positions on a polypeptide chain even though methionine is linked to these two kinds of tRNA's by the same aminoacyl-tRNA synthetase. Transformylase is the enzyme that transfers an activated formyl group from N^{10}-formytetrehydrofolate to methionyl-inititor tRNA.

143. (A)
The aim of this experiment was to ascertain the conditions that are necessary for photosynthesis. Test tube II was placed in light which allowed carbon dioxide to be absorbed by the leaves of the Elodea plant. Carbon dioxide's absence in the surrounding water solution, caused the bromthymol to turn blue. If it was present (carbon dioxide), the color of the solution would have remained yellow.

144. (D)
The biological membrane, like a 2-dimensional fluid structure, allows the constituent lipid and protein molecules certain kinds of movement in the plane of a membrane. Lateral diffusion refers to the exchange of positions among neighboring molecules. Rotational diffusion refers to rotation about a long axis perpendicular to the plane of a membrane. Transverse diffusion or flip-flop is the transition of membrane molecules from one monolayer to the other. Flip-flop is rare for lipid molecules and even unobserved for membrane proteins because of a high energy barrier posed by a hydrophobic and hydrophilic interaction. The inability for protein molecules to flip-flop helps preserve the membrane asymmetry.

145. (E)
Many terrestrial animals excrete uric acid or its salts which allows them to conserve water. These animals lay eggs which are enclosed in a relatively impermeable shell with four different membranes inside: amnion, allantois, yolk sac and chorion. Allantois is specialized in receiving and storing urinary waste since the embryo is possible to be poisoned when exposed to these wastes.

In the nitrogen metabolism of fully terrestrial animals, uric acid excretion is complementary with egg laying while urea excretion is complementary with viviparity (giving birth to live young).

146. (A)

Pavlov studied the hormone secretin in dogs. This hormone is secreted from the duodenum of the small intestine in response to the arrival of acid food. It signals the pancreas to release bicarbonate ions that neutralize this acidity.

147. (D)

At the lungs the partial pressure of oxygen is about 98 percent saturated with oxygen. It is however only 58 percent saturated at the partial pressure of oxygen typical of tissues at rest. The difference (40 percent) which is between the percent of the partial pressure of oxygen at the lungs and the percent of the partial pressure of oxygen typical of tissues at rest therefore represents the approximate percentage of oxygen carried by hemoglobin that is actually released to the tissues.

148. (D)

In the figure two different dissociation curves are given for human hemoglobin, one for the carbon dioxide concentration in arterial blood and the other for the carbon dioxide concentration in venous blood. An increase in CO_2 concentration shifts the curve to the right thus indicating that a higher partial pressure of oxygen is needed to load the hemoglobin and, that oxyhemoglobin releases oxygen more readily. The release of O_2 is facilitated by the waste of CO_2 which is picked up by the blood in the capillary beds of the tissues at the same time O_2 is released.

For questions 149–153, refer to the graph below.

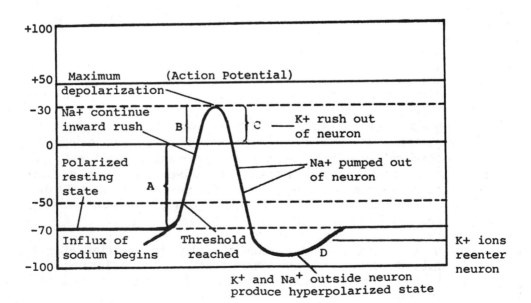

548

149. (E)
The resting potential of a nerve cell is normally about -70mV. If positive charges are added to the inside of the membrane the voltage across the membrane will become more positive, i.e. less negative than the resting membrane potential, or upon receiving a stimulus from another cell would be depolarized.

150. (B)
Nerve cells upon stimulation convert their membrane potentials to action potentials. This is offset by a stimulus of about -60 millivolts.

151. (A)
After achieving maximum depolarization the increased sodium permeability of the membrane is rapidly shut off, and an almost immediate outpour of K^+ follows, i.e. the membrane's permeability to K^+ suddenly increases.

152. (D)
Hyperpolarization is the stage at which the membrane becomes more negative than its resting level. The cell's permeability to potassium ions is increased for a longer period thus resulting in the excess loss of potassium ions.

153. (B)
Depolarization of the membrane beyond the threshold levels leads to an opening of sodium channels. The membrane's increased permeability to Na^+ enables it to diffuse into the cell. This is due to the great electrochemical gradient across the membrane. The addition of sodium further depolarizes the membrane, and so more sodium channels are opened. This positive feedback between depolarization and sodium addition leads to a very rapid and great change in membrane potential, from about -60 millivolts to +30 millivolts, in about one millisecond. The entrance of sodium ceases at +30 millivolts because this is its equilibrium level.

154. (D)
Magnesium is the inorganic component of chlorophyll. As a cofactor of enzymes it is an important accompaniment for their activity in carbohydrate metabolism.

155. (C)
More potassium is found intracellularly where it maintains the resting membrane potential of the cells. As a cofactor of enzymes it is an important accompaniment for their activity for protein synthesis and carbohydrate metabolism.

156. (A)
Nitrogen is the element in the amino group of amino acids, the subunits of proteins. As one of the four most common elements of living matter, it is found in many molecules of life.

157. (E)
Calcium is found in the channels of the cell membranes where it influences the passage of molecules through these gateways. It is found in the middle lamellae of the plant cell wall.

158. (B)
Phosphorous is the part of the phosphate group in nucleotides, the subunits of nucleic acids. A phosphate group replaces one of the three fatty acids in a triglyceride when it is changed into a phospholipid. This element is also part of some coenzymes.

159–163.
159. (E) 160. (C), 161. (D), 162. (A), 163. (B)

The answers for questions 159-163 are given in the questions themselves. [Each biome is described and categorized in the question which corresponds to it.]

164. (E)
Plants require a lot of water for their existence. This is due to their ability to rapidly evaporate water absorbed. The process used is transpiration.

165. (A)
During embryonic development, after numerous cell divisions, a stage is attained where the embryo resembles a hollow fluid filled wall that is one cell layer thick — blastula.

166. (C)
During embryonic development, the embryo is covered by four extraembryonic tissues. The chorion, amnion, yolk sac and allantois.

The amnion is a tissue from the inner cell mass that develops into the "bag" or membrane of the "bag of waters"; the fluid in which the baby develops is called amnionic fluid or "waters".

167. (B)
During fertilization male and female haploid gametes fuse to form a diploid single-celled fertilized ovum called a zygote.

168. (D)
Corona Radiata is a dense covering of follicle cells that encompass the egg at ovulation. These cells are held by hyaluronic acid until penetrated by the enzyme hyaluronidase, which allows fusion of sperm and egg to offset fertilization.

169-171.
    169. (A), 170. (B), 171. (C)
The three protein purification techniques described in questions (169) through (171) all rely on certain properties of specific proteins in order to separate them from solution.

Selective precipitation takes advantage of solubility differences among proteins. The solubility of a protein depends on the relative balance between protein-solvent interactions, which tend to keep it in solution, and protein-protein interactions, which tend to cause it to aggregate and precipitate. By increasing the ionic strength of a solution (done by adding a salt - most commonly ammonium sulfate), proteins are made to precipitate out of solution. This is commonly called salting-out.

Ion-exchange chromatography is a purification technique which takes advantage of proteins' ionic charge. Proteins of various ionic charges will become neutralized at different pH levels. A pH exists for each protein at which its negative charges equal its positive charges. This stage of neutrality is called the isoelectric point.

Ion-exchange chromatography depends on the ionic association of proteins with charged groups bound to an inert supporting material, often cellulose. The cellulose exists as either a positively charged or negatively charged resin; if positively charged, it will bind negatively charged proteins, and if negatively charged, it will attract positively charged proteins. The proteins bound to the resin are thus separated from solution, and subsequently they may be extracted from the resin by increasing the ionic strength of the buffer, thus causing the protein molecules to precipitate.

Gel filtration is a technique that separates primarily by molecular weight, and also by shape. The means for separation is a cylindrical column through which the solution of proteins slowly passes. The most commonly employed materials for gel filtration are beads made of a cross-linked polysaccharide that are perforated by holes of specific diameter These beads allow entry of molecules up to a certain molecular weight and exclude larger molecules or ones with an inappropriate shape. A column of these beads acts as a molecular sieve.

172. (C)
Both synapsis (pairing of homologous chromosomes) and crossing over (point where chromosomes are extremely condensed and form cross-bridges) occur during Prophase I meiosis.

173. (A)
Interphase is the "resting" period in the cell cycle which precedes mitosis and meiosis. This phase is segmented into three sub-phases: G_1, S, and G_2.

During the S sub-phase DNA and chromosomal proteins are replicated.

174. (B)
Cytokinesis produces daughter cells which have the haploid number of double-stranded chromosomes during the telophase I stage.

175. (D)
Phloem consists of elongated, tube-like cells, called sieve-tube members with specialized pores. It transports

carbohydrates and other products of photosynthesis to other parts of the body.

176. (E)
Interferon is a group of non-specific proteins produced by virus infected cells which diffuse to other non-infected cells via the blood. It resists intital infection while antibodies are being produced.

177. (A)
Both brain and spinal cord are wrapped in three sheets of connective tissues known as meninges (singular meninx). The first sheet (pia mater) lies on the surface of the brain and spinal cord. The second sheet, the dura mater, which is a tough membrane is fastened against the bony neural arches of the vertebrate. The third sheet, the arachnoid, is a very fragile membrane which lies between the dura mater and pia mater.

178. (B)
Humans and animals have receptors that regulate pressure in the walls of the arteries and chest. These receptors are called baroreceptors.

179. (C)
Meissner's corpuscles are nerve endings located close to the surface of the skin, that responds to light or fine touch.

180. (D)
The degree of stretch and contraction of flexor muscles induce receptors in the flexor tendons, i.e. Golgi tendon organs to fire an impulse.

181. (A)
Respiration is metabolism of chemical bond energy into one that is biologically usable, i.e. ATP. For each NADH

molecule entering the respiratory chain, 3 ATP molecules are produced.

182. (B)
Fermentation is a process by which organic compounds (carbohydrates) are catabolized to produce acids and alcohol.

183. (C)
Photosynthesis is the ability to transform the energy of the sun into chemical bond energy. This process is initiated by chlorophyll a.

The excited chlorophyll b molecule transfers light energy via high energy electrons to a chlorophyll a molecule. Once activated or elevated to a higher energy level the chlorophyll a molecule initiates the photosynthetic process.

184. (B)
The nucleus, a spherical or oblong body encompasses chromosomes and genes, the ultimate regulators of life.

It monitors changing conditions both inside the cell and in the external environment, and responds to input of information by either activating or inhibiting the appropriate genetic programs.

185. (E)
Chloroplasts are membrane bound organelles found in all photosynthetic cells of green plants. They have the ability to transform the energy of the sun into chemical energy, which is then passed to a chain of other compounds involved in light and dark reations, aided by chlorophylls.

186. (D)
Lysosomes are spherical or irregular shaped membrane bound bodies found in animal cells. All lysosomes are related, directly or indirectly, to intracellular digestion. They contain enzymes known collectively as acid hydrolases. These enzymes can quickly dissolve all the major macromolecules that comprise the cell.

Organic substances in cells are broken down to simpler forms which can be used by the cell to yield energy for its life sustaining process by these digestive enzymes.

187. (A)
Microtubules are thin hollow cylinders composed of tubulin protein. They are often distributed in cells in an arrangement that suggests their role in maintaining cell shape (a "cytoskeletal" role). They are the basic substance in the cilia and flagella of motile cells.

188. (C)
The endoplasmic reticulum is a complex of membranes that traverses the cytoplasm. The membranes form interconnecting channels that take the form of flattered sacs and tubes. It is responsible for transporting certain molecules to specific areas within the cytoplasm. Lipids and proteins are mainly transported by this system. The endoplasmic reticulum is more than a passive channel for intracellular transport. It contains a variety of enzymes playing important roles in a metabolic process.

189. (A)
During enzymatic activity, enzyme (E) combines with substrate (S) to form an intermediate enzyme-substrate complex.

When the substrate binds to the enzyme, it binds specifically with a relatively small part of the enzyme molecule; the active site.

190. (E)
Some enzymes are capable of being inhibited by their own product. These enzymes are called regulatory enzymes. They have two binding sites, an active site and a regulatory site.

191. (C)
The rising of the respiratory quotient (RQ) after eating a meal rich in carbohydrates indicates that carbon dioxide

is being used up.   Carbohydrates are used up first in
respiration.

192. (E)
   If the person who eats the meal rich in carbohydrates
does not eat in about 12 hours, the RQ will drop to near
0.70.   This indicates that he is now using his fat reserves
for respiratory needs.

193. (E)
   After not eating for a week or two, depending upon the
stored fat, the respiratory quotient will be established at
about 0.80.   This indictes that the fat reserve is used up
and that the protein of the muscles and other organs are
being used up thus causing nitrogenous wastes to be given
off.

   The average RQ should be reflective of the utilization
of some of all three classes of food if the diet is balanced.
It will generally fall between 0.80 and 0.90.

194.   (C)
   This is a simple Mendelian cross concerning the A allele.
Consider A1 to be the dominant allele, A3 to be the
recessive.   A standard cross yields 50% heterozygous (no
MS) and 50% homozygous recessive.

195.   (A)
   Penetrance is the percentage of individuals in a population
who carry a gene in the correct combination for expression
and who express the gene phenotypically.   Incomplete
penetrance occurs when genes do not express themselves
as expected.

196.   (D)
   When given the frequency of recombination for alleles as
a whole, relative allelic recombination may be determined
by adding the total recombination frequencies.   For the
given genotypes, A and C recombine at frequencies of 27%
and 14% which sum to 41%.

197. (C)

With the intestine inverted, the enzymes and permeases required for the active transport of glucose across the membrane are in contact with the substrates. Sodium ions are coupled to the glucose molecules and, together, they are transported on the carrier molecule (permease). Maltose is a glucose disaccharide. It must first be hydrolyzed to its glucose subunits before it may be transported. This takes additional time.

198. (E)

Na^+ functions in the coupled transport of glucose across the membrane. It is not, however, the carrier molecule. Instead it functions in maintaining the electrochemical gradient of the cell undergoing glucose transport.

199. (A)

The illustrated process requires energy to transport glucose molecules and maintain the Na^+ gradient. This should be obvious since glucose is being transported from an area of low concentration to one of high concentration.

200. (B)

The presence of a high concentration of lactose and a low concentration of glucose in the environment of E. coli forces the organisms to metabolize lactose, which requires the enzyme β-D-galactosidase. Under normal conditions (high concentration of glucose), the level of cyclic AMP (cAMP) concentration is low. However, whenever the glucose concentration level falls, the concentration of cAMP will increase. cAMP complexes with a catabolic gene activator protein (CAP). This complex binds to the promoter region of the lac-operon on the chromosome, and facilitates the binding of the RNA polymerase to the gene, which will then produce the necessary mRNA to begin enzyme production. In a medium in which both glucose and lactose are present, the organism will preferentially metabolize glucose, because there is not a very large difference between the concentrations of the two sugars.

201. (E)

Enzymes are inducible if they require the presence of a substrate (in this case, lactose) in order for their production to commence. Promotion, repression, and operation of enzymes refer to difference functional areas of the lac-operon. Enzyme substitution never occurs.

202. (B)
Production of the catabolite, β-D-galactosidase, is repressed due to the presence of glucose, which prevents the genetic mechanisms necessary for the production of β-D-galactosidase from acting.

203. (B)
Glucose is a reducing sugar. Any chemical entity that reduces is oxidized in the process.

204. (E)
In spectrophotometry, the wavelength is set at the point of maximum absorption, which is indirectly proportional to the point of minimum transmission of light waves.

205. (A)
Glucose is an aldose, hexose, and weak electrolyte; however, it is the first of these characteristics that is defined in the chemical reaction given. An aldose is a sugar whose carbon number 1 is a carbonyl carbon. A hexose is a six-carbon sugar in the form of a ring. Glucose is more reactive in its aldose form than it is in its cyclohexose form.

206. (E)
The ability to grow on a minimal medium showed that Salmonella strain k is a wild type. Essential amino acids are amino acids not synthesized by the organism and have to be obtained from outside sources.

207. (B)
Salmonella strain k served as a control in this

experiment. With a control, one can be sure that the inability of the mutant Salmonella bacteria to grow in the minimal medium is due to their own mutations. Any other outside influences affect the growth of both wild type and mutant bacteria. A standard wild type strain of bacteria can serve as a reference for the comparison of growth rate for other strains, but growth rate is not a concern in this experiment.

Reversion is a process through which an original mutation is converted to become normal.

208. (A)
Answer (B) is wrong because endogenous amino acids are the same structurally and chemically as the exogenously introduced amino acids. Answer (C) is also incorrect since an abnormal degradation system will also degrade the particular amino acid provided in the medium and no growth will be the result of a supplemented minimal medium. Answer (D) is proved to be wrong by the positive growth of mutant strains in LB medium. Answer (A) is a possible inference and it needs further experiment to prove whether it is correct or not.

209. (E)
All cells require amino acids for manufacturing proteins, nucleotides for making nucleic acids, and vitamins which assist in enzyme-mediated reactions. Some bacteria produce the above components and do not require external assistance (culture medium, etc.). They are able to grow on minimal media.

Fastidious organisms cannot grow on minimal media because they require unusual nutrients.

These growth requirements cannot be synthesized by all organisms, and are thus provided for in a culture medium.

210. (E)
All four strains of the bacteria, Salmonella, registered growth on the LB media because it was an enriched medium which contained the nutrients required to support their growth. Selective medium however allowed only certain bacteria to grow while inhibiting the growth of others, while

transport medium preserved viability without multiplication and showed no noticeable growth. Selective and differential medium, as well as enrichment and differential medium caused growth inhibition of certin bacteria while allowing others to grow.

Amies medium is a transport medium.

GRE

BIOLOGY APPENDIX

CLASSIFICATION OF LIVING THINGS

KINGDOM MONERA

DIVISION SCHIZOMYCETES. Bacteria*

> CLASS MYXOBACTERIA. *Myxococcus, Chondromyces*
> CLASS SPIROCHETES. *Leptospira, Cristispira, Spirocheta, Treponema*
> CLASS EUBACTERIA. *Staphylococcus, Escherichia, Salmonella, Pasteurella, Streptococcus, Bacillus, Spirillum, Caryophanon*
> CLASS RICKETTSIAE. *Rickettsia, Coxiella*
> CLASS BEDSONIA. *Chlamydia*

DIVISION CYANOPHYTA. Blue-green algae. *Gloeocapsa, Microcystis, Oscillatoria, Nostoc, Scytonema*

KINGDOM PROTISTA

DIVISION EUGLENOPHYTA. Euglenoids. *Euglena, Eutreptia, Phacus, Colacium*

DIVISION CHLOROPHYTA. Green algae

> CLASS CHLOROPHYCEAE. True green algae. *Chlamydomonas, Volvox, Ulothrix, Spirogyra, Oedogonium, Ulva*
> CLASS CHAROPHYCEAE. Stoneworts. *Chara, Nitella, Tolypella*

DIVISION CHRYSOPHYTA

> CLASS XANTHOPHYCEAE. Yellow-green algae. *Botrydiopsis, Halosphaera, Tribonema, Botrydium*
> CLASS CHRYSOPHYCEAE. Golden-brown algae. *Chrysamoeba, Chromulina, Synura, Mallomonas*
> CLASS BACILLARIOPHYCEAE. Diatoms, *Pinnularia, Arachnoidiscus, Triceratium, Pleurosigma*

DIVISION PYRROPHYTA. Dinoflagellates. *Gonyaulax, Gymnodinium, Ceratium, Gloeodinium*

DIVISION PHAEOPHYTA. Brown algae. *Sargassum, Ectocarpus, Fucus, Laminaria*

DIVISION RHODOPHYTA. Red algae. *Nemalion, Polysiphonia, Dasya, Chondrus, Batrachospermum*

*There is no generally accepted classification for bacteria at the class level. The system used here is based on R.Y. Stanier, M. Doudoroff, and E. A. Adelberg, *The Microbial World* (3rd ed.; Prentice-Hall, 1970), though these authors do not formally designate their categories as classes. Other systems usually recognize more classes, some of them very small.

KINGDOM FUNGI

DIVISION MYXOMYCOPHYTA. Slime molds

> CLASS MYXOMYCETES. True slime molds. *Physarum, Hemitrichia, Stemonitis*
> CLASS ACRASIAE. Cellular slime molds. *Dictyostelium*
> CLASS PLASMODIOPHOREAE. Endoparasitic slime molds. *Plasmodiophora*
> CLASS LABYRINTHULEAE. Net slime molds. *Labyrinthula*

DIVISION EUMYCOPHYTA. True fungi

> CLASS PHYCOMYCETES. Algal fungi. *Rhizopus, Mucor, Phycomyces*
> CLASS OÖMYCETES. Water molds, white rusts, downy mildews. *Saprolegnia, Phytophthora, Albugo*
> CLASS ASCOMYCETES. Sac fungi. *Neurospora, Aspergillus, Penicillium, Saccharomyces, Morchella, Ceratostomella*
> CLASS BASIDIOMYCETES. Club fungi. *Ustilago, Puccinia, Coprinus, Lycoperdon, Psalliota, Amanita*

KINGDOM PLANTAE

DIVISION BRYOPHYTA

> CLASS HEPATICAE. Liverworts. *Marchantia, Conocephalum, Riccia, Porella*
> CLASS ANTHOCEROTAE. Hornworts. *Anthoceros*
> CLASS MUSCI. Mosses. *Polytrichum, Sphagnum, Mnium*

DIVISION TRACHEOPHYTA. Vascular plants

Subdivision Psilopsida. *Psilotum, Tmesipteris*
Subdivision Lycopsida. Club mosses. *Lycopodium, Phylloglossum, Selaginella, Isoetes, Stylites*
Subdivision Sphenopsida. Horsetails. *Equisetum*
Subdivision Pteropsida. Ferns. *Polypodium, Osmunda, Dryopteris, Botrychium, Pteridium*
Subdivision Spermopsida. Seed plants

> CLASS PTERIDOSPERMAE. Seed ferns. No living representatives
> CLASS CYCADAE. Cycads. *Zamia*
> CLASS GINKGOAE. *Ginkgo*
> CLASS CONIFERAE. Conifers. *Pinus, Tsuga, Taxus, Sequoia*
> CLASS GNETEAE. *Gnetum, Ephedra, Welwitschia*
> CLASS ANGIOSPERMAE. Flowering plants
>> SUBCLASS DICOTYLEDONEAE. Dicots. *Magnolia, Quercus, Acer, Pisum, Taraxacum, Rosa, Chrysanthemum, Aster, Primula, Ligustrum, Ranunculus*
>> SUBCLASS MONOCOTYLEDONEAE. Monocots. *Lilium, Tulipa, Poa, Elymus, Triticum, Zea, Ophyrys, Yucca, Sabal*

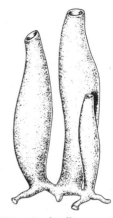

KINGDOM ANIMALIA

SUBKINGDOM PROTOZOA

PHYLUM PROTOZOA. Acellular animals

Subphylum Plasmodroma

CLASS FLAGELLATA (or Mastigophora). Flagellates. *Trypanosoma, Calonympha, Chilomonas* (also *Euglena, Chlamydomonas,* and other green flagellates included in Plantae as well)

CLASS SARCODINA. Protozoans with pseudopods. *Amoeba, Globigerina, Textularia, Acanthometra*

CLASS SPOROZOA. *Plasmodium, Monocystis*

Subphylum Ciliophora

CLASS CILIATA. Ciliates. *Paramecium, Opalina, Stentor, Vorticella, Spirostomum*

SUBKINGDOM PARAZOA

PHYLUM PORIFERA. Sponges

CLASS CALCAREA. Calcareous (chalky) sponges. *Scypha, Leucosolenia, Sycon, Grantia*

CLASS HEXACTINELLIDA. Glass sponges. *Euplectella, Hyalonema, Monoraphis*

CLASS DEMOSPONGIAE. *Spongilla, Euspongia, Axinella*

SUBKINGDOM MESOZOA

PHYLUM MESOZOA. *Dicyema, Pseudicyema, Rhopalura*

SUBKINGDOM METAZOA

SECTION RADIATA

PHYLUM COELENTERATA (or Cnidaria)

CLASS HYDROZOA. Hydrozoans. *Hydra, Obelia, Gonionemus, Physalia*

CLASS SCYPHOZOA. Jellyfishes. *Aurelia, Pelagia, Cyanea*

CLASS ANTHOZOA. Sea anemones and corals. *Metridium, Pennatula, Gorgonia, Astrangia*

PHYLUM CTENOPHORA. Comb jellies

CLASS TENTACULATA. *Pleurobrachia, Mnemiopsis, Cestum, Velamen*

CLASS NUDA. *Beroe*

SECTION PROTOSTOMIA

PHYLUM PLATYHELMINTHES. Flatworms

CLASS TURBELLARIA. Free-living flatworms. *Planaria, Dugesia, Leptoplana*

CLASS TREMATODA. Flukes. *Fasciola, Schistosoma, Prosthogonimus*

CLASS CESTODA. Tapeworms. *Taenia, Dipylidium, Mesocestoides*

PHYLUM NEMERTINA (or Rhynchocoela). Proboscis worms. *Cerebratulus, Lineus, Malacobdella*

PHYLUM ACANTHOCEPHALA. Spiny-headed worms. *Echinorhynchus, Gigantorhynchus*

PHYLUM ASCHELMINTHES

CLASS ROTIFERA. Rotifers. *Asplanchna, Hydatina, Rotaria*

CLASS GASTROTRICHA. *Chaetonotus, Macrodasys*

CLASS KINORHYNCHA (or Echinodera). *Echinoderes, Semnoderes*

CLASS NEMATODA. Round worms. *Ascaris, Trichinella, Necator, Enterobius, Ancylostoma, Heterodera*

CLASS NEMATOMORPHA. Horsehair worms. *Gordius, Paragordius, Nectonema*

PHYLUM ENTOPROCTA. *Urnatella, Loxosoma, Pedicellina*

PHYLUM PRIAPULIDA. *Priapulus, Halicryptus*

PHYLUM ECTOPROCTA (or Bryozoa). Bryozoans, moss animals

CLASS GYMNOLAEMATA. *Paludicella, Bugula*

CLASS PHYLACTOLAEMATA. *Plumatella, Pectinatella*

PHYLUM PHORONIDA. *Phoronis, Phoronopsis*

PHYLUM BRACHIOPODA. Lamp shells

CLASS INARTICULATA. *Lingula, Glottidia, Discina*

CLASS ARTICULATA. *Magellania, Neothyris, Terebratula*

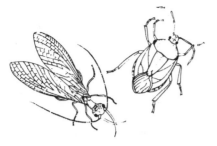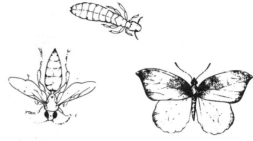

PHYLUM MOLLUSCA. Molluscs

    CLASS AMPHINEURA

        SUBCLASS APLACOPHORA. Solenogasters. *Chaetoderma, Neomenia, Proneomenia*
        SUBCLASS POLYPLACOPHORA. Chitons. *Chaetopleura, Ischnochiton, Lepidochiton, Amicula*

    CLASS MONOPLACOPHORA. *Neopilina*

    CLASS GASTROPODA. Snails and their allies (univalve molluscs). *Helix, Busycon, Crepidula, Haliotis, Littorina, Doris, Limax*

    CLASS SCAPHOPODA. Tusk shells. *Dentalium, Cadulus*

    CLASS PELECYPODA. Bivalve molluscs. *Mytilus, Ostrea, Pecten, Mercenaria, Teredo, Tagelus, Unio, Anodonta*

    CLASS CEPHALOPODA. Squids, octopuses, etc. *Loligo, Octopus, Nautilus*

PHYLUM SIPUNCULIDA. *Sipunculus, Phascolosoma, Dendrostomum*

PHYLUM ECHIURIDA. *Echiurus, Urechis, Thalassema*

PHYLUM ANNELIDA. Segmented worms

    CLASS POLYCHAETA (including Archiannelida). Sandworms, tubeworms, etc. *Nereis, Chaetopterus, Aphrodite, Diopatra, Arenicola, Hydroides, Sabella*

    CLASS OLIGOCHAETA. Earthworms and many fresh-water annelids. *Tubifex, Enchytraeus, Lumbricus, Dendrobaena*

    CLASS HIRUDINEA. Leeches. *Trachelobdella, Hirudo, Macrobdella, Haemadipsa*

PHYLUM ONYCHOPHORA. *Peripatus, Peripatopsis*

PHYLUM TARDIGRADA. Water bears. *Echiniscus, Macrobiotus*

PHYLUM PENTASTOMIDA. *Cephalobaena, Linguatula*

PHYLUM ARTHROPODA

    Subphylum Trilobita. No living representatives

    Subphylum Chelicerata

        CLASS EURYPTERIDA. No living representatives
        CLASS XIPHOSURA. Horseshoe crabs. *Limulus.*
        CLASS ARACHNIDA. Spiders, ticks, mites, scorpions, whipscorpions, daddy longlegs, etc. *Archaearanea, Latrodectus, Argiope, Centruroides, Chelifer, Mastigoproctus, Phalangium, Ixodes*
        CLASS PYCNOGONIDA. Sea spiders. *Nymphon, Ascorhynchus*

Subphylum Mandibulata

    CLASS CRUSTACEA. *Homarus, Cancer, Daphnia, Artemia, Cyclops, Balanus, Porcellio*

    CLASS CHILOPODA. Centipeds. *Scolopendra, Lithobius, Scutigera*

    CLASS DIPLOPODA. Millipeds. *Narceus, Apheloria, Polydesmus, Julus, Glomeris*

    CLASS PAUROPODA. *Pauropus*

    CLASS SYMPHYLA. *Scutigerella*

    CLASS INSECTA. Insects

        ORDER COLLEMBOLA. Springtails. *Isotoma, Achorutes, Neosminthurus, Sminthurus*

        ORDER PROTURA. *Acerentulus, Eosentomon*

        ORDER DIPLURA. *Campodea, Japyx*

        ORDER THYSANURA. Bristletails, silverfish, firebrats. *Machilis, Lepisma, Thermobia*

        ORDER EPHEMERIDA. Mayflies. *Hexagenia, Callibaetis, Ephemerella*

        ORDER ODONATA. Dragonflies, damselflies. *Archilestes, Lestes, Aeshna, Gomphus*

        ORDER ORTHOPTERA (including Isoptera). Grasshoppers, crickets, walking sticks, mantids, cockroaches, termites, etc. *Schistocerca, Romalea, Nemobius, Megaphasma, Mantis, Blatta, Periplaneta, Reticulitermes*

        ORDER DERMAPTERA. Earwigs. *Labia, Forficula, Prolabia*

        ORDER EMBIARIA (or Embiidina or Embioptera). *Oligotoma, Anisembia, Gynembia*

        ORDER PLECOPTERA. Stoneflies. *Isoperla, Taeniopteryx, Capnia, Perla*

        ORDER ZORAPTERA. *Zorotypus*

        ORDER CORRODENTIA. Book lice. *Ectopsocus, Liposcelis, Trogium*

        ORDER MALLOPHAGA. Chewing lice. *Cucletogaster, Menacanthus, Menopon, Trichodectes*

        ORDER ANOPLURA. Sucking lice. *Pediculus, Phthirius, Haematopinus*

        ORDER THYSANOPTERA. Thrips. *Heliothrips, Frankliniella, Hercothrips*

        ORDER HEMIPTERA (including Homoptera). Bugs, cicadas, aphids, leafhoppers, etc. *Belostoma, Lygaeus, Notonecta, Cimex, Lygus, Oncopeltus, Magicicada, Circulifer, Psylla, Aphis*

        ORDER NEUROPTERA. Dobsonflies, alderflies, lacewings, mantispids, snakeflies, etc. *Corydalus, Hemerobius, Chrysopa, Mantispa, Agulla*

        ORDER COLEOPTERA. Beetles, weevils. *Copris, Phyllophaga, Harpalus, Scolytus, Melanotus, Cicindela, Dermestes, Photinus, Coccinella, Tenebrio, Anthonomus, Conotrachelus*

        ORDER HYMENOPTERA. Wasps, bees, ants, sawflies. *Cimbex, Vespa, Glypta, Scolia, Bembix, Formica, Bombus, Apis*

ORDER MECOPTERA. Scorpionflies. *Panorpa, Boreus, Bittacus*

ORDER SIPHONAPTERA. Fleas. *Pulex, Nosopsyllus, Xenopsylla, Ctenocephalides*

ORDER DIPTERA. True flies, mosquitoes. *Aedes, Asilus, Sarcophaga, Anthomyia, Musca, Chironomus, Tabanus, Tipula, Drosophila*

ORDER TRICHOPTERA. Caddisflies. *Limnephilus, Rhyacophila, Hydropsyche*

ORDER LEPIDOPTERA. Moths, butterflies. *Tinea, Pyrausta, Malacosoma, Sphinx, Samia, Bombyx, Heliothis, Papilio, Lycaena*

SECTION DEUTEROSTOMIA

PHYLUM CHAETOGNATHA. Arrow worms. *Sagitta, Spadella*

PHYLUM ECHINODERMATA

CLASS CRINOIDEA. Crinoids, sea lilies. *Antedon, Ptilocrinus, Comactinia*

CLASS ASTEROIDEA. Sea stars. *Asterias, Ctenodiscus, Luidia, Oreaster*

CLASS OPHIUROIDEA. Brittle stars, serpent stars, basket stars, etc. *Asteronyx, Amphioplus, Ophiothrix, Ophioderma, Ophiura*

CLASS ECHINOIDEA. Sea urchins, sand dollars, heart urchins. *Cidaris, Arbacia, Strongylocentrotus, Echinanthus, Echinarachnius, Moira*

CLASS HOLOTHUROIDEA. Sea cucumbers. *Cucumaria, Thyone, Caudina, Synapta*

PHYLUM POGONOPHORA. Beard worms. *Siboglinum, Lamellisabella, Oligobrachia, Polybrachia*

PHYLUM HEMICHORDATA

CLASS ENTEROPNEUSTA. Acorn worms. *Saccoglossus, Balanoglossus, Glossobalanus*

CLASS PTEROBRANCHIA. *Rhabdopleura, Cephalodiscus*

PHYLUM CHORDATA. Chordates

Subphylum Urochordata (or Tunicata). Tunicates

CLASS ASCIDIACEA. Ascidians or sea squirts. *Ciona, Clavelina, Molgula, Perophora*

CLASS THALIACEA. *Pyrosoma, Salpa, Doliolum*

CLASS LARVACEA. *Appendicularia, Oikopleura, Fritillaria*

Subphylum Cephalochordata. Lancelets, amphioxus. *Branchiostoma, Asymmetron*

Subphylum Vertebrata. Vertebrates

CLASS AGNATHA. Jawless fishes. *Cephalaspis,*[*] *Pteraspis,*[*] *Petromyzon, Entosphenus, Myxine, Eptatretus*

CLASS PLACODERMI. No living representatives

CLASS CHONDRICHTHYES. Cartilaginous fishes. *Squalus, Hyporion, Raja, Chimaera*

CLASS OSTEICHTHYES. Bony fishes

SUBCLASS SARCOPTERYGII

ORDER CROSSOPTERYGII (or Coelacanthiformes). Lobe-fins. *Latimeria*

ORDER DIPNOI (or Dipteriformes). Lungfishes. *Neoceratodus, Protopterus, Lepidosiren*

SUBCLASS BRACHIOPTERYGII. Bichirs. *Polypterus*

SUBCLASS ACTINOPTERYGII. Higher bony fishes. *Amia, Cyprinus, Gadus, Perca, Salmo*

CLASS AMPHIBIA

ORDER ANURA. Frogs and toads. *Rana, Hyla, Bufo*

ORDER URODELA. Salamanders. *Necturus, Triturus, Plethodon, Ambystoma*

ORDER APODA. *Ichthyophis, Typhlonectes*

CLASS REPTILIA

ORDER CHELONIA. Turtles. *Chelydra, Kinosternon, Clemmys, Terrapene*

ORDER RHYNCHOCEPHALIA. Tuatara. *Sphenodon*

ORDER CROCODYLIA. Crocodiles and alligators. *Crocodylus, Alligator*

ORDER SQUAMATA. Snakes and lizards. *Iguana, Anolis, Sceloporus, Phrynosoma, Natrix, Elaphe, Coluber, Thamnophis, Crotalus*

CLASS AVES. Birds. *Anas, Larus, Columba, Gallus, Turdus, Dendroica, Sturnus, Passer, Melospiza*

CLASS MAMMALIA. Mammals

SUBCLASS PROTOTHERIA

ORDER MONOTREMATA. Egg-laying mammals. *Ornithorhynchus, Tachyglossus*

SUBCLASS THERIA. Marsupial and placental mammals

ORDER MARSUPIALIA. Marsupials. *Didelphis, Sarcophilus, Notoryctes, Macropus*

ORDER INSECTIVORA. Insectivores (moles, shrews, etc.). *Scalopus, Sorex, Erinaceus*

ORDER DERMOPTERA. Flying lemurs. *Galeopithecus*

ORDER CHIROPTERA. Bats. *Myotis, Eptesicus, Desmodus*

ORDER PRIMATES. Lemurs, monkeys, apes, man. *Lemur, Tarsius, Cebus, Macacus, Cynocephalus, Pongo, Pan, Homo*

[*] Extinct.

ORDER EDENTATA. Sloths, anteaters, armadillos. *Bradypus, Myrmecophagus, Dasypus*

ORDER PHOLIDOTA. Pangolin. *Manis*

ORDER LAGOMORPHA. Rabbits, hares, pikas. *Ochotona, Lepus, Sylvilagus, Oryctolagus*

ORDER RODENTIA. Rodents. *Sciurus, Marmota, Dipodomys, Microtus, Peromyscus, Rattus, Mus, Erethizon, Castor*

ORDER CETACEA. Whales, dolphins, porpoises. *Delphinus, Phocaena, Monodon, Balaena*

ORDER CARNIVORA. Carnivores. *Canis, Procyon, Ursus, Mustela, Mephitis, Felis, Hyaena, Eumetopias*

ORDER TUBULIDENTATA. Aardvark. *Orycteropus*

ORDER PROBOSCIDEA. Elephants. *Elephas, Loxodonta*

ORDER HYRACOIDEA. Coneys. *Procavia*

ORDER SIRENIA. Manatees. *Trichechus, Halicore*

ORDER PERISSODACTYLA. Odd-toed ungulates. *Equus, Tapirella, Tapirus, Rhinoceros*

ORDER ARTIODACTYLA. Even-toed ungulates. *Pecari, Sus, Hippopotamus, Camelus, Cervus, Odocoileus, Giraffa, Bison, Ovis, Bos*

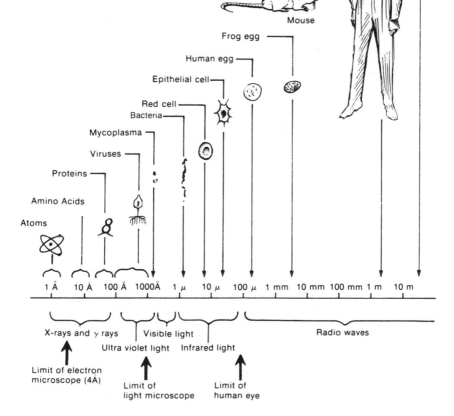

The dimensions of Biology. To accommodate the wide range of sizes in the world of living things, the unit scale is drawn logarithmically.

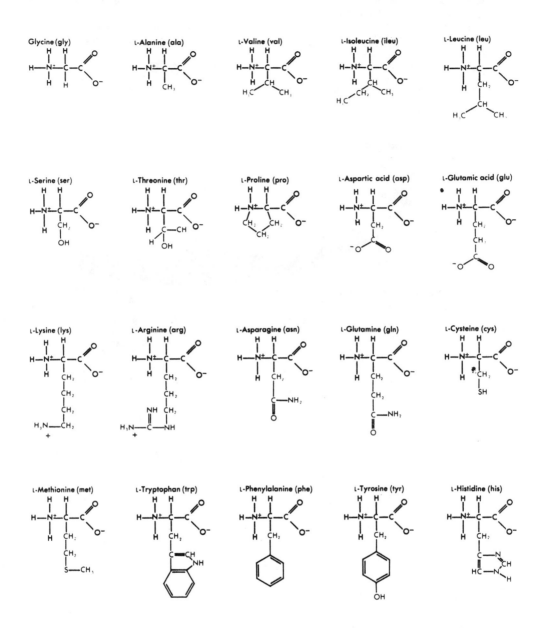

The twenty common amino acids used as buiding blocks in protein synthesis. As proteins are used to create portions of the cell membrane and enzymes which control an organism's metabolism, it is easy to see why amino acids are so important in life.

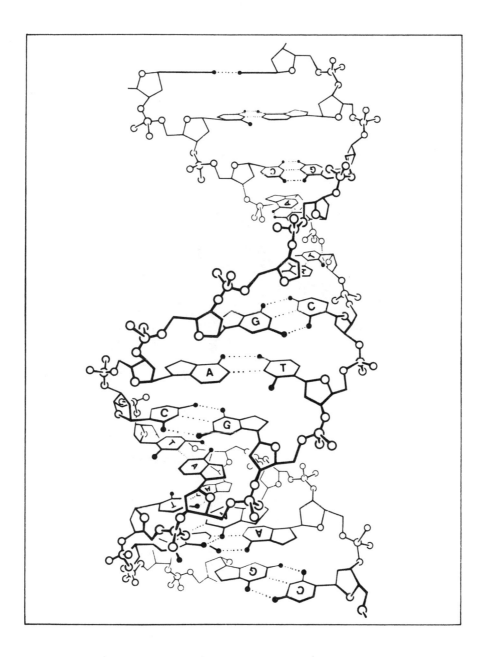

The DNA double helix. The purine (G= Guanine, A= Adenine) and pyrimidine (T= Thymine, C= Cytosine) base pairs are connected by hydrogen bonds. When these bonds are broken, the DNA can unwind and replicate (as in mitosis) or act as a template for mRNA synthesis. The ribose sugar and phosphate groups act as a backbone, helping to maintain the sequential order of the nitogenous bases. There are ten base pairs in one turn.

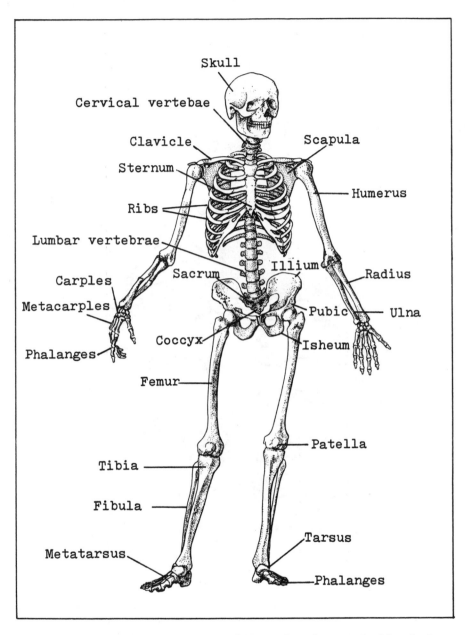

The human skeleton with major bones indicated.

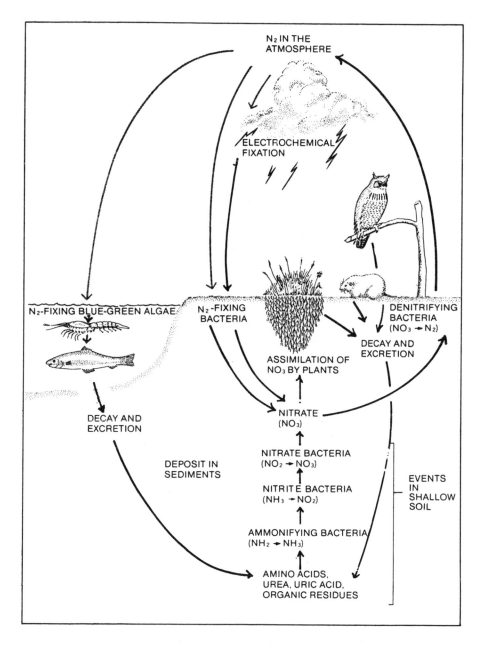

The nitrogen cycle. Bacteria, blue-green algae, or lightning can convert the nitrogen in the air into nitrates (NO_3), which can be absorbed by plants through their roots. Animals obtain nitrogen by eating these plants or other animals. This nitogen can be returned to the soil via dead organinc matter or waste products. Denitrifying bacteria convert nitrates back into atmospheric nitrogen (N_2).

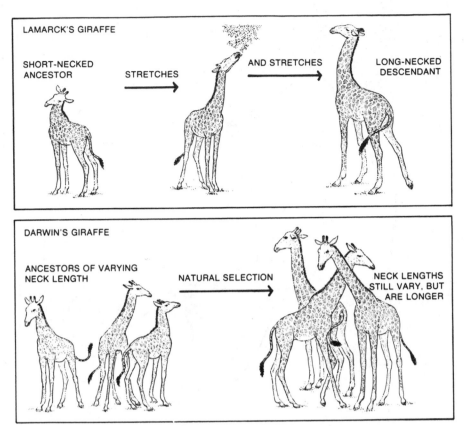

Rival explanations of evolution were advanced by
Lamarck and Darwin. According to Lamarck, the long
neck of the giraffe arose as a result of generations
of stretching to reach food. According to Darwin,
giraffes with longer necks arose as a result of
natural selection.

A comparison of the structure of the tail of the
primitive bird Archaeopteryx (right) and the tail of
a modern bird (left).

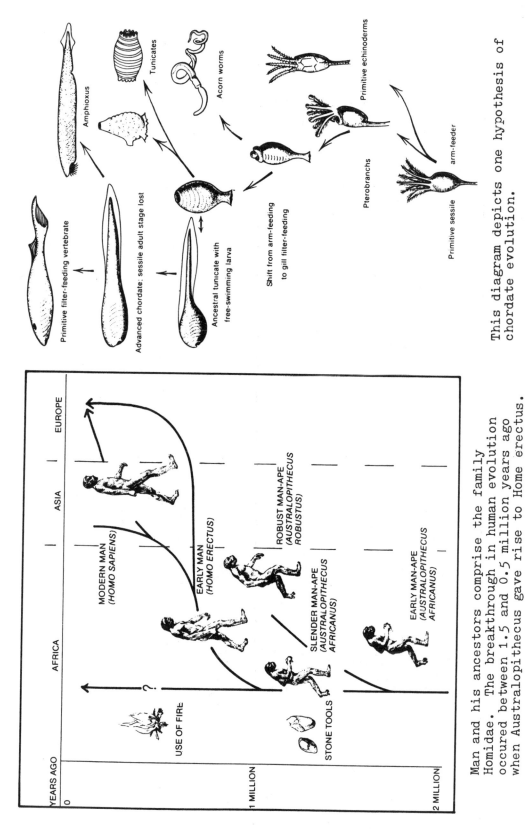

This diagram depicts one hypothesis of chordate evolution.

Man and his ancestors comprise the family Homidae. The breakthrough in human evolution occured between 1.5 and 0.5 million years ago when Australopithecus gave rise to Home erectus.

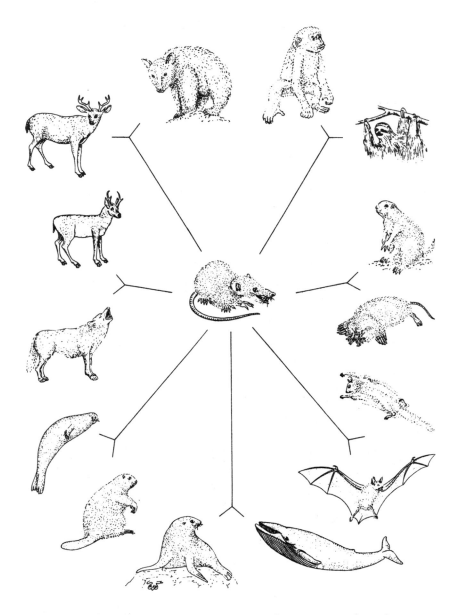

Adaptive radiation. All the various mammals shown
above have evolved from a common hypothetical
ancestor, depicted in the center. Each of the descendants has become adapted to a different type of
habitat.

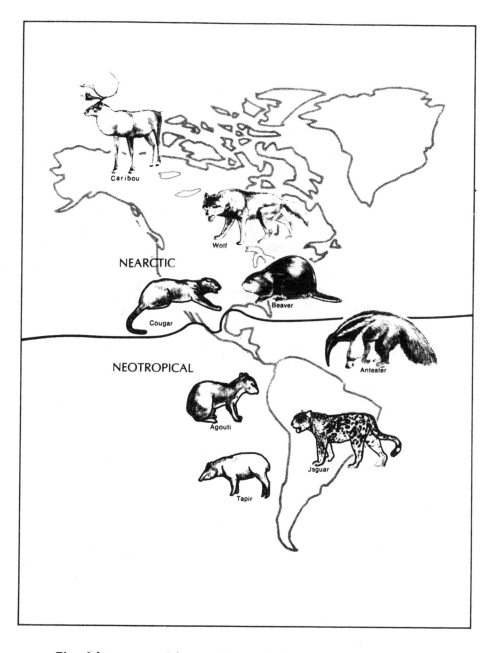

The biogeographic realms of the world with some
of their characteristic animals.